This book provides the first systematic treatment of the thermodynamic theory of site-specific effects in biological macromolecules. It describes the phenomenological and conceptual bases required to allow a mechanistic understanding of these effects from analysis of experimental data.

The thermodynamic theory also results in novel experimental strategies that enable the derivation of information on local, site-specific properties of a macromolecular system from analysis of perturbed global properties. The treatment focuses on binding phenomena, but is amenable to extension both conceptually and formally to the analysis of other cooperative processes, such as protein folding and helix–coil transitions. Much attention is devoted to the analysis of mutational perturbations used as probes of site-specific energetics.

The arrangement of various sections in the book allows its use both as a reference work on cooperative binding phenomena for any scientist involved in structure–function studies of biological macromolecules, and as a text for graduate students in biochemistry and biophysics.

THERMODYNAMIC THEORY OF SITE-SPECIFIC BINDING PROCESSES IN BIOLOGICAL MACROMOLECULES

THERMODYNAMIC THEORY OF SITE-SPECIFIC BINDING PROCESSES IN BIOLOGICAL MACROMOLECULES

ENRICO DI CERA

Washington University School of Medicine

CAMBRIDGE
UNIVERSITY PRESS

PUBLISHED BY THE PRESS SYNDICATE OF THE UNIVERSITY OF CAMBRIDGE
The Pitt Building, Trumpington Street, Cambridge, United Kingdom

CAMBRIDGE UNIVERSITY PRESS
The Edinburgh Building, Cambridge CB2 2RU, UK
40 West 20th Street, New York NY 10011–4211, USA
477 Williamstown Road, Port Melbourne, VIC 3207, Australia
Ruiz de Alarcón 13, 28014 Madrid, Spain
Dock House, The Waterfront, Cape Town 8001, South Africa

http://www.cambridge.org

First published 1995
First paperback edition 2005

A catalogue record for this book is available from the British Library

Library of Congress cataloguing in publication data

Di Cera, Enrico.
Thermodynamic theory of site-specific binding processes in
biological macromolecules / Enrico Di Cera.
p.   cm.
Includes bibliographical references.
ISBN 0 521 41659 0 (hardback)
1. Binding sites (Biochemistry)–Thermodynamics.   2. Ligand
binding (Biochemistry)–Thermodynamics.   3. Cooperative binding
(Biochemistry)–Thermodynamics.   4. Macromolecules–Thermodynamics.
I. Title.
QP517.B42D5   1995
574.192–dc20   95-15257   CIP

ISBN 0 521 41659 0 hardback
ISBN 0 521 61975 0 paperback

It is certainly not the least charm of a theory that it is refutable; it is with precisely this charm that it entices subtler minds.

Nietzsche, *Beyond Good and Evil*

To Antonella and Leonardo

Contents

Contents

Preface

Biological function in a macromolecular system often arises from the contribution of many constituent structural domains that communicate in a cooperative fashion. Formally, a *global* macromolecular property, F, can be cast as follows

$$F = f_1 + f_2 + \ldots + f_N$$

where the fs depict the contribution of individual structural domains that behave as subsystems open to interaction with the rest of the system. Relevant examples of global properties are: protein stability, where the fs represent the contribution of particular folding units to the macroscopic free energy of unfolding; helix–coil transitions, where the fs represent the helicity of individual residues and their contribution to the helix state of the peptide as a whole; binding and linkage phenomena, where the fs denote the probability of binding to individual sites and F is the average number of ligated sites. In all these cases, it is the cooperative behavior of the individual constituents that sets the behavior of the system as a whole. Also, the code for cooperativity and structure–function relationships is embodied by the *local* properties fs, rather than F. Consequently, a description of cooperative phenomena must be sought for in terms of *local* rather than *global* quantities. The limitations of a *global* description are demonstrated by the rather obvious fact that F is uniquely defined once the *local* properties fs are known, but the reverse is not true in general. Hence, the detailed aspects of the communication among constituent domains of a macromolecule that encapsulate the connection between structure and function are often swamped in the *global* picture and crucial information on mechanisms of cooperativity is lost.

The need for a description of cooperative effects in terms of *local* properties has been recognized for a long time, since the pioneering studies of Wegscheider (1895) on the ionization reactions of polybasic substances. For many years, however, our understanding of cooperative phenomena has remained limited to effects arising in the *global* picture due to the limitations imposed by experimental techniques. Likewise, theoretical treatments of cooperative effects in biology have so far been focused on the analysis of *global* effects. Recent advances in various areas, and especially in X-ray crystallography, NMR spectroscopy and recombinant DNA technology, have made it possible to access information at the *local* level and introduce a wide range of site-specific perturbations of macromolecular systems in a way never before possible. Crucial events pertaining to *global* phenomena can now be dissected in terms of the contribution of individual binding sites, folding units, amino acid residues or even atoms, thereby revealing the true and extraordinary complexity of cooperative effects in biology. These exciting new developments reinforce the notion that a predictive understanding of function and energetics from structure can only be gained from information on the *local* properties of a biological macromolecule and the network of communication among its constituent domains. A theoretical treatment of cooperative effects arising at the *local*, site-specific level is therefore both timely and important.

This monograph provides the first systematic treatment of the thermodynamic theory of site-specific effects in biological macromolecules. It describes the phenomenological and conceptual basis to gain a mechanistic understanding of these effects from analysis of experimental data. The theory also brings about novel experimental strategies of deriving information on *local* properties of a macromolecular system from analysis of perturbed *global* properties. Although the treatment focuses on binding phenomena, it is amenable of extension both conceptually and formally to the analysis of other cooperative processes, like protein folding and helix–coil transitions. Much attention is also devoted to the analysis of mutational perturbations, used as probes of site-specific energetics. Chapter 1 gives an introduction to basic thermodynamic concepts that form the backbone of the phenomenological theory of ligand binding. Cooperativity and linkage in the *global* description are dealt with in Chapter 2. Most of this chapter is a summary of the classical treatment of ligand binding and linkage effects detailed in excellent monographs (Hill, 1984; Wyman and Gill, 1990). The treatment of *local*, site-specific effects is introduced in Chapter 3 as an extension of concepts introduced in Chapters 1 and 2. The formalism and conceptual framework based on contracted partition functions are also introduced in Chapter 3. Applications of the theory of site-specific effects are given in Chapter 4. The case of hemoglobin cooperativity is discussed in Section 4.4. A novel strategy of approach to site-specific effects based on the analysis of *global* properties of structurally perturbed systems is introduced in Section 4.5. A general approach to site-specific cooperativity based on the analysis of pairwise coupling patterns derived from mutational perturbations is given in Section 4.6. Finally, Chapter 5 addresses the rather difficult problem of site-specific effects in Ising networks. These networks can be used as models for molecular recognition and serve to explore rules for site–site communication in extended systems that cannot be analyzed within the framework of the phenomenological theory. Section 5.4 describes possible new approaches to the study of the Ising problem in two and higher dimensions using site-specific thermodynamics.

The arrangement of various sections of this monograph allows for a great deal of flexibility in using it as a reference for cooperative binding phenomena, or as a textbook for graduate students in biochemistry and biophysics curricula. The reader thoroughly familiar with ensemble theory and the classical Hill–Wyman treatment of binding and linkage can start with Chapter 3, although reading of Section 1.2 and Section 2.4 is recommended. The reader interested mostly in applications of the theory to binding phenomena and mutational perturbations can start with Sections 4.4–4.6 and then move to Sections 5.1 and 5.2, although reading of the basic Sections 3.1 and 3.2 is also recommended. In using the monograph as a textbook in biochemistry and biophysics curricula, the following 'pathways' are recommended:

	Biochemistry	Biophysics
Chapter 1		Sections 1.5, 1.6
Chapter 2	Sections 2.1–2.3, 2.5	Sections 2.1–2.3, 2.5, 2.6
Chapter 3	Sections 3.1, 3.2	Sections 3.1–3.3
Chapter 4	Sections 4.2, 4.4–4.6	Sections 4.2, 4.4–4.6
Chapter 5		Sections 5.3, 5.4

In either pathway the monograph will serve its purpose of providing the conceptual tools for deciphering cooperativity codes in biological macromolecules. These tools should be acquired in conjunction with the recent developments in spectroscopic techniques and recombinant DNA technologies by those students who seek to obtain a more integrated training in biochemistry and biophysics. Knowledge of calculus and elementary differential equations is required in either pathway. The requirement can be reduced to a minimum in a short course emphasizing the practical applications of site-specific thermodynamics, which may include Sections 3.1, 3.2, 4.4–4.6 only. Such a module may be integrated in the core course of a biochemistry curriculum. A more demanding course emphasizing the statistical thermodynamic foundations can be prepared by addition of Sections 1.1–1.4 to the biophysics pathway.

Much of the inspiration to develop the theory dealt with in this monograph has come from the landmark experimental work of Gary Ackers on hemoglobin (Ackers, Doyle, Myers and Daugherty, 1992) that has led to the first, most accurate and complete dissection of the site-specific properties of a cooperative system in biology. The author is greatly indebted to Professors Gary Ackers and Timothy Lohman for reading parts of the monograph and providing helpful suggestions, and to Professor Michele Perrella for providing experimental data prior to publication and helpful comments on Section 4.4. The author is also grateful to Professors Ken Dill, Neville Kallenbach, Eaton Lattman and George Rose for helpful discussions on some of the topics covered in the monograph. A very special thanks goes to Professors John Edsall and Bruno Zimm for several comments that were invaluable in the final preparation. Professor Jeffries Wyman and the late Professor Stanley Gill provided excellent advice in the early stages of preparation of the monograph. Helpful comments also came from many graduate students and postdoctoral fellows here at Washington University and especially from Quoc Dang, Karl-Peter Hopfner, Sean Keating, Yong Kong and Murad Nayal while in the author's laboratory. The skillful secretarial assistance of Anna Goffinet was invaluable throughout the preparation of the monograph and Sophie Silverman carefully proofread various versions of the manuscript. Finally, the author gratefully acknowledges partial support from National Science Foundation Grants DMB91-04963 and MCB94-06103 during the period in which the monograph was written.

St. Louis,   E. Di Cera

1

Statistical thermodynamic foundations

We are interested in a system composed of a biological macromolecule and a number of ligands, of which the solvent can be considered one, under conditions of temperature and pressure of biological relevance. Our goal is to characterize the behavior of the macromolecule in its interaction with the various components of the system and the rules underlying the mutual interference of physical and chemical variables. In this chapter we deal with the statistical thermodynamic foundations of binding processes and the concepts that form the basis of our treatment.

1.1 Postulates and basic ensembles

The physico-chemical properties of a system are defined thermodynamically by a set of macroscopic quantities accessible to experimental measurements (Fermi, 1936; Schrödinger, 1946). If the macromolecule is taken as the system, then all observables reflect properties of the system and the way it is affected by physical and chemical driving forces. A macromolecule in the presence of multiple ligands can be seen as a system existing in a number of distinct energy states, E_1, E_2, ... E_r, totally analogous to the energy states of a quantum mechanical system. Each energy state may be coupled to a conformational state of the macromolecule, i.e., a specific arrangement of its secondary, tertiary or quaternary structure. In addition, E_j is specified by the number of ligands bound, or by a particular ligated configuration. Different energy states may group together in the same energy level according to their degeneracy. We consider a prototypic system composed of the macromolecule and the solvent, in contact with a heat bath at constant temperature T, and focus our attention on a particular configuration of the macromolecule (our system) with N water molecules (the solvent) bound to it, as shown in

Figure 1.1. We use the term *bound* in a thermodynamic sense, to distinguish the water molecules in any form of interaction with the system from those belonging to the bulk solvent. In what follows, we make no attempt to discriminate among the various binding modes that characterize protein hydration, whether specific or non-specific. We use the macromolecule in solution to illustrate the salient thermodynamic features of the simplest system of interest. If V is the volume of the macromolecule, then any energy state can be written as $E_j(N, V)$ and the spectrum of energy values $E_1, E_2, \ldots E_r$ denotes the possible alternative states accessible to

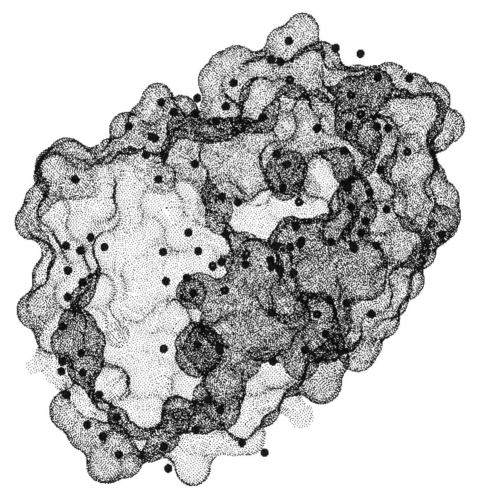

Figure 1.1 Connolly surface of lysozyme, constructed from the crystal structure (Weaver and Matthews, 1987). Water molecules are indicated by solid dots. Some molecules are present as components of the bulk solvent. Other molecules are 'bound' to the accessible surface of the enzyme and make polar contacts with specific residues.

the macromolecule with specified values of V and N. The question that arises at this point concerns the probability of finding the macromolecule in a particular energy state, E_j, when the temperature of the heat bath is held constant. To solve this problem we use the ensemble method introduced by Gibbs (Gibbs, 1902, 1928; Hill, 1960).

Consider an ensemble of a very large number of identical replicas of our system, as depicted in Figure 1.2, with the same value of V and N, and in thermal contact with each other. Each system is in contact with a heat bath at constant T and is allowed to exchange heat, but not water molecules, with the surrounding systems. After equilibrium is reached and T is uniform everywhere, the entire ensemble of systems, or supersystem, is thermally insulated by an adiabatic wall. If Γ is the total number of replicas in the supersystem, then each system can be thought of as being in contact with a heat bath at temperature T formed by the remaining $\Gamma - 1$ systems. The supersystem artificially generated by our replication process is an *isolated* system that cannot exchange heat or matter with the environment, as opposed to each constituent system of the ensemble that is *closed* and isothermal, since it is allowed to exchange heat, but not matter, with the environment. The properties of our original system can be derived from the properties of the supersystem characterized by a

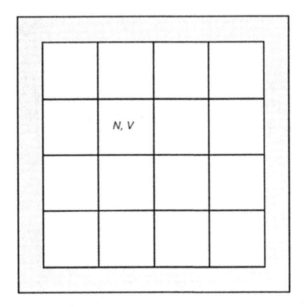

Figure 1.2 Schematic representation of a canonical ensemble. A number of elementary systems with fixed values of V and N are in thermal contact with each other at constant temperature. The supersystem formed by Γ elementary systems is insulated by an adiabatic wall from the environment.

volume ΓV, ΓN water molecules and a total energy E_t. In the thermodynamic limit $\Gamma \to \infty$, the supersystem is composed of an extremely large number of replicas and the properties of each system separately can be derived as 'ensemble averages' taken over the entire collection of replicas. The validity of this assertion is supported by two postulates of Gibbs.

The first postulate states that:

The long time average of a variable in the system at equilibrium is equal to its ensemble average in the thermodynamic limit $\Gamma \to \infty$, provided the ensemble of systems replicates the system of interest and its environment.

This postulate is intuitively obvious. It simply states the equivalence between averages obtained by monitoring the behavior of the system at equilibrium over a long time scale and those obtained over the ensemble artificially constructed. The time evolution of a variable of interest, say the energy of the system, is plotted in Figure 1.3 as a trajectory whose

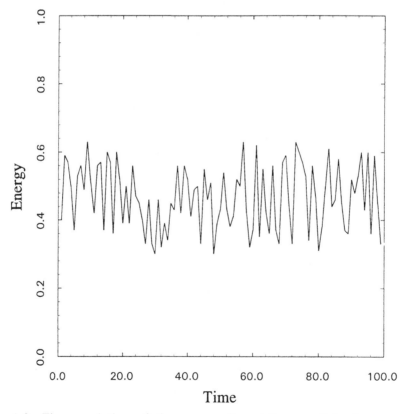

Figure 1.3 Time evolution of the energy (in arbitrary units) of a canonical ensemble at equilibrium. The time average of the energy values gives the thermodynamic measure of the energy E.

time average gives the thermodynamic measure of $E(N, V)$. The contact
with the heat bath makes the energy fluctuate and assume any of the
allowable values in the spectrum. If we were to monitor the energy for our
system at equilibrium, we would obtain a plot similar to that shown in
Figure 1.3. The average value of the energy must, of course, be independent of the particular time at which the observation is started. Hence, we
could make repeated observations about our system and obtain a bundle
of trajectories such as the one shown in Figure 1.3, each of which would
yield the same value of the average energy $E(N, V)$. We may think of
each trajectory as belonging to a replica of our original system. In the limit
where this number becomes arbitrarily large, we can take a snapshot of
the bundle of trajectories at any given time and display the energy values
for all individual systems, as shown in Figure 1.4. The average of these

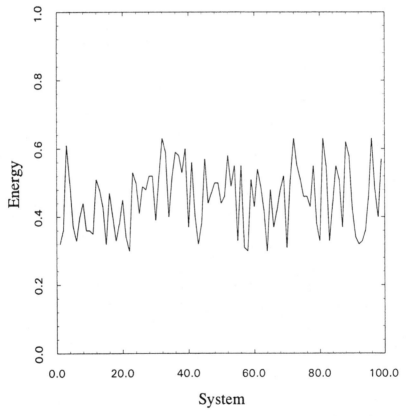

Figure 1.4 Spectrum of energy values for a bundle of trajectories such as the one
shown in Figure 1.3, taken at a fixed time, for a canonical ensemble at equilibrium. The abscissa enumerates a representative sample of systems in the
supersystem in Figure 1.2. The ensemble average of the energy values is equivalent to the time average derived from Figure 1.3 and gives the thermodynamic
measure of the energy E.

values must be independent of the particular time chosen. Sections of the bundle taken at different times must yield energy profiles with the same average value, and this value is evidently the same as that obtained in the long time average of a single trajectory. Hence the first postulate, which establishes a complete equivalence between time averages over the system and ensemble averages independent of time over a supersystem constructed of a large number of replicas of the original system.

In order to determine the properties of our system specifically we need a second postulate, also known as the 'principle of equal *a priori* probabilities' for an isolated system, that states:

In an isolated system for which N, V and the energy E are specified, all energy states have equal a priori probabilities.

All allowable energy states, consistent with the specified values of N, V and E, occur with the same frequency if the behaviour of the system is followed over time or, equivalently, the system spends equal amounts of time in any of the allowable energy states. This important consequence of the two postulates is known as the 'ergodic hypothesis' and plays a central conceptual role in statistical mechanics and its mathematical foundations (Tolman, 1938; Khinchin, 1949; Kubo, 1965; de Groot and Mazur, 1984). In an ensemble representative of an isolated system, the probability of occurrence of any energy state is exactly the same for all systems and if any system is chosen at random, it will be in any of the allowable energy states with equal probability. The second postulate as stated above applies only to *isolated* systems, while the first postulate holds quite generally for any equilibrium system.

We are now in the position to find the properties of our original system composed of the macromolecule with a volume V, a number N of water molecules bound, and in contact with a heat bath at constant temperature T. The ensemble of elementary systems shown in Figure 1.2 is, by construction, an isolated system which must obey the Gibbs postulates. The energy of the supersystem is

$$E_t = \sum_{j=1}^{r} n_j E_j \tag{1.1}$$

where n_j is the number of systems in the energy state E_j. Also,

$$\Gamma = \sum_{j=1}^{r} n_j \tag{1.2}$$

is an obvious conservation relationship. There is a very large number of distributions, D, of n_j values consistent with eqs (1.1) and (1.2), and specifically there are as many as

$$\Omega(D) = \frac{\Gamma!}{n_1(D)!n_2(D)! \ldots n_r(D)!} \tag{1.3}$$

possible ways of realizing the particular distribution, D, of such values. Here $n_j(D)$ denotes the number of systems in state E_j for the particular distributions D. The probability, ψ_j, of finding the macromolecule in a given energy state E_j is the weighted mean of all possible $n_j(D)$ values, divided by the total number of systems, i.e.,

$$\psi_j = \frac{\sum\limits_{n} n_j(D)\Omega(D)}{\Gamma\sum\limits_{n}\Omega(D)} \tag{1.4}$$

The distribution of n_j values, $\Omega(D)$, is multinomial and goes into a Gaussian for large Γ by virtue of the De Moivre–Laplace theorem (Wilson, 1911; Feller, 1950). In the limit $\Gamma \to \infty$, the Gaussian goes into a completely sharp Dirac δ-function centered about the most probable distribution $D = D^*$. Since the δ-function vanishes everywhere except for $D = D^*$, then the simple result $\psi_j = n_j^*/\Gamma$ is obtained from eq (1.4), where $n_j^* = n_j(D^*)$ is the value of n_j in the most probable distribution.

The value of n_j^* can be found by maximizing Ω, or equivalently $\ln \Omega$, subject to the constraints in eqs (1.1) and (1.2). Application of the method of Lagrange's undetermined multipliers (Wilson, 1911) gives

$$\frac{\partial}{\partial n_j}[\ln \Omega(D) - \alpha\Gamma - \beta E_t] = 0 \tag{1.5}$$

where α and β are the undetermined multipliers. Using the Stirling approximation $\ln x! \approx x \ln x - x$ for the factorial terms of $\Omega(D)$ in eq (1.3), and recalling that Γ and E_t are functions of n_j, yields

$$\ln \Gamma - \ln n_j^* - \alpha - \beta E_j = 0 \tag{1.6}$$

Hence,

$$\psi_j = \frac{n_j^*}{\Gamma} = \exp(-\alpha - \beta E_j) \tag{1.7}$$

gives the probability that the macromolecule exists in the energy state E_j defined by V and N. The value of β is $(k_B T)^{-1}$, where k_B is the Boltzmann constant (Tolman, 1938; Hill, 1960). The value of α is obtained from the conservation relationship (1.2), which embodies the obvious fact that the sum of all ψ_j must equal unity, so that

$$\alpha = \ln \sum_{j=1}^{r} \exp\left(-\frac{E_j}{k_B T}\right) \tag{1.8}$$

Hence,

$$\psi_j = \frac{\exp\left[-\dfrac{E_j(N, V)}{k_B T}\right]}{\displaystyle\sum_{j=1}^{r} \exp\left[-\dfrac{E_j(N, V)}{k_B T}\right]} \tag{1.9}$$

gives the explicit expression for the probability of the macromolecule existing in the energy state E_j. This quantity decreases exponentially with the value of E_j. The ensemble average over all possible energy states

$$E = \sum_{j=1}^{r} \psi_j E_j = \langle E \rangle \tag{1.10}$$

defines the energy of the system in the thermodynamic sense.

The closed isothermal system considered in the foregoing analysis is called a *canonical ensemble* and the probability distribution in eq (1.9) is the Boltzmann distribution for a canonical ensemble (Tolman, 1938). The properties of the macromolecule considered as a closed isothermal system, or a canonical ensemble, are completely specified by its volume V, the number of water molecules bound N and the temperature of the heat bath T. Thermodynamically, such a system is an $N-V-T$ ensemble. We now introduce an important quantity that allows us to compute all thermodynamic functions of interest for the system. This is the canonical *partition function* defined as the sum

$$Z(N, V, T) = \sum_{j=1}^{r} \exp\left[-\frac{E_j(N, V)}{k_B T}\right] \tag{1.11}$$

The partition function enumerates all possible energy states of the system as they appear in the Boltzmann distribution. The relative contribution of each term is weighted exponentially by the Boltzmann factor associated with it. The probability of any energy state, E_j, is given by the ratio between the term containing the value of E_j and the partition function, as seen in eq (1.9). Mathematically, this can be expressed in compact form with the use of the partial derivative

$$\psi_j = -k_B T \frac{\partial \ln Z}{\partial E_j} = \frac{\partial F}{\partial E_j} \tag{1.12}$$

where

$$F(N, V, T) = -k_B T \ln Z(N, V, T) \tag{1.13}$$

is the Helmholtz free energy of the macromolecule and plays the role of the potential associated with the partition function Z. The energy of the macromolecule can be derived in an analogous way as

$$E = k_B T^2 \left(\frac{\partial \ln Z}{\partial T} \right)_{N,V} = -T^2 \left(\frac{\partial \frac{F}{T}}{\partial T} \right)_{N,V} = \left(\frac{\partial \frac{F}{T}}{\partial \frac{1}{T}} \right)_{N,V} \tag{1.14}$$

Since Z is also a function of V and N, it is important to derive the quantities associated with these independent variables. The derivative

$$P = k_B T \left(\frac{\partial \ln Z}{\partial V} \right)_{N,T} = -\left(\frac{\partial F}{\partial V} \right)_{N,T} \tag{1.15}$$

gives the pressure of the system and

$$\mu = -k_B T \left(\frac{\partial \ln Z}{\partial N} \right)_{V,T} = \left(\frac{\partial F}{\partial N} \right)_{V,T} \tag{1.16}$$

is the chemical potential of water.

The change of Z or F with respect to the external conditions subject to experimental control yields information on the quantities associated with them. The energy of the system is obtained as the 'response function' to a change in temperature. Likewise, the pressure and chemical potential of water are obtained as responses to changes of V and N respectively. From the definition of F in eq (1.13) it also follows that

$$\left(\frac{\partial F}{\partial T} \right)_{N,V} = -k_B \ln Z - k_B T \left(\frac{\partial \ln Z}{\partial T} \right)_{N,V} = \frac{F - E}{T} = -S \tag{1.17}$$

The entropy, S, of the macromolecule is derived from the change of the Helmholtz free energy as a function of temperature. An explicit expression for S is obtained from the definition of ψ_j as follows. Taking the logarithm of ψ_j one has

$$\ln \psi_j = -\frac{E_j}{k_B T} - \ln Z \tag{1.18}$$

Multiplying both members by ψ_j and summing over all values of j leads to

$$\frac{F - E}{k_B T} = \sum_{j=1}^{r} \psi_j \ln \psi_j \tag{1.19}$$

Hence,

$$S = -k_B \sum_{j=1}^{r} \psi_j \ln \psi_j = -\frac{F - E}{T} \qquad (1.20)$$

provides an important definition of the entropy in terms of the probability of occurrence of the energy states, and the thermodynamic potentials E and F. Eqs (1.15)–(1.17) also allow for a definition of the thermodynamic potential F in differential, or Pfaffian form (de Heer, 1986) as follows

$$dF = \left(\frac{\partial F}{\partial N}\right)_{V,T} dN + \left(\frac{\partial F}{\partial V}\right)_{N,T} dV + \left(\frac{\partial F}{\partial T}\right)_{N,V} dT$$
$$= \mu\, dN - P\, dV - S\, dT \qquad (1.21)$$

This Pfaffian form summarizes the properties of the macromolecule as a canonical ensemble. The integral form of F is

$$F = E - TS \qquad (1.22)$$

and follows directly from eq (1.20).

Having defined the properties of our original closed isothermal system, we turn to the supersystem itself which is an isolated system for which N, V and the total energy E are fixed. Consider all systems in the supersystem that belong to the same energy state E_j, group and surround them by an adiabatic wall. This yields an ensemble of systems with the same value of N, V and $E = E_j$. Such an ensemble is called *microcanonical* and differs from the canonical ensemble insofar as all systems in the ensemble have the same energy. Application of the same arguments developed for the canonical ensemble leads to the conclusion that ψ_j must be independent of j in the microcanonical ensemble. This is because the microcanonical ensemble is a degenerate canonical ensemble where all systems have the same energy $E = E_1 = E_2 = \ldots = E_r$. We also know, by construction, that the microcanonical ensemble is a subensemble of the canonical ensemble in Figure 1.2, where all systems with the same energy have been grouped together and isolated. Hence, the degeneracy $\Omega(D^*)$ of the most probable distribution D^* in a microcanonical ensemble with energy $E = E_j$ coincides with the value of n_j^* in the canonical ensemble. The value of the probability ψ for the energy E is simply $1/\Omega(D^*)$ and is the same for all systems of the ensemble since they belong to the same energy level and there are a total of $\Omega(D^*)$ such systems. This follows directly from the Boltzmann distribution in eq (1.9) by letting $E_j = E$ for all $j = 1$, $2, \ldots \Omega(D^*)$. The partition function of the macromolecule as a microcanonical ensemble is given by Ω and is completely defined once N, V

and E are fixed. The thermodynamic potential associated with Ω is the entropy S and is obtained from eq (1.20) by letting $\psi_j = \psi = 1/\Omega$ for all $j = 1, 2, \ldots \Omega$, i.e.,

$$S(N, V, E) = k_B \ln \Omega(N, V, E) \tag{1.23}$$

The microcanonical ensemble is therefore an $N-V-E$ ensemble. The Pfaffian form associated with the potential S is

$$dS = \left(\frac{\partial S}{\partial N}\right)_{V,E} dN + \left(\frac{\partial S}{\partial V}\right)_{N,E} dV + \left(\frac{\partial S}{\partial E}\right)_{N,V} dE$$
$$= -\frac{\mu}{T} dN + \frac{p}{T} dV + \frac{1}{T} dE \tag{1.24}$$

and is a consequence of eq (1.22). In fact, taking the differential of eq (1.22) yields

$$dF = dE - T\,dS - S\,dT = -S\,dT - P\,dV + \mu\,dN \tag{1.25}$$

and hence eq (1.24). The response functions of the macromolecule treated as a microcanonical ensemble are derived from eq (1.24) as follows

$$1 = k_B T \left(\frac{\partial \ln \Omega}{\partial E}\right)_{N,V} = T\left(\frac{\partial S}{\partial E}\right)_{N,V} \tag{1.26}$$

$$P = k_B T \left(\frac{\partial \ln \Omega}{\partial V}\right)_{N,E} = T\left(\frac{\partial S}{\partial V}\right)_{N,E} \tag{1.27}$$

$$\mu = -k_B T \left(\frac{\partial \ln \Omega}{\partial N}\right)_{V,E} = -T\left(\frac{\partial S}{\partial N}\right)_{V,E} \tag{1.28}$$

They provide an alternative way of characterizing the pressure and chemical potential of water using the response of the macromolecule to variables controlled externally. The Pfaffian form of the energy of the macromolecule is implicit in the differential form of the entropy in eq (1.24) and is given by

$$dE = T\,dS - P\,dV + \mu\,dN \tag{1.29}$$

This form will be very useful in the following sections. Note that the energy, like the entropy, is a function of extensive quantities only and therefore scales linearly with the dimensions of the system. We will analyze the important consequences of this fact in Section 1.3.

The microcanonical and canonical ensembles provide two important conceptual frameworks for characterizing the properties of the macromolecule in our original system. When we look at the macromolecule as a microcanonical ensemble, we focus our attention on the properties of a configuration of volume V, with N water molecules bound and a particular energy state E, which is fixed. On the other hand, the macromolecule as a canonical ensemble is a system of fixed volume V, and N water molecules bound, but an energy E that can assume any value out of a spectrum of discrete energy states compatible with the preassigned values of N and V. The canonical partition function (1.11) encapsulates all the energy states of the system. If multiple states belong to the same energy level E_j, then E_j is said to be degenerate and $\Omega(N, V, E_j)$ is the number of configurations belonging to E_j, for N and V preassigned. The canonical partition function can be cast in two equivalent forms, depending on whether the summation is carried out over energy *states* or energy *levels* with their appropriate degeneracy taken into account, i.e.,

$$Z(N, V, T) = \sum_{j=1}^{r} \exp\left[-\frac{E_j(N, V)}{k_B T}\right] = \sum_{j=1}^{s} \Omega(N, V, E_j) \exp\left[-\frac{E_j(N, V)}{k_B T}\right]$$

$$(1.30)$$

The partition function cast in terms of energy levels shows the contribution of microcanonical components to the overall sum, as embodied by the microcanonical partition function Ω. This term reflects the properties of a macromolecule of fixed volume V, with N water molecules bound and belonging to a fixed energy level E_j. The canonical partition function is the sum of several microcanonical partition functions weighted according to a Boltzmann factor. This is consistent with the fact that a canonical ensemble is merely a collection of microcanonical ensembles. The partition function enumerates all possible configurations consistent with the constraints imposed upon the system. In the case of the microcanonical ensemble we deal with the degeneracy of the energy level E assigned to the macromolecule of the volume V and with N water molecules bound. In the case of the canonical ensemble, we deal with the entire spectrum of energy values allowed by preassigned values of V and N.

Analogous considerations can be applied to the case where different possible values of N are considered. In this case the macromolecule is studied as a system for which V is fixed, but the number of water molecules N and the energy E can both change. Since the energy level $E_j(N, V)$ depends on N and V, it is expected to vary when N changes from zero to its upper bound M. Likewise, the entire energy spectrum is expected to change with N. In this case we seek to define the average value of the energy E in the system, as well as the average number of water molecules bound to the macromolecule, $\langle N \rangle$. Intuitively, we expect the partition function of the macromolecule to be the sum of a number of canonical partition functions over the discrete variable N, weighted by a proper term that needs to be determined. This is by analogy with the transition between the microcanonical and canonical ensembles, where the Boltzmann factors containing the energy terms E_j were introduced as weighting factors of the microcanonical contributions to the canonical partition function.

Consider the supersystem in Figure 1.5 constructed as in the case of the canonical ensemble by grouping together a very large number of replicas of our system of interest. Notice that the wall between adjacent systems has been modified to allow systems to exchange not only heat, but also matter in the form of water molecules. The system of interest is now open and isothermal and defines a *grand canonical ensemble*. The heat bath sets the temperature T of the system at equilibrium and a reservoir of water molecules (the solvent) sets the chemical potential of water μ. We allow a long time for equilibrium of the supersystem and then compute the most probable distribution of energy and N values for our system as in the case of the canonical ensemble. Likewise, we maximize $\ln \Omega[D(N)]$ using the method of undetermined multipliers. Here $D(N)$ is the most probable

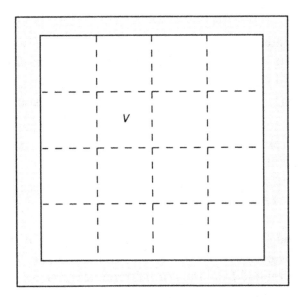

Figure 1.5 Schematic representation of a grand canonical ensemble. A number of elementary systems with fixed values of V are in thermal contact with each other at constant temperature. Each system can also exchange molecules with its neighbors. The supersystem formed by Γ elementary systems is insulated by an adiabatic wall from the environment.

distribution of systems with energy states $E_1(N, V)$, $E_2(N, V)$, ... $E_r(N, V)$ when N is assigned, and we have to consider all possible values of N. The constraints to be taken into account are evidently

$$\Gamma = \sum_{N=0}^{M} \sum_{j=1}^{r} n_j(N) \tag{1.31}$$

$$E_t = \sum_{N=0}^{M} \sum_{j=1}^{r} n_j(N) E_j(N, V) \tag{1.32}$$

$$N_t = \sum_{N=0}^{M} \sum_{j=1}^{r} n_j(N) N \tag{1.33}$$

where $n_j(N)$ is the number of systems in the energy state $E_j(N, V)$ and N_t is the total number of water molecules in the supersystem. This is the only additional constraint in the grand canonical ensemble compared to the canonical ensemble. A condition analogous to eq (1.5) states that

$$\frac{\partial}{\partial n_j(N)}[\ln \Omega[D(N)] - \alpha\Gamma - \beta E_t - \gamma N_t] = 0 \tag{1.34}$$

with

$$\Omega[D(N)] = \frac{\Gamma!}{n_1[D(N)]! n_2[D(N)]! \ldots n_r[D(N)]!} \tag{1.35}$$

Here $n_j[D(N)]$ indicates the number of systems with N water molecules bound and in the energy state $E_j(N, V)$, in the particular distribution $D(N)$. Solution of eq (1.34) is straightforward and leads to

$$\ln \Gamma - \ln n_j(N)^* - \alpha - \beta E_j(N, V) - \gamma N = 0 \tag{1.36}$$

where $n_j(N)^*$ is the most probable value of $n_j(N)$. The probability of finding the macromolecule with N water molecules bound and in the energy state $E_j(N, V)$ is

$$\psi_j(N) = \frac{n_j(N)^*}{\Gamma} = \exp\left[-\alpha + \frac{N\mu - E_j(N, V)}{k_B T}\right] \tag{1.37}$$

The undetermined multipliers β and γ have been expressed in terms of the appropriate driving forces as $\beta = 1/k_B T$ and $\gamma = -\mu/k_B T$. Again, the value of α is determined from the normalization condition for the probabilities ψs as

$$\alpha = \ln \sum_{N=0}^{M} \sum_{j=1}^{r} \exp\left[\frac{N\mu - E_j(N, V)}{k_B T}\right] \tag{1.38}$$

Hence,

$$\psi_j(N) = \frac{\exp\left[\dfrac{N\mu - E_j(N, V)}{k_B T}\right]}{\displaystyle\sum_{N=0}^{M}\sum_{j=1}^{r}\exp\left[\dfrac{N\mu - E_j(N, V)}{k_B T}\right]} \tag{1.39}$$

The energy and the number of water molecules bound to the macro-molecule in the grand canonical ensemble are fluctuating quantities and can only be defined as ensemble averages once the definition of $\psi_j(N)$ is available. The relevant expressions are

$$E = \sum_{N=0}^{M}\sum_{j=1}^{r}\psi_j(N)E_j(N, V) \tag{1.40}$$

$$W = \sum_{N=0}^{M}\sum_{j=1}^{r}\psi_j(N)N \tag{1.41}$$

The double sum in the logarithm of (1.38) and in the denominator of (1.39) is the grand canonical partition function

$$\Xi(V, T, \mu) = \sum_{N=0}^{M}\sum_{j=1}^{r}\exp\left[\frac{N\mu - E_j(N, V)}{k_B T}\right] \tag{1.42}$$

The grand canonical partition function enumerates all possible states with respect to the number of water molecules bound and the allowable energy states.

Two alternative forms of eq (1.42) deserve particular attention. Upon rewriting Ξ in terms of energy levels and their degeneracy as follows

$$\Xi(V, T, \mu) = \sum_{N=0}^{M}\sum_{j=1}^{s}\Omega(N, V, E_j)\exp\left[\frac{N\mu - E_j(N, V)}{k_B T}\right] \tag{1.43}$$

the contribution of microcanonical and canonical components becomes clear. Starting with a microcanonical ensemble for which N, V and E_j are fixed, the partition function Ω is nested in a series expansion containing all possible values of E_j subject to the fixed values of V and N. Each term of this series, defining the canonical partition function, is weighted according to the Boltzmann factor. Next, the canonical partition function is nested in another series expansion over all possible values of N to generate the grand canonical partition function. The alternative form for the grand canonical partition function is a polynomial expansion in the water activity $w = \exp(\mu/k_B T)$, i.e.,

$$\Xi(V, T, w) = \sum_{N=0}^{M} Z(N, V, T)w^N \tag{1.44}$$

The coefficients of this expansion are merely the canonical partition functions for a given set of values of V, N and T.

The thermodynamic potential associated with Ξ is

$$J(\mu, V, T) = -k_B T \ln \Xi(\mu, V, T) \tag{1.45}$$

and shows that the grand canonical ensemble is a $\mu-V-T$ ensemble. The basic properties of the macromolecule can be derived from differentiation once Ξ or J are known. The number of water molecules bound is

$$W = k_B T \left(\frac{\partial \ln \Xi}{\partial \mu}\right)_{V,T} = -\left(\frac{\partial J}{\partial \mu}\right)_{V,T} \tag{1.46}$$

Note that W is the same as N interpreted in the thermodynamic sense as an ensemble average. The pressure is

$$P = k_B T \left(\frac{\partial \ln \Xi}{\partial V}\right)_{\mu,T} = -\left(\frac{\partial J}{\partial V}\right)_{\mu,T} \tag{1.47}$$

Finally, the derivative with respect to T yields

$$E - W\mu = k_B T^2 \left(\frac{\partial \ln \Xi}{\partial T}\right)_{\mu,V} = -T^2 \left(\frac{\partial \frac{J}{T}}{\partial T}\right)_{\mu,V} = \left(\frac{\partial \frac{J}{T}}{\partial \frac{1}{T}}\right)_{\mu,V} \tag{1.48}$$

The derivative of the potential J with respect to T is again related to the entropy of the macromolecule in a direct way, as seen for the canonical ensemble. In fact,

$$\left(\frac{\partial J}{\partial T}\right)_{\mu,V} = -k_B \ln \Xi - k_B T \left(\frac{\partial \ln \Xi}{\partial T}\right)_{\mu,V} = \frac{J - E + W\mu}{T} = -S \tag{1.49}$$

Arguments similar to those used for the canonical ensemble show that the definition of S for a grand canonical ensemble is

$$S = -k_B \sum_{N=0}^{M} \sum_{j=1}^{r} \psi_j(N) \ln \psi_j(N) \tag{1.50}$$

The Pfaffian form associated with J is

$$dJ = \left(\frac{\partial J}{\partial \mu}\right)_{V,T} d\mu + \left(\frac{\partial J}{\partial V}\right)_{\mu,T} dV + \left(\frac{\partial J}{\partial T}\right)_{\mu,V} dT = -W\, d\mu - P\, dV - S\, dT \tag{1.51}$$

and the integral form derived from eq (1.49) is

$$J = E - TS - W\mu = F - W\mu \tag{1.52}$$

The results of the foregoing analysis obtained for the microcanonical, canonical and grand canonical ensembles are summarized in Table 1.1.

Table 1.1. *Summary of ensembles.*

Ensemble	Partition function	Potential		Pfaffian form
Microcanonical	$\Omega(E, V, N)$	$S = k_B \ln \Omega$		$T\,dS = -\mu\,dN + P\,dV + dE$
Canonical	$Z(N, V, T)$	$F = -k_B T \ln Z$		$dF = -S\,dT - P\,dV + \mu\,dN$
Grand canonical	$\Xi(\mu, V, T)$	$J = -k_B T \ln \Xi$		$dJ = -S\,dT - P\,dV - N\,d\mu$

Analysis of the properties of the macromolecule as a thermodynamic system entails the definition of a set of independent variables that can be controlled by external constraints. In the microcanonical ensemble the energy, volume and number of water molecules are fixed. In the canonical ensemble the energy fluctuates and can be defined only as an ensemble average over many distinct energy states or levels. In the grand canonical ensemble both the energy and the number of water molecules bound are subject to fluctuations. The fluctuating variable is defined in thermodynamic terms by computing the response of the partition function of the system, or equivalently the thermodynamic potential associated with it, with respect to the variables held fixed by external constraints. For instance, the change of F or J with T gives information on the entropy of the macromolecule. The change of J with μ provides the average number of water molecules bound. The properties of the macromolecule are thus derived from the response of the system to changes of temperature and the chemical potential of water.

The properties of most interest in biochemical and biophysical studies are extensive quantities such as S, E, V or N. These quantities are derived as response functions of the system to conjugate intensive quantities such as T for S and E, P for V and μ for N. Investigation of the properties of the macromolecule thus coincides with setting some external constraints that can be controlled experimentally, such as T, P or μ, and then measuring the response of the system. The basic ensembles dealt with in the foregoing analysis can be considered in this perspective. In the microcanonical ensemble all the properties of interest, i.e., the extensities E, V and N, are fixed. In this case it is not possible to measure any of these properties from fluctuations arising from changes in thermodynamic driving forces. The partition function is merely a constant that enumerates the configurations compatible with the preassigned values of E, V and N. In the canonical ensemble only V and N are fixed. In this case it is possible to derive the energy of the macromolecule from the response of F to temperature changes. The canonical ensemble provides the simplest case of a system where a basic extensive property of the macromolecule is obtained as an ensemble average. In the grand canonical ensemble only V is fixed. Here both the energy, E, and the number of water molecules bound, N or W, are derived from ensemble averages. Both E and N are inherently fluctuating quantities to be derived as response functions to changes in T and μ. Clearly, the grand canonical ensemble is the system that most closely approximates the conditions usually encountered in practice. In general, it is very difficult to control extensive quantities pertaining to the macromolecule, while it is relatively easy to control temperature, pressure and the chemical potential of various components present in solution. For this reason, a thermodynamic description of the properties of the macromolecule must be based on suitable thermodynamic potentials, or partition functions, that are defined in terms of intensive quantities only.

1.2 The generalized ensemble

The basic ensembles dealt with so far provide a conceptual framework for the derivation of any potential of interest. Let us rewrite the integral definitions of the potentials F and J as follows

$$F = E - TS \tag{1.53}$$

$$J = F - N\mu = E - TS - N\mu \tag{1.54}$$

The potential F for the canonical ensemble is obtained from the energy E of the microcanonical ensemble by subtracting the term TS. Since E is a function of N, V and S only, the transformation (1.53) makes F a function of N, V and T. If a similar transformation is applied to F by subtracting the term $N\mu$, then the potential J is obtained which is a function of μ, V and T. This transformation is equivalent to subtracting TS and $N\mu$ from the energy E. Hence, any potential can be constructed from the energy E by replacing a given extensive quantity, S, V or N, with the conjugate intensity, T, P or μ. The replacement is known as a Legendre transformation. The pair of quantities to be exchanged is subtracted from, or added to, the energy E to yield the new potential, depending on the sign of the pair in the Pfaffian form dE. Consider the differential of the energy

$$\mathrm{d}E = T\,\mathrm{d}S - P\,\mathrm{d}V + \mu\,\mathrm{d}N \tag{1.55}$$

which applies to the microcanonical ensemble. Assume that a new potential is to be derived from E where T replaces S. Upon subtracting the differential $d(TS)$ from both sides of eq (1.55) one has

$$dE - d(TS) = T\,dS - P\,dV + \mu\,dN - T\,dS - S\,dT$$
$$= -S\,dT - P\,dV + \mu\,dN \quad (1.56)$$

The right-hand side of eq (1.56) is now a function of N, V and T as desired, while the left-hand side is the differential of a new potential $E - TS$, which gives the Helmholtz free energy of a canonical ensemble. The transformation operated on E is, in fact, the same transformation that generates the canonical from the microcanonical ensemble. The potential for the grand canonical ensemble is derived from E by subtracting TS and $N\mu$.

In general, a total of $2^3 = 8$ potentials can be generated with three independent variables. The first member of this set of potentials is the energy E, which is a function of extensive quantities only. As we have seen, some of the remaining potentials reflect properties of basic ensembles. Some have very little practical interest, considering the independent variables to be controlled experimentally. An example is the potential $X = E - N\mu$, which is a function of S, V and μ. There are two potentials, however, of considerable practical interest since they refer to the properties of the macromolecule when pressure and temperature are set by external constraints. The first potential is the Gibbs free energy of the macromolecule obtained from E by interchanging P and V, and T and S, i.e.,

$$G = E - TS + PV \quad (1.57)$$

The Gibbs free energy is a function of N, P and T and the N–P–T ensemble associated with it is called *mechano-thermal*. The Pfaffian form associated with G is evidently

$$dG = -S\,dT + V\,dP + \mu\,dN \quad (1.58)$$

The partition function of the mechano-thermal ensemble can be derived in a way analogous to that considered for the canonical and grand canonical ensemble. The transformation which produces the desired potential from the energy E maps into a corresponding transformation involving the partition function of the system. The canonical partition function is derived from the microcanonical partition function by summing over all possible energy states. Likewise, summing over all possible energy states and values of N yields the grand canonical partition function. Hence, the change of a given extensive quantity with its conjugate intensity is achieved by summing over all possible values allowable to the extensive quantity of interest. This turns out to be the result of finding the most probable distribution using undetermined multipliers, as shown in Section 1.1. To obtain the partition function for the mechano-thermal ensemble from the microcanonical ensemble, we have to sum over all energy states at constant T, and over all possible volume states at constant P. Due to the continuous nature of the V variable, the sum over V must be replaced by an integral from the lowest (V_{min}) to the highest (V_{max}) value of V. Mathematically, the same result is obtained by summing from zero to infinity if we assume that macromolecular configurations with $V < V_{min}$ or $V > V_{max}$ have vanishingly small probabilities of existence and make no contribution to the integral. The partition function $Y(P, T, N)$ for the mechano-thermal ensemble is

$$Y(P,\ T,\ N) = \int_0^\infty \sum_{j=1}^r \exp\left[-\frac{PV + E_j(N,\ V)}{k_B T}\right] dV$$
$$= \int_0^\infty Z(V,\ T,\ N) \exp\left(-\frac{PV}{k_B T}\right) dV \quad (1.59)$$

and is mathematically identical to the Laplace transform of the canonical partition function in V space. Again, the connection between Y and G is of the form

$$G(N,\ P,\ T) = -k_B T \ln Y(N,\ P,\ T) \qquad (1.60)$$

The macromolecule as a mechano-thermal ensemble is indeed a mechano-thermal device that can assume a number of different energy and volume states. This is exactly the scenario of interest in biology. The Gibbs free energy refers to the ensemble of such states when the number of water molecules bound is held fixed.

The set of response functions derived from the potential G, or equivalently from the partition function Y, gives information on some important extensive properties of the macromolecule, such as V and E. The relevant expressions are

$$V = -k_B T \left(\frac{\partial \ln Y}{\partial P}\right)_{N,T} = \left(\frac{\partial G}{\partial P}\right)_{N,T} \qquad (1.61)$$

$$E + PV = k_B T^2 \left(\frac{\partial \ln Y}{\partial T}\right)_{N,P} = -T^2 \left(\frac{\partial \frac{G}{T}}{\partial T}\right)_{N,P} = \left(\frac{\partial \frac{G}{T}}{\partial \frac{1}{T}}\right)_{N,P} \qquad (1.62)$$

$$\mu = -k_B T \left(\frac{\partial \ln Y}{\partial N} \right)_{P,T} = \left(\frac{\partial G}{\partial N} \right)_{P,T} \qquad (1.63)$$

The important new feature in the mechano-thermal ensemble is that the volume of the macromolecule can be derived from the response of the system to pressure. The response to temperature is $E + PV$. Using the foregoing arguments this response can be considered as a new potential where V has been replaced by its conjugate P as an independent variable. The potential

$$H = E + PV \qquad (1.64)$$

is the enthalpy of the macromolecule and describes the properties of an N–P–S ensemble. The relationship in eq (1.62) is an important thermodynamic expression known as the Gibbs–Helmholtz equation, which relates the enthalpy to a change in the Gibbs free energy of the system. Combination of eqs (1.57) and (1.64) yields the key thermodynamic relation

$$G = H - TS \qquad (1.65)$$

from which any of the three thermodynamic quantities G, H and S can be derived from knowledge of the other two. We also have the property, analogous to eqs (1.17) and (1.49), that

$$\left(\frac{\partial G}{\partial T} \right)_{N,P} = -k_B \ln Y - k_B T \left(\frac{\partial \ln Y}{\partial T} \right)_{N,P} = \frac{G - E - PV}{T} = -S \quad (1.66)$$

which complements the Gibbs–Helmholtz equation.

The second potential of interest is obtained from E by interchanging all extensive variables with their conjugate intensities, i.e.,

$$\Pi = E - TS + PV - N\mu \qquad (1.67)$$

We shall see that this potential, to be referred to as the generalized potential Π, coincides with the chemical potential of the macromolecule μ_m. The potential Π is a function of all intensive quantities μ, P and T and the μ–P–T ensemble associated with it is called *generalized*. The Pfaffian form associated with Π is

$$d\Pi = -S\,dT + V\,dP - N\,d\mu \qquad (1.68)$$

The partition function $\Psi(P, T, \mu)$ of the generalized ensemble can be derived from that of the microcanonical ensemble by summing over all possible energy and volume states, as well as over all possible values of N. The solution is evidently

$$\Psi(P, T, \mu) = \int_0^\infty \sum_{j=1}^r \sum_{N=0}^M \exp \left[\frac{N\mu - PV - E_j(N, V)}{k_B T} \right] dV \qquad (1.69)$$

with

$$\Pi(\mu, P, T) = -k_B T \ln \Psi(\mu, P, T) \qquad (1.70)$$

Again, it is quite instructive to consider two alternative forms of Ψ. One is obtained by rewriting Ψ in terms of energy levels and their degeneracy as follows

$$\Psi(P, T, \mu) = \int_0^\infty \sum_{j=1}^s \sum_{N=0}^M \Omega(N, V, E_j) \exp\left[\frac{N\mu - PV - E_j(N, V)}{k_B T}\right] dV$$

$$(1.71)$$

This expression reveals the 'anatomy' of a generalized ensemble. The original microcanonical kernel Ω is subject to a series of transformations leading to stepwise substitution of each extensive quantity with its conjugate intensity. Each substitution is brought out by filtering Ω with an exponential term containing the thermodynamic pair of interest. Grouping terms in eq (1.71) according to a particular choice of independent variables generates various equivalent forms of Ψ. The form of considerable practical interest is a polynomial expansion in the water activity w, as seen for the grand canonical ensemble, i.e.,

$$\Psi(P, T, w) = \sum_{N=0}^M Y(N, P, T) w^N \qquad (1.72)$$

The coefficients of this expansion are the mechano-thermal partition functions related to the Gibbs free energy of the macromolecule. In the expansion of the generalized ensemble, each intermediate state of the macromolecule defined by a given number of water molecules bound is a mechano-thermal device that can assume a number of different energy and volume states. Each intermediate enters the generalized partition function with a weight set by the Gibbs free energy of the corresponding mechano-thermal device. The whole ensemble is generated by expanding the number of mechano-thermal devices along the N coordinate.

The generalized ensemble provides the ideal system for studying the properties of the macromolecule, or any other component of interest. The importance of this ensemble was first pointed out by Stockmayer (1950). All the intensive quantities are controlled by external constraints and the extensive quantities reflecting the behavior of the macromolecule are determined from the response of Π to thermodynamic driving forces such as P, T and μ. We have for these quantities the set of relationships

$$V = -k_B T \left(\frac{\partial \ln \Psi}{\partial P}\right)_{\mu, T} = \left(\frac{\partial \Pi}{\partial P}\right)_{\mu, T} \qquad (1.73)$$

$$E + PV - N\mu = k_B T^2 \left(\frac{\partial \ln \Psi}{\partial T}\right)_{\mu,P} = -T^2 \left(\frac{\partial \frac{\Pi}{T}}{\partial T}\right)_{\mu,P} \quad (1.74)$$

$$N = k_B T \left(\frac{\partial \ln \Psi}{\partial \mu}\right)_{P,T} = -\left(\frac{\partial \Pi}{\partial \mu}\right)_{P,T} \quad (1.75)$$

The change of Π with T gives the entropy of the macromolecule, as expected

$$\left(\frac{\partial \Pi}{\partial T}\right)_{\mu,P} = -k_B \ln \Psi - k_B T \left(\frac{\partial \ln \Psi}{\partial T}\right)_{\mu,P}$$

$$= \frac{\Pi - E - PV + N\mu}{T} = -S \quad (1.76)$$

The entropy is related to the probability $\psi_j(N, V)$ of observing a particular configuration of the macromolecule with preassigned values of N, V and E_j. The expression for $\psi_j(N, V)$ derived from application of the method of undetermined multipliers to the generalized ensemble is analogous to that already obtained for the canonical and grand canonical ensembles. The analytical form of $\psi_j(N, V)$ is given by the ratio

$$\psi_j(N, V) = \frac{\Omega(N, V, E_j) \exp\left[\dfrac{N\mu - PV - E_j(N, V)}{k_B T}\right]}{\displaystyle\int_0^\infty \sum_{j=1}^{s} \sum_{N=0}^{M} \Omega(N, V, E_j) \exp\left[\dfrac{N\mu - PV - E_j(N, V)}{k_B T}\right] dV} \quad (1.77)$$

Of particular importance is to cast eq (1.77) in terms of the probability ψ_N of finding the Nth ligated intermediate with N water molecules bound to it, i.e.,

$$\psi_N = \frac{Y(N, P, T)w^N}{\displaystyle\sum_{N=0}^{M} Y(N, P, T)w^N} = \frac{Y(N, P, T)w^N}{\Psi(w, P, T)} \quad (1.78)$$

The entropy of the macromolecule assumes a relatively simple form in terms of the ψs

$$S = -k_B \sum_{N=0}^{M} \psi_N \ln \psi_N \quad (1.79)$$

This relationship and eqs (1.73)–(1.75) provide the analytical tools to define the important extensive quantities of the macromolecule under consideration.

The importance of the generalized ensemble and the potential Π in the description of the properties of the macromolecule becomes even more evident once the exact nature of Π is clarified. This demands an explicit definition of the energy E. We have given the Pfaffian form of the energy in eq (1.29), but the definition of Π, or G, H, F and any other potential, requires the definition of E in its integral form. Let us rewrite eq (1.29) as follows

$$dE = T\,dS - P\,dV + \mu\,dN + \mu_m\,dN_m \qquad (1.80)$$

where the suffix m refers to the macromolecule. Since our system is the macromolecule itself, N_m is constant and its differential is zero. The term $\mu_m\,dN_m$ added to eq (1.29) to obtain eq (1.80) makes no contribution to dE and the results of the foregoing sections remain perfectly valid, as long as N_m is kept constant throughout. Specifically $N_m = 1$, since our system contains only one macromolecule. Assume now that the value of N_m is multiplied by an arbitrary constant $\lambda > 1$. This results in an expansion of our system. All other extensive quantities increase by the same amount λ, since we have to take into account λN water molecules bound, a volume λV, an entropy λS and an energy λE. The expansion, however, is inconsequential on the properties of our system if we define a new set of variables scaled by a factor λ with respect to the original extensities in eq (1.80). This leads to the important condition for the integral form of the energy E (Stanley, 1971; Hankey and Stanley, 1972; Weinhold, 1975a,b)

$$E(\lambda S, \lambda V, \lambda N, \lambda N_m) = \lambda E(S, V, N, N_m) \qquad (1.81)$$

The energy of a system where all extensities have been scaled by a factor λ is λ times the energy of the original system with unscaled variables. Mathematically, eq (1.81) states that the energy E is a first-order homogeneous function. It then follows from Euler's theorem on homogeneous functions (Pearson, 1990) that

$$E = \sum_{i=1}^{t} \left(\frac{\partial E}{\partial X_i}\right)_{\{X_{j \neq i}\}} X_i \qquad (1.82)$$

where X_i denotes one of the t independent extensive quantities defining E. The partial derivatives in eq (1.82) can be calculated from the Pfaffian form (1.80) so that

$$E = TS - PV + N\mu + N_m\mu_m \qquad (1.83)$$

which gives the integral expression for the energy E. It is now easy to compute any other potential from eq (1.83) to obtain

$$H = E + PV = TS + N\mu + \mu_\mathrm{m} \tag{1.84}$$

$$F = E - TS = -PV + N\mu + \mu_\mathrm{m} \tag{1.85}$$

$$J = E - TS - N\mu = -PV + \mu_\mathrm{m} \tag{1.86}$$

$$G = E - TS + PV = N\mu + \mu_\mathrm{m} \tag{1.87}$$

$$\Pi = E - TS + PV - N\mu = \mu_\mathrm{m} \tag{1.88}$$

where it should be remembered that $N_\mathrm{m} = 1$ by construction. Hence, from the integral definition of E, we have proved that the potential of the generalized ensemble, Π, is the chemical potential of the macromolecule itself.

The Pfaffian form involving all intensive quantities of the system is

$$S\,\mathrm{d}T - V\,\mathrm{d}P + N\,\mathrm{d}\mu + \mathrm{d}\mu_\mathrm{m} = 0 \tag{1.89}$$

Unlike the energy E, which is a function of all extensive quantities, the potential generated from E in eq (1.83) by replacing all extensities with their conjugate intensities, including the variables referring to the macromolecule, does not exist. Since P, T, μ and μ_m are all the variables to be taken into account in the system we have constructed, it follows that it is not possible to define a potential, and hence the state of a system, in terms of intensive quantities only. The relationship (1.89) that encapsulates this fundamental principle is the Gibbs–Duhem equation. Generalization of eq (1.89) to the case involving an arbitrary number of components is straightforward. The physical significance of the Gibbs–Duhem equation is as follows. Our two-component system cannot be uniquely defined in terms of P, T and the activity of water and the macromolecule. A solution of the macromolecule of a given activity (or concentration), at atmospheric pressure and 298.15 K does not define a thermodynamic system to which a potential can be associated. Knowledge of these conditions by no means allows us to derive how many water molecules are bound to the macromolecule (N), how many macromolecules are present (N_m), or the volume and energy of the system. To obtain this information we need to fix at least one of the extensive quantities involved, which is equivalent to specifying the size of the system. All the potentials defined in the foregoing analysis, except Π, contain the term $\mathrm{d}N_\mathrm{m}$ in their Pfaffian form. Since N_m is fixed, $\mathrm{d}N_\mathrm{m}$ makes no contribution and all potentials are functions of three independent variables. However, the importance of fixing N_m is revealed by the fact that its conjugate variable, μ_m, itself becomes a member of the set of potentials, and is itself a function of three independent variables. Thus, fixing one extensity results in the generation of a set of potentials that is iso-homogeneous. If one of the extensities is not fixed, then there are potentials such as E containing four independent variables and a null potential corresponding to the Gibbs–Duhem equation. In a system containing N extensive variables and a corresponding number of conjugate intensities, there can only be $N - 1$ independent intensities, as dictated by the Gibbs–Duhem equation. Hence, fixing one of the extensities is a necessary condition to define the state of the system. This operation yields a total of $N - 1$ independent variables, extensive or intensive.

1.3 The laws of thermodynamics

The generalized ensemble depicts the macromolecule as an open, isothermal system at constant pressure. These are the conditions in which macromolecules function *in vivo* and for this reason they are of most interest in practical applications. As a result of the properties of the generalized ensemble, thermodynamic driving forces act upon the system and modify its state in a way that can be predicted from the analytical behavior of a set of response functions. These driving forces may be expected to interfere with each other if the response of the macromolecule, say to P, modifies its response to T or μ. Thermodynamics provides a number of rules for the interference to take place. These rules are at the basis of many important regulatory phenomena in biology and prove useful for understanding and developing basic mechanisms of action of macromolecules under equilibrium conditions. The Pfaffian form of the energy for the system of interest composed of the macromolecule and the solvent embodies the results of the first two laws of thermodynamics. The first law establishes the conservation of energy and the fact that for any reversible transformation the change in the energy E depends solely on the initial and final states, as implied for a state function. The second law establishes that, when all other extensive variables are fixed, the energy of a system at equilibrium is at a minimum and that any perturbation of the equilibrium state triggers a response from the system to restore the initial conditions. We shall now examine the important consequences of these two laws.

Mathematically, the first law states that dE is an exact differential, since E is a state function. Let us write the familiar Pfaffian form

$$dE = T\,dS - P\,dV + \mu\,dN \qquad (1.90)$$

in terms of the three independent variables S, V and N, as done in Section 1.1. We maintain that N_m is constant and that the macromolecule is our system by construction. Euler's theorem on exact differentials (Wilson, 1911; Pearson, 1990) gives rise to a number of relationships between different independent variables. In the case of dE the reciprocity relationships are

$$\left(\frac{\partial T}{\partial V}\right)_{N,S} = -\left(\frac{\partial P}{\partial S}\right)_{N,V} \tag{1.91}$$

$$\left(\frac{\partial T}{\partial N}\right)_{V,S} = \left(\frac{\partial \mu}{\partial S}\right)_{N,V} \tag{1.92}$$

$$\left(\frac{\partial \mu}{\partial V}\right)_{N,S} = -\left(\frac{\partial P}{\partial N}\right)_{S,V} \tag{1.93}$$

These relationships are also known as Maxwell's relationships. A similar set of relationships exists for any of the potentials derived in the foregoing sections, since all of them are state functions like E and have exact differentials. Here we choose Π due to the importance of the generalized ensemble in the description of the properties of the macromolecule. The Pfaffian form can be rewritten in a particularly convenient form as follows

$$-d\Pi = S\,dT - V\,dP + N\,d\mu \tag{1.94}$$

The exact differential $-d\Pi$ yields a form similar to dE, where all extensive independent quantities are replaced by their conjugate intensities. The Maxwell's relationships associated with $-\Pi$ are

$$\left(\frac{\partial S}{\partial P}\right)_{\mu,T} = -\left(\frac{\partial V}{\partial T}\right)_{P,\mu} \tag{1.95}$$

$$\left(\frac{\partial S}{\partial \mu}\right)_{P,T} = \left(\frac{\partial N}{\partial T}\right)_{P,\mu} \tag{1.96}$$

$$\left(\frac{\partial N}{\partial P}\right)_{\mu,T} = -\left(\frac{\partial V}{\partial \mu}\right)_{P,T} \tag{1.97}$$

We will discuss these relationships in order. The first relationship connects the entropy and the volume of the macromolecule. The change in entropy due to a change in pressure, while keeping μ and T constant, is the same as the change in volume (with opposite sign) due to a change in temperature, while μ and P are constant. Hence, if pressure affects the entropy of the macromolecule at constant T and μ, then temperature must affect the volume of the macromolecule at constant μ and P. In this case entropy and volume are linked. On the other hand, if pressure has no effect on the entropy, then temperature cannot affect the volume of the macro-molecule. The second relationship connects entropy and the number of water molecules bound. If a change of the entropy of the macromolecule is observed upon a change of the chemical potential of water, then temperature must affect the amount of water bound to the macro-molecule. Entropy and N are linked. The last relationship applies to linkage effects involving N and V. If pressure affects the number of water molecules bound to the macromolecule, then a change in the chemical potential of water must affect the volume of the macromolecule.

We begin to perceive a rather basic and important aspect of the thermodynamics of macromolecular systems from the Maxwell relation-ships derived for the potential Π. Different driving forces acting on the macromolecule, such as P, T or μ, may bring about changes in the properties as revealed by specific response functions. These changes are not arbitrary, but obey some important reciprocity rules. Consider an extensive property such as N, the number of water molecules bound to the macromolecule. The response of N to driving forces such as T or P, that are not conjugate to N, provides information on how μ, the driving force conjugate to N, will affect S or V, the extensive quantities conjugate to T and P. Therefore, one can use a particular response function to predict others that are difficult to access experimentally. For example, if N is measured as a function of P and a change is observed, then it can be concluded that the binding of water will affect the volume of the macro-molecule. This is a very useful result, considering that measurements of V are not as direct and feasible as measurements of N. Likewise, assume that we are interested in the effect of μ on the entropy of the macro-molecule, but there is no direct way of measuring the effect. Measure-ments of N as a function of T will reveal whether or not a change of S due to a change of μ is to be expected.

The second law establishes a sort of variational principle of substantial importance for thermodynamic response functions. The first law establi-shes a connection between transformations involving pairs of conjugate variables through the Maxwell relationships, but it says nothing about the direction of the change. According to the first law, any transformation is allowed to take place, as long as conservation of energy is not violated. The second law imposes further restrictions and adds predictive power to the thermodynamic approach. This law states that, for a given set of extensive variables held fixed, only those transformations that minimize energy can take place in a thermodynamic system at equilibrium. Mathe-matically, the first two laws can be expressed in a simple and most elegant way using matrix algebra (Gibbs, 1928; Weinhold, 1975a,b). Consider the definition of the energy function E in terms of a set of independent extensities $\{X_i\} = X_1, X_2, \ldots X_r$ and let f_i be the intensive variable conjugate to X_i. Consider the $r \times r$ matrix $\mathbf{G}$ with elements

$$g_{ij} = \left(\frac{\partial f_i}{\partial X_j}\right)_{\{X_{k \neq j}\}} \qquad (i, j = 1, 2, \ldots r) \qquad (1.98)$$

where the suffix $\{X_{k \neq j}\}$ denotes that all Xs but X_j are kept fixed. The first law states that the matrix $\mathbf{G}$ is symmetric. In other words, the equality

$$g_{ij} = g_{ji} \qquad (1.99)$$

gives the Maxwell relationships in terms of the elements of $\mathbf{G}$. The second law provides additional information on the properties of $\mathbf{G}$. The energy E is a minimum if, and only if, the matrix $\mathbf{G}$ is positive-definite (Gantmacher, 1959). This implies that all principal minors of $\mathbf{G}$ are positive. A principal minor is a determinant obtained from $\mathbf{G}$ by deleting rows and columns crossing over a diagonal element. In the specific case of our system, we have $r = 3$ and we can identify S with X_1, V with X_2 and N with X_3. Hence, $f_1 \equiv T$, $f_2 \equiv -P$ and $f_3 \equiv \mu$. The matrix $\mathbf{G}$ is

$$\mathbf{G} = \begin{pmatrix} g_{11} & g_{12} & g_{13} \\ g_{21} & g_{22} & g_{23} \\ g_{31} & g_{32} & g_{33} \end{pmatrix} \qquad (1.100)$$

The positive-definite nature of $\mathbf{G}$ implies that *all* of the following conditions are satisfied

$$g_{11} > 0 \qquad (1.101a)$$

$$g_{22} > 0 \qquad (1.101b)$$

$$g_{33} > 0 \qquad (1.101c)$$

$$g_{11}g_{22} - g_{12}g_{21} > 0 \qquad (1.102a)$$

$$g_{11}g_{33} - g_{13}g_{31} > 0 \qquad (1.102b)$$

$$g_{22}g_{33} - g_{23}g_{32} > 0 \qquad (1.102c)$$

$$g_{11}g_{22}g_{33} + g_{12}g_{23}g_{31} + g_{13}g_{21}g_{32} - g_{13}g_{22}g_{31} - g_{12}g_{21}g_{33} - g_{11}g_{23}g_{32} > 0 \qquad (1.103)$$

The inequalities (1.101a)–(1.101c) establish that the diagonal elements of $\mathbf{G}$ are always positive. A diagonal element is by definition the response of a given intensive variable to a change in its conjugate extensity. Hence, an increase in entropy implies that temperature has increased, since $g_{11} = \partial T/\partial S > 0$. Likewise, an increase in the number of water molecules bound implies an increase in the chemical potential of water, since $g_{33} = \partial \mu/\partial N > 0$. Finally a decrease in pressure is linked to an increase in the volume of the macromolecule, since $g_{22} = -\partial P/\partial V > 0$.

While the diagonal elements of **G** involve pairs of conjugate variables, the off-diagonal terms reflect the mutual interference, or linkage, of different independent variables. In addition to the criterion involving diagonal elements, the second law defines constraints also for the off-diagonal elements, or linkage coefficients. The inequalities (1.102a)–(1.102c) demand that for any two pairs of conjugate variables the product of the diagonal elements be larger than the product of the linkage coefficients. Similar restrictions apply to all other principal minors, including the determinant of **G** in eq (1.103), which is itself a principal minor.

Other conditions embodied by the second law are derived from consideration of the potential Π as a function of the independent intensities $\{f_1\} = f_1, f_2, \ldots f_r$. The $r \times r$ matrix **G**$'$ associated with $-\Pi$ has elements

$$g'_{ij} = \left(\frac{\partial X_i}{\partial f_j}\right)_{\{f_{k \neq j}\}} \qquad (i, j = 1, 2, \ldots r) \qquad (1.104)$$

For our system of interest **G**$'$ is given by

$$\mathbf{G}' = \begin{pmatrix} g'_{11} & g'_{12} & g'_{13} \\ g'_{21} & g'_{22} & g'_{23} \\ g'_{31} & g'_{32} & g'_{33} \end{pmatrix} \qquad (1.105)$$

Again, the first law states that the matrix **G**$'$ is symmetric, so that $g'_{ij} = g'_{ji}$, while the second law imposes on **G**$'$ a set of conditions analogous to eqs (1.101)–(1.103), i.e.,

$$g'_{11} > 0 \qquad (1.106a)$$

$$g'_{22} > 0 \qquad (1.106b)$$

$$g'_{33} > 0 \qquad (1.106c)$$

$$g'_{11}g'_{22} - g'_{12}g'_{21} > 0 \qquad (1.107a)$$

$$g'_{11}g'_{33} - g'_{13}g'_{31} > 0 \qquad (1.107b)$$

$$g'_{22}g'_{33} - g'_{23}g'_{32} > 0 \qquad (1.107c)$$

$$g'_{11}g'_{22}g'_{33} + g'_{12}g'_{23}g'_{31} + g'_{13}g'_{21}g'_{32} - g'_{13}g'_{22}g'_{31} - g'_{12}g'_{21}g'_{33} - g'_{11}g'_{23}g'_{32} > 0 \qquad (1.108)$$

As in the case of **G**, we have for **G**$'$ conditions involving diagonal elements and off-diagonal linkage coefficients. The diagonal elements of **G**$'$ play a fundamental role in the thermodynamic description of the system. The response of S to T, at constant P and μ

$$g'_{11} = \left(\frac{\partial S}{\partial T}\right)_{P,\mu} = \frac{C_{P,\mu}}{T} \qquad (1.109)$$

gives the *heat capacity* of the macromolecule, divided by the absolute temperature. The heat capacity is a measure of the heat absorbed reversibly by the macromolecule due to an increase in temperature. The second law demands that the heat capacity is always positive. Hence, the macromolecule absorbs heat when T increases and releases heat when T decreases. A response function analogous to the heat capacity is the *binding capacity* (Di Cera, Gill and Wyman, 1988a)

$$g'_{33} = \left(\frac{\partial N}{\partial \mu} \right)_{P,T} = B_{P,T} \tag{1.110}$$

which represents the change in the number of water molecules bound due to a change in the chemical potential of water. The second law states that the binding capacity is always positive and that an increase/decrease of the chemical potential of water always increases/decreases the number of water molecules bound. The third diagonal element gives the change of V with respect to P, at constant T and μ

$$g'_{22} = -\left(\frac{\partial V}{\partial P} \right)_{\mu,T} = \kappa_{\mu,T} V \tag{1.111}$$

and represents the *compressibility* of the macromolecule times the volume. The compressibility is always positive, since an increase/decrease in pressure always decreases/increases the volume of the macromolecule.

The second law guarantees the stability of the equilibrium state against small, adiabatic perturbations. Assume, for instance, that the equilibrium of the system is perturbed by a slight increase in pressure. This change causes the volume of the macromolecule to decrease. The response of the macromolecule relieves the external 'stress' induced by the perturbation and restores the original conditions. The pressure is, in fact, inversely proportional to the volume of the system not occupied by the macromolecule and the solvent (outer volume), so that a decrease in the volume of the macromolecule opposes the increase in pressure by increasing the outer volume. If the second law were violated and the macromolecule were to react by increasing its volume, the outer volume would shrink and produce a further increase in pressure. The sequence of reactions triggered by the anomalous response of the macromolecule would then make the pressure grow without bounds in a spontaneous fashion, thereby leading to explosion of the system. Similar arguments apply in the case of the response of the macromolecule to T and μ. An increase in temperature causes the macromolecule to absorb heat, thereby minimizing the rise of temperature. An increase in the chemical potential of water causes water molecules to bind to the macromolecule. If the number of water molecules bound were to decrease with increasing μ, thereby violating the second law, a small increase of μ arising from a spontaneous fluctuation would eventually cause all bound water molecules to dissociate from the macromolecule. A small decrease of μ would cause all water molecules in the bulk solvent to condense on the macromolecule. Thermodynamic stability, on the other hand, makes the macromolecule act as a buffer that absorbs water when μ increases and releases water when μ decreases, in either case opposing the effect of the perturbation acting on the system. The response to T is analogous, so that heat is absorbed/released when T increases/decreases. The natural tendency of the system to react to the perturbation and restore the original conditions of equilibrium is the mark of thermodynamic stability and forms the basis of Le Chatelier's principle on the restorative tendency of the equilibrium state (Prigogine and Defay, 1954).

In addition to the diagonal elements of **G** and **G**′, other principal minors are subject to the constraints imposed by the second law. It is quite instructive to show that these minors are analogous to the response functions defined above. Consider the set of exact differentials involving the extensive quantities of the macromolecule under consideration

$$dS = \left(\frac{\partial S}{\partial T}\right)_{P,\mu} dT + \left(\frac{\partial S}{\partial P}\right)_{T,\mu} dP + \left(\frac{\partial S}{\partial \mu}\right)_{P,T} d\mu \qquad (1.112)$$

$$dV = \left(\frac{\partial V}{\partial T}\right)_{P,\mu} dT + \left(\frac{\partial V}{\partial P}\right)_{T,\mu} dP + \left(\frac{\partial V}{\partial \mu}\right)_{P,T} d\mu \qquad (1.113)$$

$$dN = \left(\frac{\partial N}{\partial T}\right)_{P,\mu} dT + \left(\frac{\partial N}{\partial P}\right)_{T,\mu} dP + \left(\frac{\partial N}{\partial \mu}\right)_{P,T} d\mu \qquad (1.114)$$

This set is equivalent to the transformation

$$dX_1 = g'_{11} df_1 + g'_{12} df_2 + g'_{13} df_3 \qquad (1.115)$$

$$dX_2 = g'_{21} df_1 + g'_{22} df_2 + g'_{23} df_3 \qquad (1.116)$$

$$dX_3 = g'_{31} df_1 + g'_{32} df_2 + g'_{33} df_3 \qquad (1.117)$$

that yields the set of extensies from the set of intensies through the matrix **G**′. The transformation exists and is unique since the second law guarantees that the determinant of the matrix **G**′ is positive (Gibbs, 1928; Shaw, 1935; Crawford, 1950, 1955; Weinhold, 1975a, b). Assume now that the variable X_3 is held fixed, so that $dX_3 = 0$. This allows df_3 to be substituted in eqs (1.115) and (1.116) using eq (1.117) to yield a new transformation

$$dX_1 = {}^3g'_{11}\,df_1 + {}^3g'_{12}\,df_2 \tag{1.118}$$

$$dX_2 = {}^3g'_{21}\,df_1 + {}^3g'_{22}\,df_2 \tag{1.119}$$

The coefficients of this new transformation define a 'contracted' form ${}^3\mathbf{G}'$ of the original matrix $\mathbf{G}'$, and denote response functions subject to the condition $X_3 =$ constant. For example, ${}^3g'_{11}$ denotes the change of S with T when P and N are kept constant, as opposed to g'_{11} that represents the analogous change when P and μ are constant. Likewise, ${}^3g'_{22}$ is the change of V with $-P$ at constant T and N, as opposed to g'_{22} that defines the same change at constant T and μ. The contraction of the matrix $\mathbf{G}'$ along the variable $X_3 = N$ has operated the substitution $f_3 \to X_3$ in the definition of the response functions involving the thermodynamic pairs f_1, X_1 and f_2, X_2. The connection between the elements of ${}^3\mathbf{G}'$ and those of $\mathbf{G}$ is

$$ {}^3g'_{ij} = g'_{ij} - \frac{g'_{i3}g'_{3j}}{g'_{33}} \quad (i, j = 1, 2) \tag{1.120}$$

Since the matrix ${}^3\mathbf{G}'$ has been derived from the matrix $\mathbf{G}'$ by fixing the state of one thermodynamic variable in the system, it is itself subject to the first two laws and is itself symmetric and positive-definite. It follows from the first law that ${}^3g'_{12} = {}^3g'_{21}$, or that

$$\left(\frac{\partial S}{\partial P}\right)_{N,T} = -\left(\frac{\partial V}{\partial T}\right)_{P,N} \tag{1.121}$$

The only difference between the Maxwell relationship above and eq (1.95) is in the independent variable N which replaces μ. The following set of inequalities is a consequence of the second law

$$ {}^3g'_{11} = \left(\frac{\partial S}{\partial T}\right)_{P,N} = \frac{C_{P,N}}{T} > 0 \tag{1.122}$$

$$ {}^3g'_{22} = -\left(\frac{\partial V}{\partial P}\right)_{T,N} = \kappa_{T,N} V > 0 \tag{1.123}$$

$$ {}^3g'_{11}{}^3g'_{22} - {}^3g'_{12}{}^3g'_{21} > 0 \tag{1.124}$$

The heat capacity and compressibility of the macromolecule retain their fundamental property of being positive. This result is a straightforward consequence of the properties of $\mathbf{G}'$. In fact, eq (1.120) implies that ${}^3g'_{11}$ is the ratio between two principal minors of $\mathbf{G}'$, i.e.,

$$ {}^3g'_{11} = \frac{g'_{11}g'_{33} - g'_{13}g'_{31}}{g'_{33}} \tag{1.125}$$

that are themselves always positive. The same argument applies to ${}^3g'_{22}$. It also follows from eq (1.125) that $g'_{11} \geqslant {}^3g'_{11}$, since the product $g'_{13}g'_{31}$ is always positive due to the Maxwell relationships. Hence,

$$C_{P,\mu} \geqslant C_{P,N} \tag{1.126}$$

The heat capacity of the macromolecule computed at a constant chemical potential of water is never smaller than the heat capacity computed at a constant number of water molecules bound. If the temperature is increased/decreased at constant μ the heat absorbed/released by the macromolecule is never smaller than that absorbed/released while keeping N constant. Likewise,

$$\kappa_{T,\mu} \geqslant \kappa_{T,N} \tag{1.127}$$

The compressibility of the macromolecule computed at constant μ is never smaller than that computed at constant N. Had we chosen the condition $X_2 = $ constant in (1.116) we would have obtained

$$C_{P,\mu} \geqslant C_{V,\mu} \tag{1.128}$$

$$B_{P,T} \geqslant B_{V,T} \tag{1.129}$$

Likewise, letting $X_1 = $ constant in eq (1.115) leads to

$$\kappa_{T,\mu} \geqslant \kappa_{S,\mu} \tag{1.130}$$

$$B_{P,T} \geqslant B_{P,S} \tag{1.131}$$

The inequalities (1.128) and (1.130) provide more information on the heat and volume changes of the macromolecule. The binding capacity is subject to similar constraints and at constant T or P is never smaller than the binding capacity computed at constant S or V. In general, for any response function expressing the change in a given extensity due to a change in the conjugate intensity, the change computed at constant f, where f is an arbitrary intensive quantity, is never smaller than that computed at constant X, where X is the extensity conjugate to f, or

$$g_{ii}' \geqslant {}^j g_{ii}' \tag{1.132}$$

where $j \neq i$. It is easy to verify by induction that the same inequality holds for the case where an arbitrary number of extensities is kept constant.

1.4 Fluctuations

It is quite instructive to note that the elements of the matrix $\mathbf{G}'$, as well as those of $\mathbf{G}$, are related to the fluctuations of important thermodynamic quantities. Consider the polynomial expression of the partition function Ψ for the generalized ensemble (1.72). The first derivative of $\ln \Psi$ with respect to $\ln w$, at constant P and T, is equivalent to differentiation with respect to μ. The result is the average number of water molecules bound, W, as follows

$$W = \left(\frac{\partial \ln \Psi}{\partial \ln w}\right)_{P,T} = \frac{\sum\limits_{j=0}^{M} j A_j w^j}{\sum\limits_{j=0}^{M} A_j w^j} = \sum\limits_{j=0}^{M} j \psi_j = \langle j \rangle \qquad (1.133)$$

Here j replaces the index N in eq (1.72), while $A_j = Y(j, P, T)$ is the mechano-thermal partition function for the configuration of the macromolecule with j water molecules bound. The derivative of W with respect to $\ln w$, $\partial W/\partial \ln w$, at constant P and T, gives the binding capacity B. This is a simpler definition than that given in eq (1.110), since the $k_B T$ term in the denominator has been omitted. This term is, of course, inconsequential when T is constant. The relevant expression written in terms of the partition function of the generalized ensemble is obtained by differentiation of eq (1.133) as follows

$$B = \left(\frac{\partial^2 \ln \Psi}{\partial \ln w^2}\right)_{P,T} = \frac{\sum\limits_{j=0}^{M} j^2 A_j w^j}{\sum\limits_{j=0}^{M} A_j w^j} - \left(\frac{\sum\limits_{j=0}^{M} j A_j w^j}{\sum\limits_{j=0}^{M} A_j w^j}\right)^2$$

$$= \sum\limits_{j=0}^{M} j^2 \psi_j - \left(\sum\limits_{j=0}^{M} j \psi_j\right)^2 = \langle j^2 \rangle - \langle j \rangle^2 \quad (1.134)$$

The binding capacity is therefore the variance of the distribution of ligated intermediates around the mean value W defined by eq (1.133), or a measure of the fluctuations associated with the average value W.[†]

Similar arguments show that the compressibility is related to the fluctuations of the volume. Upon expressing Ψ as the Laplace transform of the grand canonical partition function of the system

$$\Psi(P, T, \mu) = \int_0^\infty \exp(-PV/k_B T) \Xi(V, T, \mu) \, dV \qquad (1.135)$$

[†] The connection between the derivative of W and the variance of the fluctuations of the molecules bound was presented by Linderstrøm-Lang at a lecture at Harvard in 1939 (Cohn and Edsall, 1943; Wyman and Gill, 1990) and believed by many to be derived by him originally. Tolman (1938) points out that eq (1.134) was first derived by Pauli in 1927 (Pauli, 1927). It turns out, however, that eq (1.134) is as old as statistical mechanics and was first published by Gibbs in 1902 (Gibbs, 1902). The derivation of eq (1.134) is also implicit in Gibbs' monumental paper on the equilibrium of heterogeneous substances (Gibbs, 1875).

the volume is derived as

$$V = -k_{\mathrm{B}}T\left(\frac{\partial \ln \Psi}{\partial P}\right)_{T,\mu} = \frac{\int_0^\infty V \exp\left(-PV/k_{\mathrm{B}}T\right)\Xi(V, T, \mu)\,\mathrm{d}V}{\int_0^\infty \exp\left(-PV/k_{\mathrm{B}}T\right)\Xi(V, T, \mu)\,\mathrm{d}V} = \langle V \rangle$$

(1.136)

The compressibility at constant T and μ is

$$\kappa_{T,\mu} = -\frac{1}{V}\left(\frac{\partial V}{\partial P}\right)_{T,\mu} = \frac{1}{Vk_{\mathrm{B}}T}\left\{\frac{\int_0^\infty V^2 \exp\left(-\frac{PV}{k_{\mathrm{B}}T}\right)\Xi(V, T, \mu)\,\mathrm{d}V}{\int_0^\infty \exp\left(-\frac{PV}{k_{\mathrm{B}}T}\right)\Xi(V, T, \mu)\,\mathrm{d}V}\right.$$

$$\left. - \left[\frac{\int_0^\infty V \exp\left(-\frac{PV}{k_{\mathrm{B}}T}\right)\Xi(V, T, \mu)\,\mathrm{d}V}{\int_0^\infty \exp\left(-\frac{PV}{k_{\mathrm{B}}T}\right)\Xi(V, T, \mu)\,\mathrm{d}V}\right]^2\right\} = \frac{\langle V^2 \rangle - \langle V \rangle^2}{Vk_{\mathrm{B}}T}$$

(1.137)

and depends on the variance of the distribution of volume values. Likewise, the heat capacity is the variance of the fluctuations of the energy.

The foregoing relationships identify important response functions of the macromolecule as fluctuations of thermodynamic extensive quantities and are equilibrium analogues of 'fluctuation–dissipation relationships' (Kubo, 1965; de Groot and Mazur, 1984). The conditions imposed by thermodynamic stability on the sign of these response functions are readily understood in terms of statistical considerations, since the variance of a distribution is necessarily a positive quantity. The linkage coefficients, on the other hand, are related to the covariance of the fluctuations of extensive quantities. For example, the change of W with respect to P is derived from eq (1.133) and Ψ given in eq (1.72) as

$$-\left(\frac{\partial W}{\partial P}\right)_{\mu,T} = \frac{\langle jV \rangle - \langle j \rangle\langle V \rangle}{k_{\mathrm{B}}T}$$

(1.138)

The change of V with respect to μ yields

$$\left(\frac{\partial V}{\partial \mu}\right)_{P,T} = \frac{\langle Vj \rangle - \langle V \rangle\langle j \rangle}{k_{\mathrm{B}}T}$$

(1.139)

The Maxwell relationship (1.97) has a straightforward interpretation in terms of the covariance of the fluctuations in the number of water molecules bound and the volume of the macromolecule. The covariance of two variables a and b is, in fact, the same as the covariance of b and a (Feller, 1950). The matrix $\mathbf{G}'$, as well as its inverse $\mathbf{G}$, is the variance–covariance matrix, or Hessian, of the system defined in terms of the independent thermodynamic variables (Feller, 1950). Thermodynamic stability imposes on $\mathbf{G}'$ the quite obvious restriction of being positive-definite as is necessarily the case for a Hessian matrix. Alternatively, the matrix $\mathbf{G}'$, or its inverse, can be seen as the Gram matrix (Gantmacher, 1959) of a Euclidean hyperspace spanned by t independent vectors, each of which is associated with the differential of an intensive quantity. In this metric picture, elaborated by Weinhold in a series of most elegant papers (Weinhold, 1975a, b, c), each response function becomes the projection of a given vector onto itself (diagonal elements of the Gram matrix), or another vector of the basis set (off-diagonal elements of the Gram matrix). Thermodynamic laws translate into rules of Euclidean geometry. The projection of any vector onto itself is necessarily a positive quantity, hence the positive nature of the binding capacity, compressibility and heat capacity. On the other hand, the projection of a vector a onto a vector b equals the projection of b on a, hence the Maxwell reciprocity relationships involving linkage coefficients. The positive-definite nature of $\mathbf{G}'$ is a direct consequence of the positive-definite metric of the Euclidean space associated with it. The determinant of $\mathbf{G}'$ is related to the volume of the hyperspace and must be positive. Any principal minor of $\mathbf{G}'$ represents the volume of a subspace orthogonal to a given vector of the basis set and must be positive for the same reason.

1.5 Simple models for energy transition

We shall now examine a number of specific models that are relevant to macromolecular interactions. These models will be useful for our subsequent analysis of cooperative transitions. We start with the simplest model of interest by considering the macromolecule as a canonical ensemble, with the volume and number of water molecules bound fixed to arbitrary values. The canonical partition function of the system is given by eq (1.30)

$$Z(T) = \sum_{j=1}^{s} \Omega(E_j) \exp\left(-\frac{E_j}{k_B T}\right) \qquad (1.140)$$

where N and V have been omitted since they are assumed to be constant. We assume that the macromolecule can exist in two energy levels, $-\varepsilon$ and

$+\varepsilon$, each of which has a degeneracy of one. The partition function is in this case

$$Z(T) = \exp\left(\frac{\varepsilon}{k_{\mathrm B}T}\right) + \exp\left(-\frac{\varepsilon}{k_{\mathrm B}T}\right) = 2\cosh\left(\frac{\varepsilon}{k_{\mathrm B}T}\right) \qquad (1.141)$$

The Helmholtz free energy, the energy E and the entropy S of the macromolecule are given by

$$F = -k_{\mathrm B}T\ln 2 - k_{\mathrm B}T\ln\cosh\left(\frac{\varepsilon}{k_{\mathrm B}T}\right) \qquad (1.142)$$

$$E = -\varepsilon\tanh\left(\frac{\varepsilon}{k_{\mathrm B}T}\right) \qquad (1.143)$$

$$S = k_{\mathrm B}\ln 2 + k_{\mathrm B}\ln\cosh\left(\frac{\varepsilon}{k_{\mathrm B}T}\right) - \frac{\varepsilon\tanh\left(\varepsilon/k_{\mathrm B}T\right)}{T} \qquad (1.144)$$

The energy E is derived from F using eq (1.14). The important response function of the system is the heat capacity at constant V obtained from E by differentiation with respect to T, i.e.,

$$C_V = \frac{\varepsilon^2}{k_{\mathrm B}T^2\cosh^2\left(\varepsilon/k_{\mathrm B}T\right)} = \frac{k_{\mathrm B}\Theta^2}{\cosh^2\Theta} \qquad (1.145)$$

where $\Theta = \varepsilon/k_{\mathrm B}T$ is a dimensionless variable. The heat capacity is illustrated in Figure 1.6 as a function of Θ. The function is positive everywhere and tends to zero for $\Theta \to 0$ and $\Theta \to \infty$, corresponding to $T \to \infty$ and $T \to 0$ respectively. Since C_V is a measure of the fluctuations of E, the value of C_V goes to zero whenever E approaches asymptotically a constant value. As $T \to 0$, the macromolecule freezes into the lowest energy level $-\varepsilon$. As $T \to \infty$, both energy levels become equally populated and the system freezes into another regime that is the average of the two possible energy levels. In either case, no fluctuations of the energy occur. It is intuitively obvious that at some intermediate values of T the macromolecule makes a transition between these two limiting regimes. The maximum value of C_V is observed for a value of $\Theta = \Theta_{\max}$ such that

$$\Theta_{\max}\tanh\Theta_{\max} = 1 \qquad (1.146)$$

thereby indicating that the macromolecule is making a transition between two energy levels. The transition occurs at a value of the energy

$$E_{\max} = -k_{\mathrm B}T_{\max} \qquad (1.147)$$

A peak in the heat capacity profile of the system thus indicates the existence of at least two energy levels accessible to the macromolecule. The levels are differentially populated as the temperature increases. At low temperature only the level with minimum energy is populated, while, as temperature increases, levels with higher energy also become pop-

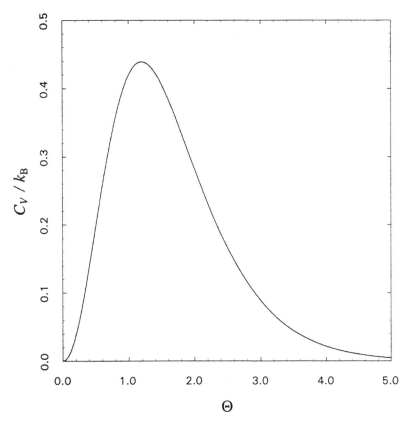

Figure 1.6 Plot of the heat capacity, in k_B units, versus the dimensionless variable $\Theta = \varepsilon/k_B T$ for the two-level model for energy transition. The peak in the function C_V is known as a Schottky transition and reveals the presence of a transition between two distinct energy levels.

ulated. A fundamental property of the system, i.e., the existence of multiple energy levels, is revealed by the analysis of the fluctuations of the energy using thermodynamic response functions. The heat capacity of the form (1.145) is also called the Schottky specific heat and is characteristic of any substance having an 'excitation' energy 2ε (Kubo, 1965).

The next model of interest is the case of a quantized energy spectrum. In this case the macromolecule modeled as a canonical ensemble can exist in an infinite number of equally degenerate energy levels

$$E_j = E_0 + j\varepsilon \tag{1.148}$$

with $j = 0, 1, \ldots \infty$. The canonical partition function is given in this case by

$$Z(T) = \sum_{j=0}^{\infty} \exp\left(-\frac{E_j}{k_B T}\right) = \frac{\exp\left(-E_0/k_B T\right)}{1 - \exp\left(-\varepsilon/k_B T\right)} \tag{1.149}$$

Hence,

$$F = E_0 + k_B T \ln\left[1 - \exp\left(-\frac{\varepsilon}{k_B T}\right)\right] \tag{1.150}$$

$$E = E_0 + \frac{\varepsilon}{\exp\left(\varepsilon/k_B T\right) - 1} \tag{1.151}$$

$$S = \frac{\varepsilon}{T[\exp\left(\varepsilon/k_B T\right) - 1]} - k_B \ln\left[1 - \exp\left(-\frac{\varepsilon}{k_B T}\right)\right] \tag{1.152}$$

$$C_V = \frac{\varepsilon^2 \exp\left(-\varepsilon/k_B T\right)}{k_B T^2 [1 - \exp\left(-\varepsilon/k_B T\right)]^2} = \frac{k_B \Theta^2 \exp\left(-\Theta\right)}{[1 - \exp\left(-\Theta\right)]^2} = \frac{k_B \Theta^2}{4 \sinh^2\left(\Theta/2\right)}$$
$$\tag{1.153}$$

The heat capacity profile is shown in Figure 1.7. A comparison with the

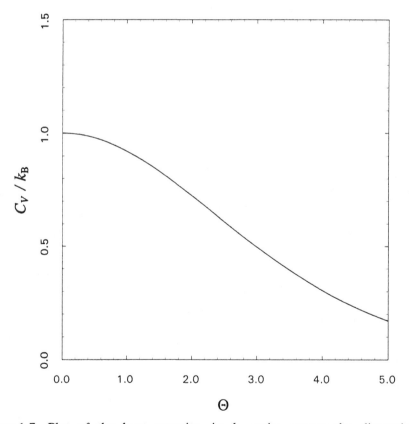

Figure 1.7 Plot of the heat capacity, in k_B units, versus the dimensionless variable $\Theta = \varepsilon/k_B T$ for the quantized energy model for energy transition. Unlike the two-level model, there is no peak in the function C_V which changes monotonically from zero to k_B.

heat capacity profile in Figure 1.6 obtained for the two-level model shows the absence of a peak. The heat capacity for the quantized energy levels changes from a value of zero, obtained for $T \to 0$, to a value of k_B for $T \to \infty$. This indicates that for $T \to 0$ the macromolecule freezes into a constant energy level, as in the case of the two-level model, but at higher temperature the energy increases in a linear fashion with T. In fact, from eq (1.151) it follows that

$$E \approx E_0 + k_B T \tag{1.154}$$

for $T \gg \varepsilon/k_B$. The presence of fluctuations of the energy at high temperature values for the quantized energy model is to be understood in terms of the availability of arbitrarily high energy levels for the system. These levels, which are not populated at low temperature, become progressively stabilized as T increases, thereby causing the energy of the system to fluctuate constantly.

1.6 Simple models for binding

We now consider the grand canonical analogue of the models treated in the previous section. These simplified models are conceptually useful in the discussion of binding phenomena. The macromolecule is modeled as a generalized ensemble for which P and T are fixed, while the number of water molecules bound is subject to fluctuations under the effect of the driving force μ. We start from eq (1.72) written as a polynomial in the water activity w as follows

$$\Psi(w) = \sum_{N=0}^{M} Y_N w^N \tag{1.155}$$

which incorporates the mechano-thermal partition functions for given values of N. We now assume that the macromolecule has a single binding site for water, which can exist in two states, free ($N = 0$) or bound ($N = M = 1$). This situation parallels the two-level model for energy transitions. Likewise, we assume that the ligation state of the macromolecule uniquely specifies its Gibbs free energy level, G_0 for the site free and G_1 for the site bound. The degeneracy of these levels is obtained from the dispositions allowed for the water molecule in all possible ligated configurations. Evidently, there is only one way to obtain the site, free or bound, and the microcanonical components of the partition function (1.71) are equal to one. Hence,

$$\Psi(w) = \exp\left(-\frac{G_0}{k_B T}\right)(1 + Kw) \tag{1.156}$$

where

$$K = \exp\left(\frac{G_0 - G_1}{k_B T}\right) = \exp\left(-\frac{\Delta G}{k_B T}\right) \qquad (1.157)$$

is the equilibrium constant for the reaction free $\rightarrow$ bound of the site and is a function of T and P. This is a consequence of the *law of mass action* (Prigogine and Defay, 1954) through which water binds to the site on the macromolecule in a specific and 'saturable' way that can be predicted from knowledge of the water activity. The standard free energy of ligation for the site is

$$\Delta G = -k_B T \ln K \qquad (1.158)$$

The partition function (1.156) is a binomial expression in the variable w, times a factor that depends only on temperature. This factor is inconsequential for the properties of the macromolecule if our discussion remains confined to an isothermal system. The quantities of interest in this system are the average number of water molecules bound, W, and its fluctuations or binding capacity B. Both quantities are obtained from differentiations of $\ln \Psi$ as already shown in the foregoing sections. The explicit expressions for W and B are

$$W = \frac{\partial \ln \Psi}{\partial \ln w} = \frac{Kw}{1 + Kw} \qquad (1.159)$$

$$B = \frac{\partial W}{\partial \ln w} = \frac{Kw}{(1 + Kw)^2} \qquad (1.160)$$

The functions W and B are plotted in Figure 1.8 versus the logarithm of the water activity. The value of W goes from 0 to 1, two asymptotic values obtained for $w = 0$ and $w = \infty$. The binding capacity is positive everywhere and shows a peak for $w_{max} = K^{-1}$. This is the value of w for which $W = \frac{1}{2}$ and the macromolecule is equally partitioned between free and bound sites. The two-state model for the ligation of the site echoes the two-level model for energy transitions examined in the previous section. The driving force for binding is $\ln w$, while, in the case of energy transitions, it is the temperature T. The function W parallels E, while the binding capacity parallels the heat capacity. The physical significance of the binding capacity can be understood as a Schottky-like effect arising as a consequence of the transition between two ligation states. A peak in the binding capacity profile indicates the existence of such a transition reflecting the ligation of a site.

The quantized analogue for binding is provided by a binding site that can accommodate an arbitrary number of water molecules. This situation,

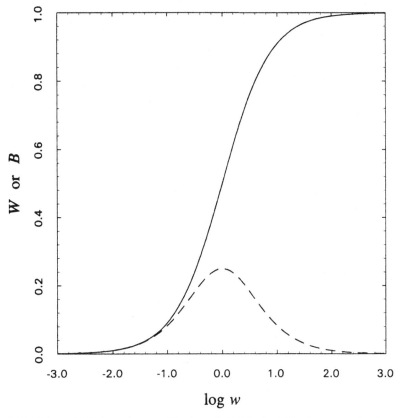

Figure 1.8 Plot of W (continuous line) and B (dashed line) versus the logarithm of water activity for the case of water binding to a single site of the macromolecule according to the law of mass action. Curves were drawn according to eqs (1.159) and (1.160) with $K = 1 \text{ M}^{-1}$.

which is of interest when dealing with protein hydration phenomena, entails binding of a water molecule to the site much like what is observed in the two-state model, followed by subsequent binding of additional water molecules in a piggy-back fashion, as illustrated in Figure 1.9. This case, to be referred to as *condensation* (Mayer and Mayer, 1940), is quite different from mass law binding since it does not show saturation. We assume that the Gibbs free energy of the configuration of the site with N water molecules bound is given by

$$G_N = G_0 + NG \tag{1.161}$$

where $N = 0, 1, \ldots \infty$. The analogy with eq (1.148) is obvious. The degeneracy of each free energy level is equal to 1, since there is only one way of ligating the first water molecule, or any water molecule thereafter

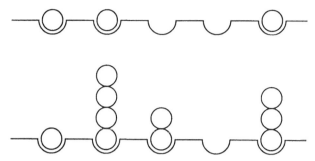

Figure 1.9　Schematic representation of mass law binding (top) and condensation (bottom), for the paradigmatic example of water binding to a macromolecule. In the case of saturable binding, each site is occupied by at most one water molecule. In the case of condensation, multiple layers are formed as vertical piles of molecules for each site. Saturation is the landmark property of mass law binding.

in a piggy-back fashion. The partition function can be written as

$$\Psi(w) = \frac{\exp\left(-G_0/k_B T\right)}{1 - Kw} \tag{1.162}$$

where $K = \exp\left(-\varepsilon/k_B T\right)$ and again the constant term $\exp\left(-G_0/k_B T\right)$ is inconsequential. The relevant response functions are

$$W = \frac{\partial \ln \Psi}{\partial \ln w} = \frac{Kw}{1 - Kw} \tag{1.163}$$

$$B = \frac{\partial W}{\partial \ln w} = \frac{Kw}{(1 - Kw)^2} \tag{1.164}$$

These expressions are the same as those derived for the case of mass law binding, except for the minus sign in the denominator. The replacement has the nontrivial consequence that both W and B are defined only in a finite range of w such that

$$K^{-1} = w_c > w \geq 0 \tag{1.165}$$

The critical value w_c defines the water activity at which all water molecules present in solution condense on the macromolecule, with a consequent divergence of W and B. The behavior of W and B is illustrated in Figure 1.10. The value of W grows without bound as $w \to w_c$, and so does B. The saturation seen in the function W for the two-level model in Figure 1.8, which is the mark of mass law binding, is never reached in the model for condensation. Accordingly, the binding capacity does not peak for any value of w, since the system never settles down to a definite configuration upon binding and the fluctuations grow without bound as w increases.

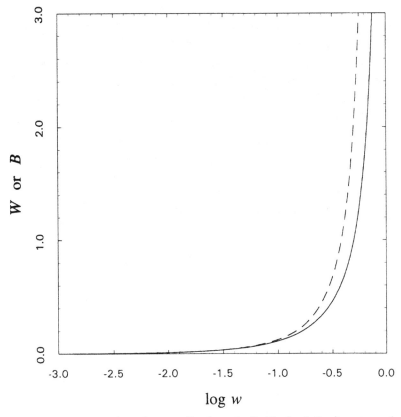

Figure 1.10 Plot of W (continuous line) and B (dashed line) versus the logarithm of water activity, for the case of water binding in a nonsaturable way (condensation) to a single site of the macromolecule. Curves were drawn according to eqs (1.163) and (1.164) with $K = 1 \, \mathrm{M}^{-1}$. The critical value of the water activity is $w_\mathrm{c} = 1 \, \mathrm{M}$.

The two models analyzed above provide important examples of interactions. In the case of the two-state model, the statistics hinge on the transition between the free and bound states of the macromolecule. This situation is completely analogous to that arising in the Fermi–Dirac statistics of quantum mechanics, where any given quantum state can be occupied by at most one particle (Tolman, 1938; Hill, 1960; Kubo, 1965). In our case, the binding site of the macromolecule plays the role of the quantum state that can be either free or occupied by a particle to be identified with a water molecule. The macromolecule binding according to the law of mass action behaves like a 'fermion'. In the condensation model, there is no upper bound for the variable W and one or more water molecules can occupy the binding site. This case is completely analogous to that arising in the Bose–Einstein statistics of quantum mechanics, where any given quantum state can be occupied by an arbitrary number of particles (Tolman, 1938; Hill, 1960; Kubo, 1965). The macromolecule binding according to the condensation model behaves like a 'boson'. The two models can be merged to describe high-affinity mass law binding to a given site, followed by low-affinity condensation. This hybrid model is known as the Brunauer–Emmett–Teller theory of adsorption and yields the simple Langmuir theory as a special case in the absence of condensation (Hill, 1960). Interesting extensions of this theory to the case of fractal layers of adsorption have been derived (Avnir, Farin and Pfeifer, 1984; Fleischmann, Tildesley and Ball, 1990). In the rest of this monograph, we will deal exclusively with mass law binding.

2

Global binding processes

In this chapter we start our discussion of binding processes. Here we deal explicitly with processes that involve the system as a whole. The properties of the macromolecule in this global description can be derived in a straightforward manner from the principles outlined in the previous sections.

2.1 The reference system

The simplest system of interest in the discussion of binding processes is that composed of a macromolecule M containing a single site for a ligand X. The term 'ligand' indicates a molecule that interacts specifically with M according to the law of mass action as discussed in Section 1.6. In the previous chapter we have considered water as ligand. Here we are interested in the interaction of M with a 'specific' ligand and regard the solvent as an 'inert' component. A treatment of linked hydration effects is given in Section 2.6. For the sake of simplicity in the following we consider temperature and pressure constant. We also assume that the macromolecule does not change its aggregation state. A rather extensive and elegant treatment of aggregation effects linked to binding processes has been developed by Hermans and Premilat (1975). The macromolecule is modeled as a generalized ensemble for which P and T are fixed, while the number of ligands bound is subject to fluctuations under the effect of the driving force $\mu = k_B T \ln x$, where x is the ligand activity. The relevant partition function of this system has the same form as eq (1.155), i.e.,

$$\Psi(x) = Y_0 + Y_1 x \qquad (2.1)$$

The mechano-thermal partition functions, Y_0 and Y_1, incorporate the properties of the two possible ligated intermediates and specify the values of the related Gibbs free energies G_0 and G_1. The standard free energy difference between the bound and free forms of the site is

$$\Delta G = G_1 - G_0 = -k_B T \ln K \tag{2.2}$$

where K is the equilibrium constant for the reaction

$$M + X \Leftrightarrow MX \tag{2.3}$$

or, according to the law of mass action

$$K = \frac{[MX]}{[M]x} \tag{2.4}$$

We implicitly assume that the solution is dilute enough to allow for the concentrations of the macromolecular forms M and MX to be used in place of their activities. The equilibrium constant K is a function of T and P, as expected for a mechano-thermal ensemble, and the change of K with T and P provides information on enthalpy and volume changes associated with the binding reaction (2.3), as implied by eqs (1.61) and (1.62). Detailed treatments of these effects are given elsewhere (Wyman and Gill, 1990; Weber, 1992). The degeneracy of G_0 and G_1 is set by the dispositions allowed for the ligand in all possible ligated configurations. Evidently, there is only one way to obtain the site free or bound, which sets the microcanonical components of the partition function (2.1) equal to one. We now rewrite the partition function as

$$\Psi(x) = \exp\left(-\frac{G_0}{k_B T}\right)(1 + Kx) \tag{2.5}$$

At constant T and P the prefactor makes no contribution to the binding properties of the system and can be set equal to one without loss of generality. This is equivalent to setting $G_0 = 0$, or taking the unligated form of the macromolecule as reference. Under this assumption, the partition function assumes the rather simple form

$$\Psi(x) = 1 + Kx \tag{2.6}$$

The average number of ligated sites per macromolecule, X, and the fluctuations of X reflecting the binding capacity B, are obtained from differentiation of $\ln \Psi$ as shown in Section 1.6. The result is

$$X = \frac{d \ln \Psi}{d \ln x} = \frac{Kx}{1 + Kx} \tag{2.7}$$

$$B = \frac{dX}{d \ln x} = \frac{Kx}{(1 + Kx)^2} \tag{2.8}$$

The partial derivative has been replaced with the total derivative for the sake of simplicity, since T and P are constant. The functions X and B plotted versus the logarithm of the ligand activity give the results already shown in Figure 1.8. Binding to the macromolecule shows saturation at $X = 1$ as $x \to \infty$. The binding capacity shows a peak for $x = K^{-1}$, which is the value of the ligand activity at half saturation ($X = \frac{1}{2}$).

Consider the case of a macromolecule containing N sites that are identical and independent. This is a *reference system* whose properties will be of great importance in our discussion. The reference system is a scaled replica of the simpler case $N = 1$ and as such the relevant thermodynamic quantities X and B are expected to show no substantial difference other than scaling factors. There are essentially two ways of deriving the properties of the reference system. One is simple and intuitive. Since the N sites are identical and independent, the potential $\Pi = -k_B T \ln \Psi$ of the macromolecule is N times that of a single site. It follows from (1.70) that

$$\Psi(x) = (1 + Kx)^N \tag{2.9}$$

$$X = \frac{d \ln \Psi}{d \ln x} = \frac{NKx}{1 + Kx} \tag{2.10}$$

$$B = \frac{dX}{d \ln x} = \frac{NKx}{(1 + Kx)^2} \tag{2.11}$$

The functions X and B scale with N, as expected. The alternative way of deriving the partition function (2.9) is more instructive. Consider the polynomial expansion for the system containing N sites

$$\Psi(x) = \sum_{j=0}^{N} Y_j x^j \tag{2.12}$$

The jth mechano-thermal partition function in eq (2.12) contains the Gibbs free energy of the jth ligated intermediate. Since the sites are all identical and independent, the Gibbs free energy level G_j can be written in the form

$$G_j = jG \tag{2.13}$$

where G is the Gibbs free energy level of a ligated site and the unligated configuration is assumed to have a free energy level of zero. The degeneracy, or microcanonical component, associated with each Y_j is computed from the number of distinct ways j sites out of N can be ligated, since any of these possible dispositions is linked to the same value of G_j. The number of such dispositions is given by the binomial coefficient $C_{j,N} = N!/j!(N-j)!$. Hence,

$$Y_j = C_{j,N} \exp(-G_j/k_B T) = C_{j,N} \exp(-jG/k_B T) = C_{j,N} K^j \tag{2.14}$$

where K has already been defined in the simpler case $N = 1$. Substitution of (2.14) into (2.12) yields

$$\Psi(x) = \sum_{j=0}^{N} C_{j,N} K^j x^j = \sum_{j=0}^{N} \binom{N}{j} K^j x^j = (1 + Kx)^N \qquad (2.15)$$

which is identical to Eq (2.9). The jth mechano-thermal partition function Y_j in the expansion of Ψ is related to the Gibbs free energy for the formation of the jth ligated intermediate from the reference unligated form of the macromolecule, or the reaction

$$M + jX \Leftrightarrow MX_j \qquad (2.16)$$

The standard free energy change associated with this reaction is

$$\Delta G_j = -k_B T \ln C_{j,N} - j k_B T \ln K \qquad (2.17)$$

The degeneracy, $C_{j,N}$, accounts for the different dispositions of j ligands among N sites. This term contributes to the free energy change in an additive manner and must be taken into account when dealing with binding to multiple sites ($N > 1$).

The probability of finding the jth ligated configuration in the reference system is derived from (1.78) as

$$\psi_j = \frac{C_{j,N} K^j x^j}{(1 + Kx)^N} = C_{j,N} N^{-N} X^j (N - X)^{N-j} = C_{j,N} \theta^j (1 - \theta)^{N-j} \quad (2.18)$$

where $\theta = X/N$ is the fractional saturation, or the probability of finding a ligated site in the system. The values of ψ_j as a function of θ are given in Figure 2.1 for the case $N = 4$. The functions ψ_0 and ψ_N are monotonic, with ψ_0 decreasing and ψ_N increasing as a function of X or θ. On the other hand, the functions relative to the intermediates of ligation such that $0 < j < N$ show a distinct peak occurring at intermediate values of saturation. It follows from eq (2.18) that the peak for the function ψ_j occurs at $X = j$, or $\theta = j/N$. The jth ligated intermediate has a maximum of occurrence when the system contains j ligated sites on the average. The two limiting configurations with 0 and N ligated sites are no exceptions to this rule, since they are peaked at $X = 0$ and $X = N$ respectively. The functions ψs contain all information necessary to define the key properties of the system. In fact, the partition function can be rewritten as

$$\Psi(x) = \sum_{j=0}^{N} C_{j,N} K^j x^j = (1 + Kx)^N \sum_{j=0}^{N} \psi_j = \psi_0^{-1} \qquad (2.19)$$

and equals the inverse of the probability of finding the unligated form of the macromolecule. This result is valid in general, regardless of the

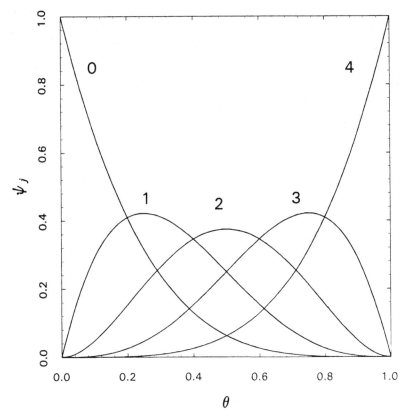

Figure 2.1   Plot of the probability of occurrence of ligation intermediates as a function of fractional saturation for a reference system. Curves were drawn according to eq (2.18) with $N = 4$. The jth ligated intermediate, as indicated, shows a maximum in the plot for a value of $X = j$, or $\theta = j/N$.

specific form of ψ_0 (Hill, 1960, 1984; d'A. Heck, 1971; Schellman, 1975; Wyman and Gill, 1990). Likewise, the functions X and B can be expressed in terms of the ψs using eqs (1.133) and (1.134) as follows

$$X = \sum_{j=0}^{N} j\psi_j \tag{2.20}$$

$$B = \sum_{j=0}^{N} j^2\psi_j - \left(\sum_{j=0}^{N} j\psi_j\right)^2 \tag{2.21}$$

Hence, for a reference system, we have the following important relationship between X and B

$$B = \frac{X(N - X)}{N} = B_0 \tag{2.22}$$

The suffix denotes the particular value of B in a reference system, which will be used in our subsequent analysis. The binding capacity is plotted versus X in Figure 2.2 for the case $N = 4$. The fluctuations of ligated intermediates around the average ligation value X are reflected by B_0. It is a unimodal function of X and goes to zero for $X = 0$ and $X = N$, as expected. For $X \to 0$ the binding capacity tends to vanish due to the low concentration of ligand molecules, while for $X \to N$ all sites become bound and B tends to vanish as no sites are available for ligation. Under these limiting conditions the possible configurations of the system 'freeze' around the average value of X and no fluctuations are possible. The binding capacity is a maximum for $X = N/2$, where the dispersion of ligated intermediates around the average value is a maximum. The distribution of intermediates in Figure 2.1 provides the physical basis for the behavior of B. The intermediates are all significantly populated as X

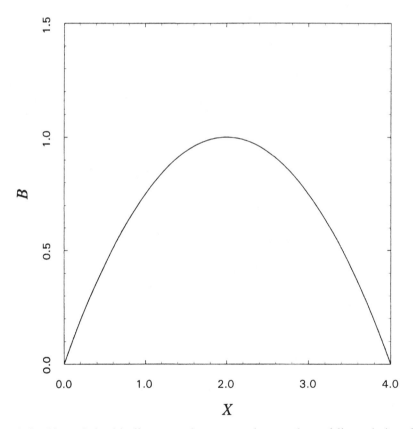

Figure 2.2   Plot of the binding capacity versus the number of ligated sites for a reference system. The curve was drawn according to eq (2.22) with $N = 4$.

approaches its mid-point value $N/2$, while only one configuration is present at the limiting values of $X = 0$ and $X = N$. The maximum value of B for a reference system is $N/4$ and provides a measure of the maximum dispersion of ligated intermediates around X.

Another important property of the reference system is derived from consideration of the 'affinity function' ^xK. This function gives the average binding affinity of a site as a function of the ligand activity. If an 'average' site in the system is considered, then the binding reaction

$$\text{site} \cdot \text{free} + \text{ligand} \Leftrightarrow \text{site} \cdot \text{bound} \tag{2.23}$$

can be associated with an apparent equilibrium constant

$$^xK = \frac{[\text{site} \cdot \text{bound}]}{[\text{site} \cdot \text{free}][\text{ligand}]} \tag{2.24}$$

The concentration of bound sites is Xm_t, where m_t denotes the total concentration of the macromolecule. Likewise, the concentration of free sites is $(N - X)m_t$ and therefore

$$^xK = \frac{X}{(N - X)x} \tag{2.25}$$

This is the equilibrium constant for the reaction (2.23) and gives the affinity function of the system as a function of the ligand activity. The importance of ^xK will become evident when dealing with cooperative systems. The free energy of binding associated with ^xK is

$$^x\Delta G = -k_B T \ln \frac{X}{(N - X)x} \tag{2.26}$$

The affinity function of a reference system can be computed from the definition (2.25) using the expression for X given by (2.10) to obtain

$$^xK = K \tag{2.27}$$

The fact that ^xK does not change with x for a reference system is intuitively obvious. A particularly useful plot is the logarithm of ^xK versus the logarithm of x. This 'affinity plot' transforms the data shown in Figure 2.3 for Na^+ binding to thrombin into those shown in Figure 2.4 (Wells and Di Cera, 1992; Ayala and Di Cera, 1994). The affinity function can be constructed directly from experimental data once X is known as a function of x. For $N = 1$, or in the case of a reference system for arbitrary N, a plot of $\ln {}^xK$ versus $\ln x$ should yield a straight line with zero slope. The intercept in this plot gives $\ln K$, from which the value of K is determined directly.

A familiar transformation of the function X is the Hill plot (Hill, 1910)

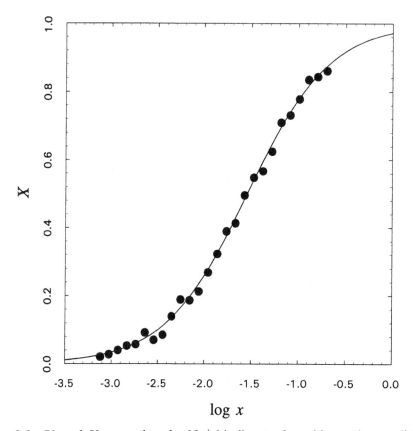

$$\log x$$

Figure 2.3   Plot of X versus $\ln x$ for Na$^+$ binding to thrombin, under conditions of: 5 mM Tris, 0.1% PEG, I = 0.2 M, pH 8.0 at 25 °C (Ayala and Di Cera, 1994). The binding reaction involves a single site (Dang, Vindigni and Di Cera, 1995) and obeys eq (2.7) for a reference system, with $K = 35\ \mathrm{M}^{-1}$.

where the logarithm of the ratio $X/(N - X)$, or equivalently $\theta/(1 - \theta)$, is plotted versus the logarithm of the ligand activity. In the case of the reference system, the Hill plot is a straight line with unit slope, as shown in Figure 2.5. The value of the abscissa where the ordinate vanishes is $-\ln K$. The slope of the Hill plot defines the Hill coefficient n_H. For a reference system $n_\mathrm{H} = 1$, independent of x. The Hill plot is a represent-ation of the function

$$\ln \frac{X}{N - X} = \ln {}^x\!K + \ln x \tag{2.28}$$

and is merely a plot of $\ln {}^x\!K$ rotated by 45° counterclockwise. Hence, the Hill coefficient is the slope of the Hill plot, or the slope of the affinity plot plus one. Other transformations of the function X have little thermody-

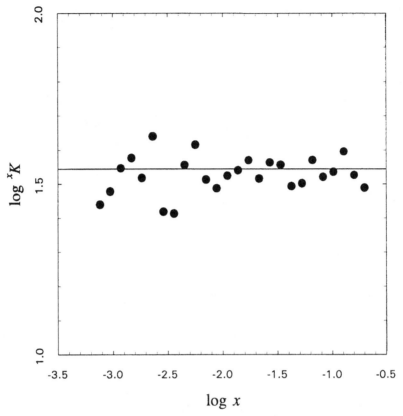

Figure 2.4   Affinity plot for Na$^+$ binding to thrombin. The data are the same as those shown in Figure 2.3, but plotted as ln$^x K$ versus ln x. The affinity plot is a straight line with zero slope, as implied by eq (2.27) for a reference system, and yields the value of ln K as the intercept of the ln$^x K$ axis, with $K = 35$ M^{-1}.

namic significance, but are worth mentioning for historical reasons. In the popular Scatchard plot (Scatchard, 1949) given in Figure 2.6, the ratio X/x is plotted versus X. In the case of a reference system the slope in this plot is equal to $-K$ and the intercept on the ordinate is NK, as can be easily verified from eq (2.10). In the equally popular double-reciprocal plot given in Figure 2.7 the function $1/X$ is plotted versus $1/x$. The slope in this plot equals $1/NK$ and the intercept is equal to $1/N$. The Scatchard and double-reciprocal transformations are mentioned solely for the sake of completeness and will not be pursued any further in our treatment. These linear transformations were invaluable in those days when nonlinear least-squares analysis of experimental data could not be performed routinely on a computer. These plots, however, convey very little information in thermodynamic terms and should be used with extreme caution in

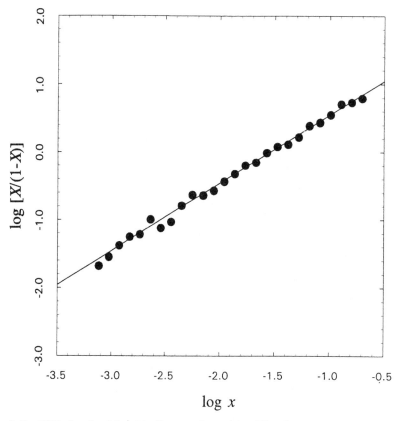

Figure 2.5   Hill plot for Na$^+$ binding to thrombin. The data are the same as those shown in Figure 2.3, but plotted as $\ln[X/(1-X)]$ versus $\ln x$. The Hill plot is a straight line with a slope of one, as implied by eq (2.28) for a reference system. The intercept on the abscissa where the ordinate vanishes gives the value of $-\ln K$, with $K = 35$ M^{-1}.

practical applications. The Scatchard plot may grossly underestimate the number of binding sites (Klotz, 1985), while the double-reciprocal plot may yield severely biased estimates of binding parameters (Di Cera, 1992a). These pitfalls, which were very well known to Scatchard (1949), continue to be ignored by many experimentalists in their data analyses. As a general rule, data should be plotted and analyzed in the form taken experimentally, with a distribution of errors in the weighting scheme for parameter estimation that reflects the one derived from actual measurements. The problem of correct parameter estimation in data analysis is of considerable importance and complexity. The interested reader is referred to the excellent monographs by Bard (1974) and by Box and Jenkins (1976) for a detailed treatment.

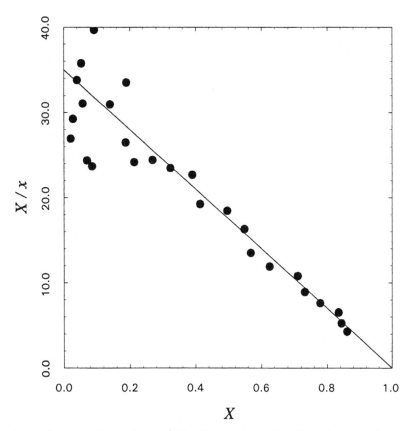

Figure 2.6 Scatchard plot for Na^+ binding to thrombin. The data are the same as those shown in Figure 2.3, but plotted as X/x versus X. The plot is linear, with an intercept on the ordinate of $K = 35 \, M^{-1}$. The error of data points at low saturation is emphasized in the Scatchard plot and complicates the resolution of binding parameters (Klotz, 1985).

2.2 Cooperative binding

In the reference system all sites are alike and independent. The partition function assumes a simple binomial form raised to the Nth power, where N equals the number of sites in the system. In general, one can expect the sites to be different and/or influence each other's binding reactions. In this case, of considerable interest in biology, the simple rules examined for the reference system no longer apply. The partition function of a macromolecule containing N sites is

$$\Psi(x) = \sum_{j=0}^{N} Y_j x^j \qquad (2.29)$$

The jth mechano-thermal partition function in eq (2.29) contains the

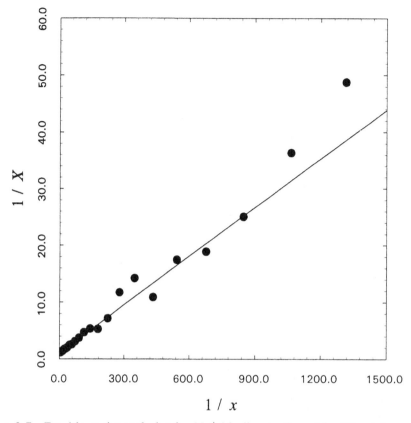

Figure 2.7 Double-reciprocal plot for Na^+ binding to thrombin. The data are the same as those shown in Figure 2.3, but plotted as $1/X$ versus $1/x$. The plot is linear with a slope equal to $1/K$, with $K = 35\ M^{-1}$. Note how the data for x small significantly deviate from the straight line, unlike those obtained for large x. This is a consequence of the serious bias introduced by the transformation (Di Cera, 1992a).

Gibbs free energy of the jth ligated intermediate. Since there are $N + 1$ such intermediates, and one can be taken as reference, there are N independent mechano-thermal partition functions in eq (2.29). This is the fundamental difference with the reference system that can be fully described in terms of the free energy of binding to any of the N equivalent and independent sites. Consider the reaction

$$M + jX \Leftrightarrow MX_j \qquad (2.30)$$

where the jth ligated intermediate is generated from the unligated form. The standard free energy change associated with this reaction is (d'A. Heck, 1971)

$$\Delta G_j = k_B T \ln \frac{Y_0}{Y_j} = -k_B T \ln A_j \tag{2.31}$$

where A_j is the jth *overall* equilibrium binding constant. There are N such independent constants that define the partition function of the system. The degeneracy associated with each Y_j is again given by $C_{j,N}$, or by the number of dispositions of j ligands among N sites. Hence,

$$Y_j = C_{j,N} \exp\left(-\frac{G_j}{k_B T}\right) \tag{2.32}$$

and $G_0 = 0$ can be assumed without loss of generality. If the sites are all alike and independent, as in the case of the reference system, the free energy level G_j is just jG, where G is the same as the free energy change for binding to any of the sites. The partition function in this case factors out in terms of N identical binomials. If the conditions required by the reference system are not met, no such simplification is possible for G_j and N independent binding free energies must be considered. Also, the partition function does not factor out in terms of identical binomials. This is the mark of cooperative binding phenomena.

In order to understand the onset of cooperativity better, it is useful to define a new set of equilibrium constants, the *stepwise* constants, k, that reflect the binding affinity of sequential ligation steps. Unlike the As, the ks have the same dimension (M^{-1}) and can be compared directly. Consider the free energy level G_j in eq (2.32) split into j terms as follows

$$G_j = G_1' + G_2' + \ldots + G_j' \tag{2.33}$$

Since G_j is a state function defining the free energy level of the jth ligated intermediate with respect to the reference, unligated configuration, it makes no difference how many intermediate free energy terms we use to define it. Hence, we choose a set of free energy values such that G_j' defines the free energy change due to adding one ligand to the $(j-1)$th ligated configuration to yield the jth ligated one. The equilibrium constant associated with G_j' is

$$k_j = \exp\left(-\frac{G_j'}{k_B T}\right) \tag{2.34}$$

and defines the *stepwise* equilibrium binding constant for the jth ligation step. The connection between overall and stepwise constants is

$$A_j = C_{j,N} k_1 k_2 \ldots k_j \tag{2.35}$$

$$k_j = \frac{C_{j-1,N}}{C_{j,N}} \frac{A_j}{A_{j-1}} = \frac{j}{N-j+1} \frac{A_j}{A_{j-1}} \tag{2.36}$$

where it should be recalled that $C_{0,N} = C_{N,N} = 1$ and $A_0 = 1$. In the case of the reference system, $A_j = C_{j,N}K^j$ and $k_1 = k_2 = \ldots = k_N = K$. All ligation steps have the same affinity. A cooperative system is characterized by a change in affinity with ligation. If the affinity increases with ligation we have positive cooperativity, otherwise cooperativity is negative. For $N \geqslant 3$ cooperativity can also give rise to mixed patterns, where increases in affinity alternate with decreases and *vice versa*. For a practical point of view, the definition of positive and negative cooperativity in a macroscopic sense refers explicitly to the comparison of the first and last steps of ligation. Specifically, positive macroscopic cooperativity demands $k_N > k_1$, while negative macroscopic cooperativity demands $k_N < k_1$.

Regardless of the nature of the binding process under investigation, the relevant response functions X, B and ^xK can be determined in a straightforward manner from the partition function

$$\Psi(x) = \sum_{j=0}^{N} A_j x^j \tag{2.37}$$

cast in terms of the overall equilibrium constants for the sake of simplicity. The average number of ligated sites is given by the expression

$$X = \frac{\mathrm{d}\ln\Psi}{\mathrm{d}\ln x} = \frac{\sum_{j=0}^{N} jA_j x^j}{\sum_{j=0}^{N} A_j x^j} = \langle j \rangle \tag{2.38}$$

and is bounded from zero to N. The binding capacity is given by

$$B = \frac{\mathrm{d}X}{\mathrm{d}\ln x} = \frac{\sum_{j=0}^{N} j^2 A_j x^j}{\sum_{j=0}^{N} A_j x^j} - \left(\frac{\sum_{j=0}^{N} jA_j x^j}{\sum_{j=0}^{N} A_j x^j}\right)^2 = \langle j^2 \rangle - \langle j \rangle^2 \tag{2.39}$$

and is bounded from zero to $X(N - X)$. The affinity function is given by

$$^xK = \frac{X}{(N - X)x} = \frac{\sum_{j=0}^{N} jA_j x^{j-1}}{\sum_{j=0}^{N} (N - j)A_j x^j} \tag{2.40}$$

It is straightforward to show that ^xK reaches two limiting values for $x \to 0$ and $x \to \infty$ respectively. These limits are

$$^0K = \frac{A_1}{N} = k_1 \tag{2.41a}$$

$$^\infty K = \frac{NA_N}{A_{N-1}} = k_N \tag{2.41b}$$

and are equal to the first and last stepwise binding constants. This is why the stepwise constants are important in defining macroscopic cooperativity. The affinity function for $N = 2$ is plotted in Figure 2.8 for a negatively $(k_2 < k_1)$ and positively $(k_2 > k_1)$ cooperative system. It is quite instructive to note from eq (2.40) that the affinity function approaches its limiting values with a zero slope. Hence, at either low or high saturation a cooperative system behaves like a reference system, with an affinity $K = k_1$ for $x \to 0$ and $K = k_N$ for $x \to \infty$. The physical basis of this effect is as follows. At low saturation, binding to the macromolecule produces ligation of the first site, while at high saturation binding produces ligation of the last site. In either case, the cooperative nature of the system is not reflected in its macroscopic behavior since ligation involves only one site.

The properties of ^xK are of particular relevance in understanding the physical basis of cooperativity. The average number of ligated sites can be written in a rather simple way by using the definition of ^xK, i.e.,

$$X = \frac{N^xKx}{1 + {}^xKx} \tag{2.42}$$

In a reference system $^xK = K$, independent of x, and X assumes the familiar form (2.10). In general, a cooperative system can be regarded as a reference system with variable affinity. There is always a unique value of ^xK for a given x and, for this particular value of x, the function X of a cooperative system assumes the same value as the function X of a reference system with $K = {}^xK$. This is illustrated in Figure 2.9 for both positive and negative cooperativity. In making its transition from $^0K = k_1$ to $^\infty K = k_N$, a cooperative system explores a continuous spectrum of reference systems. Also, the function X is the same as X for a reference system with $K = k_1$ at low saturation and $K = k_N$ at high saturation. When the system makes a transition from low to high affinity, as in the case of positive cooperativity, the function X becomes steeper than any intermediate reference curve crossed over at intermediate saturation values, as shown in Figure 2.9(b). Actual experimental data are shown in Figure 2.10(b) for carbon monoxide binding to hemoglobin (Perrella *et al.*, 1990). The slope in the $\ln {}^xK$ plot is positive at intermediate saturation, as shown in Figure 2.8(b), and the binding capacity is higher and

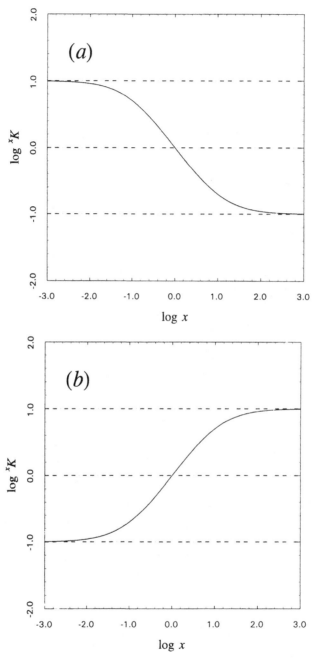

Figure 2.8   Affinity plot for $N = 2$ in the case of (a) negative or (b) positive cooperativity. Curves were drawn according to eq (2.40) with (a) $k_1 = 10\,\mathrm{M}^{-1}$, $k_2 = 0.1\,\mathrm{M}^{-1}$, and (b) $k_1 = 0.1\,\mathrm{M}^{-1}$, $k_2 = 10\,\mathrm{M}^{-1}$. Chain lines depict the values of the $\ln k_1$ and $\ln k_2$ asymptotes in the plot (see eqs (2.41a) and (2.41b)), and the value of $\ln {}^x\!K$ at half saturation, i.e., $\ln \sqrt{k_1 k_2}$. These values correspond to the value of ${}^x\!K$ for reference systems with $K = k_1$, $K = k_2$ or $K = \sqrt{k_1 k_2}$.

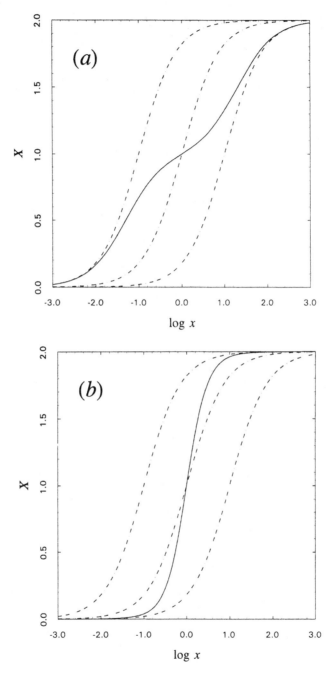

Figure 2.9 Binding curve for $N = 2$ in the case of (a) negative or (b) positive cooperativity. Curves were drawn according to eq (2.38) with (a) $k_1 = 10\,\mathrm{M}^{-1}$, $k_2 = 0.1\,\mathrm{M}^{-1}$, and (b) $k_1 = 0.1\,\mathrm{M}^{-1}$, $k_2 = 10\,\mathrm{M}^{-1}$. Alternatively, the function X can be derived from eq (2.42) using ^xK as shown in Figure 2.8. Chain lines depict binding curves of reference systems for which $K = k_1$, $K = \sqrt{k_1 k_2}$, or $K = k_2$.

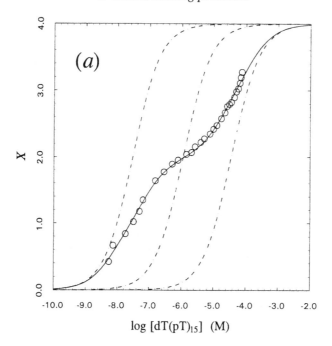

X

$\log \, [\mathrm{dT(pT)_{15}}] \;\; (\mathrm{M})$

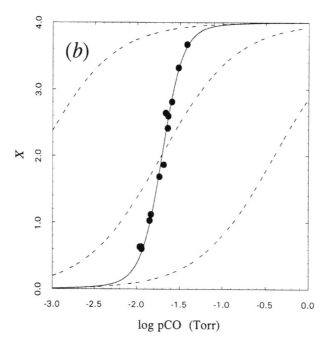

X

$\log \, p\mathrm{CO} \;\; (\mathrm{Torr})$

more narrowly peaked, as shown in Figure 2.11(b). On the other hand, when the system makes a transition from high to low affinity, as in the case of negative cooperativity, the function X becomes less steep than any intermediate reference curve crossed over at intermediate saturation values, as shown in Figure 2.9(a) and in the case of dT(pT)$_{15}$ binding to *E. coli* single strand binding protein (Bujalowski and Lohman, 1989) depicted in Figure 2.10(a). The slope in the ln$^x K$ plot is negative at intermediate saturation, as shown in Figure 2.8(a), and the binding capacity is smaller and more broadly distributed, as shown in Figure 2.11(a). The onset of cooperativity is also marked by a drastic change in the distribution of ligated intermediates, as shown in Figure 2.12. The probability of finding the jth ligated intermediate is derived from eq (1.78) as

$$\psi_j = \frac{A_j x^j}{\displaystyle\sum_{j=0}^{N} A_j x^j} \tag{2.43}$$

A landmark feature of positive cooperativity is the significant suppression of intermediates, with the shape of the binding curve, or any other response function, being determined mostly by the unligated and fully-ligated species. The relative insensitivity of positively cooperative binding isotherms to intermediates of ligation makes resolution of binding constants for such systems a rather nontrivial task. This also complicates significantly the formulation of specific molecular mechanisms of interaction.

Figure 2.10   (*a*) Binding curve of dT(pT)$_{15}$ to *E. coli* SSB tetramer under conditions of: 10 mM Tris, 0.1 mM Na$_3$EDTA, 50 mM NaCl, pH 8.1 at 25 °C (Bujalowski and Lohman, 1989). The continuous line was drawn according to eq (2.38) with the four stepwise constants: $k_1 = 3.4 \times 10^7$ M^{-1}, $k_2 = 7.3 \times 10^6$ M^{-1}, $k_3 = 8.9 \times 10^4$ M^{-1}, $k_4 = 2.8 \times 10^4$ M^{-1}. The values of k_1 and k_4 were used to construct the binding curves for the reference systems, given by chain lines ($K = k_1$ for the curve at left, $K = k_4$ for the curve at right). The reference curve in the middle corresponds to $K = (k_1 k_2 k_3 k_4)^{1/4}$. (*b*) Carbon monoxide binding curve of hemoglobin under conditions of: 3.125 mM heme, 0.1 M KCl, pH 7.0 at 20 °C (Perrella *et al.*, 1990). The continuous line was drawn according to eq (2.38) with the four stepwise constants: $k_1 = 2.5$ Torr^{-1}, $k_2 = 8.5$ Torr^{-1}, $k_3 = 225$ Torr^{-1}, $k_4 = 1460$ Torr^{-1}. The values of k_1 and k_4 were used to construct the binding curves for the reference systems, given by the chain lines ($K = k_1$ for the curve at right, $K = k_4$ for the curve at left). The reference curve in the middle corresponds to $K = (k_1 k_2 k_3 k_4)^{1/4}$. Note the different scale on the abscissa in the two cases. The binding curve for the negatively cooperative case (*a*) spans nearly eight orders of magnitude in the ligand activity, as opposed to only two for the positively cooperative case (*b*).

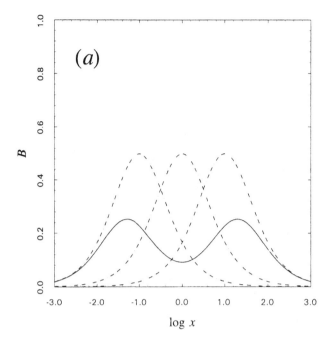

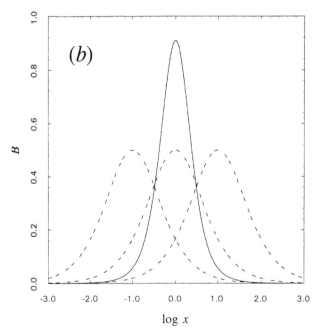

Figure 2.11 Binding capacity for $N = 2$ in the case of (a) negative or (b) positive cooperativity. Curves were drawn according to eq (2.39) with (a) $k_1 = 10\,\mathrm{M}^{-1}$, $k_2 = 0.1\,\mathrm{M}^{-1}$, and (b) $k_1 = 0.1\,\mathrm{M}^{-1}$, $k_2 = 10\,\mathrm{M}^{-1}$. Chain lines depict the binding capacity of reference systems for which $K = k_1$, $K = \sqrt{k_1 k_2}$, or $K = k_2$.

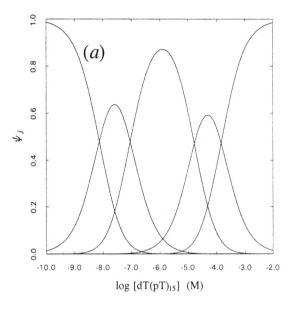

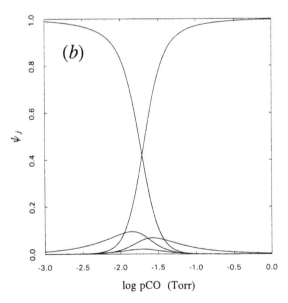

Figure 2.12 Plot of the probability of occurrence of ligation intermediates as a function of the logarithm of ligand activity for (*a*) the binding of dT(pT)₁₅ to *E. coli* SSB tetramer (Bujalowski and Lohman, 1989), and (*b*) carbon monoxide binding to hemoglobin (Perrella *et al.*, 1990). The continuous curves were drawn according to eq (2.43), with parameter values given in the legend to Figure 2.10. The distribution of intermediates should be compared with that of a reference system for $N = 4$ given in Figure 2.1. In the negatively cooperative case (*a*), intermediates make a larger contribution at intermediate saturation, while in the positively cooperative case (*b*) the contribution is almost negligible.

The slope of the affinity plot is related to a classical measure of cooperativity. The Hill coefficient (Hill, 1910; Cantor and Schimmel, 1980; Wyman and Gill, 1990)

$$n_H = \frac{d\ln\left(\dfrac{X}{N-X}\right)}{d\ln x} = \frac{d\ln{}^xK}{d\ln x} + 1 = \Delta X + 1 \tag{2.44}$$

is the slope of the Hill plot, which is the affinity plot rotated 45° counterclockwise. The value of n_H is equal to the value of the slope of the affinity plot, $\Delta X = d\ln{}^xK/d\ln x$, plus one. The significance of ΔX as an important linkage property of the system will become evident in Chapter 3. The properties of the Hill coefficient can be derived from those of the affinity function in a straightforward manner. The Hill plot approaches asymptotically two straight lines of unit slope (Wyman, 1964), $\ln k_1 + \ln x$ for $x \to 0$ and $\ln k_N + \ln x$ for $x \to \infty$, as shown in Figure 2.13. Positive cooperativity in the Hill plot corresponds to $n_H > 1$ and negative cooperativity to $n_H < 1$. The Hill coefficient has nothing to do with 'the number of interacting sites in the system', a rather misleading definition which is often used. It represents a measure of the change of X with $\ln x$ and, as such, it is related to the binding capacity. It is this connection that makes the Hill coefficient of thermodynamic interest in the discussion of binding phenomena. From the definition of n_H in eq (2.44) it follows that (Schellman, 1990a)

$$n_H = \frac{NB}{X(N-X)} = \frac{B}{B_0} \tag{2.45}$$

The Hill coefficient is the ratio between the binding capacity of the system and that of a reference system with the same number of sites. As such, it represents the ratio of the fluctuations of the ligation intermediates around the average value defined by the binding isotherm, relative to that of the reference system. A value of $n_H = 3$, as found for oxygen binding to the four sites of hemoglobin (Chu, Turner and Ackers, 1982), means that the binding capacity of hemoglobin is three times that of a reference system of four sites. A more interesting interpretation of the Hill coefficient will be given in Chapter 3 when we consider site-specific binding effects.

2.3 Properties of the binding curve

We now examine the properties of the binding curve X. Wyman was the first to recognize a fundamental property of X that can be exploited from analysis of experimental data (Wyman, 1964). If measurements of X are

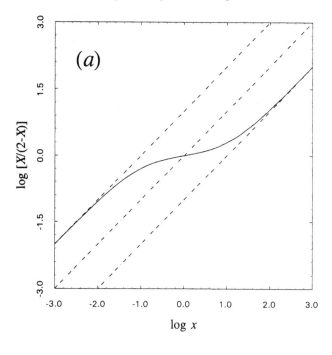

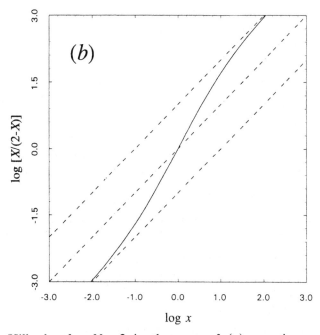

Figure 2.13 Hill plot for $N = 2$ in the case of (*a*) negative or (*b*) positive cooperativity. Curves were drawn using parameter values (*a*) $k_1 = 10 \text{ M}^{-1}$, $k_2 = 0.1 \text{ M}^{-1}$, and (*b*) $k_1 = 0.1 \text{ M}^{-1}$, $k_2 = 10 \text{ M}^{-1}$. Chain lines depict the Hill plot of reference systems for which $K = k_1$, $K = \sqrt{k_1 k_2}$, or $K = k_2$.

available, it is always possible to obtain the coefficient A_N of the partition function by direct integration. The Wyman integral equation

$$\ln x_{\mathrm{m}} = \frac{1}{N} \int_0^N \ln x \, \mathrm{d}X \tag{2.46}$$

defines the mean value of $\ln x$. The value of x_{m} associated with eq (2.46) is hereby defined as the 'mean' ligand activity[†] to indicate the value of x for which $\ln x$ assumes its mean value. Alternatively, the definition (2.46) can be interpreted as the mean value of the ligand chemical potential in $k_{\mathrm{B}}T$ units, or the average work spent on ligating one site of the system. The importance of the quantity x_{m} stems from its connection with the last coefficient of the partition function, A_N. The mean ligand activity is, in fact, the value of x where the unligated and fully ligated configurations of the macromolecule are equally populated. This follows directly from the definition of A_N and solution of the integral equation (2.46). Integration by parts yields

$$\ln x_{\mathrm{m}} = \ln(\infty) - \frac{1}{N} \int_{-\infty}^{\infty} X \, \mathrm{d}\ln x = \ln(\infty) - \frac{1}{N} \ln \frac{\Psi(\infty)}{\Psi(0)} = -\frac{1}{N} \ln A_N \tag{2.47}$$

But from the law of mass action one has

$$A_N = \frac{[\mathrm{MX}_N]}{[\mathrm{M}]x^N} \tag{2.48}$$

Hence, for $x = x_{\mathrm{m}}$, the concentration of the fully ligated species equals that of the reference, unligated configuration. The value of x_{m} uniquely defines the last coefficient of the partition function and, *vice versa*, A_N uniquely defines the value of the mean ligand activity. The connection with the total work of ligating the macromolecule should be emphasized. This quantity is expressed by the standard free energy change

$$\Delta G_N = -k_{\mathrm{B}}T \ln A_N \tag{2.49}$$

as implied by eq (2.31). The average work spent in ligating one site of the macromolecule is the value of ΔG_N divided by N, or the logarithm of x_{m} in $k_{\mathrm{B}}T$ units. This important property of the system can be derived from analysis of X in a direct way, without knowing the detailed form of the partition function. The mean ligand activity should not be confused with $x_{1/2}$, which gives the value of x at half saturation where $X = N/2$, although it is identical to this quantity in some special cases. The import-

---

[†] We employ for x_{m} the term 'mean' in place of 'median' ligand activity, as originally defined by Wyman (1964), to be consistent with the mathematical definition in eq (2.46).

ance of x_m in the analysis of linkage effects involving multiple ligands will be illustrated in Section 2.6.

In practice, calculation of x_m from measurements of X is extremely simple. Wyman pointed out that eq (2.46) can be rewritten as

$$\frac{1}{N}\int_0^N \ln \frac{x}{x_m} dX = 0 \tag{2.50}$$

whereby integration by parts yields

$$\int_{-\infty}^{\ln x_m} X \, d\ln x = \int_{\ln x_m}^{\infty} (N - X) \, d\ln x \tag{2.51}$$

The value of $\ln x_m$ defines the value of $\ln x$ where the area under the curve X, Q_u, at the left-hand side of eq (2.51), equals the area above the curve X up to the asymptote $X = N$, Q_a, as shown in Figure 2.14. Wyman's equal-area rule is an elegant property of X that is widely used in practical

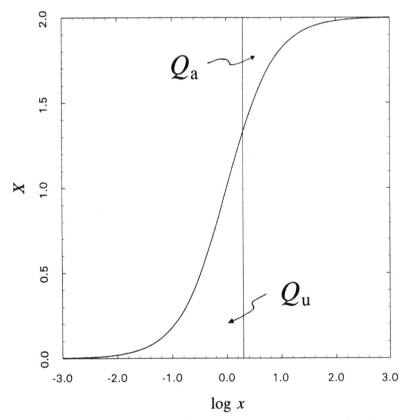

Figure 2.14 Binding curve of a two-site macromolecule. Q_u is the area under the curve, from $-\infty$ up to the point considered, while Q_a is the area above the curve up the asymptote $X = N$, from the point considered to $+\infty$.

applications. There is, however, an important property of x_m that should be emphasized for its operational significance (Di Cera, 1994a). Let us define the functions Q_u and Q_a as follows

$$Q_u(x) = \int_{-\infty}^{\ln x} X \, d \ln x' \tag{2.52a}$$

$$Q_a(x) = \int_{\ln x}^{\infty} (N - X) \, d \ln x' \tag{2.52b}$$

It follows immediately from eq (2.51) that

$$\ln x_m = \ln x + \frac{1}{N}[Q_a(x) - Q_u(x)] \tag{2.53}$$

for any value of x. Therefore, $\ln x_m$ is an invariant property of the binding curve X, since for any two points x_1 and x_2 one has

$$\ln x_1 + \frac{1}{N}[Q_a(x_1) - Q_u(x_1)] = \ln x_2 + \frac{1}{N}[Q_a(x_2) - Q_u(x_2)] = \ln x_m \tag{2.54}$$

As a result of this invariant property, the mean ligand activity can be measured from any point along X. Given the arbitrary point $\ln x_1$, the value of $Q_u(x_1)$ is given by the area under the curve X from $-\infty$ to $\ln x_1$ and likewise the value of $Q_a(x_1)$ is given by the area above X up to the asymptote $X = N$, from $\ln x_1$ to ∞. The difference $Q_a(x_1) - Q_u(x_1)$, divided by N and added to $\ln x_1$, gives $\ln x_m$, regardless of the point x_1. Wyman's equal-area rule is embodied by the invariant property (2.53) as a special case.

Another property of the binding curve X arises in connection with the analysis of its shape. The binding curve X plotted versus $\ln x$ is symmetric when it can be reproduced exactly upon a rotation of $180°$ around a point of coordinates $\ln x_s$, $X = N/2$ (Wyman, 1948; Whitehead, 1980a; Weber, 1982; Di Cera, Hopfner and Wyman, 1992). The point x_s is the center of symmetry of X. Symmetry introduces mathematical constraints among the otherwise independent coefficients of the partition function Ψ. From a mathematical point of view, a function $F(x)$, with $-\infty \leqslant x \leqslant \infty$, can show a number of symmetry properties. If F is an even function, then $F(x) = F(-x)$ and any point in the positive half plane at x has an image at $-x$. Symmetry in this case is *specular* since F in one half plane is the specular image of F in the other half plane. If F is an odd function, then $F(x) = -F(-x)$ and symmetry is *rotational* since F is reproduced exactly when rotated $180°$ around $x = 0$. Specular and rotational symmetries are related by differentiation. It is easy to prove that if F has rotational symmetry, then its derivative dF/dx must have specular symmetry. Hence, if X is symmetric around $\ln x_s$, then the binding capacity B must have specular symmetry around $\ln x_s$. The condition of rotational symmetry for a generic function $F(x)$, where $-\infty \leqslant \ln x \leqslant \infty$ and x_s is the value of x at the center of symmetry, is given by

$$F(x_s\lambda) + F(x_s\lambda^{-1}) = F(0) + F(\infty) \tag{2.55}$$

for any $\lambda \geqslant 0$. At the center of symmetry the value of F is by definition half way between the limiting values $F(0)$ and $F(\infty)$, i.e.,

$$F(x_s) = \frac{F(0) + F(\infty)}{2} \tag{2.56}$$

so that

$$F(x_s\lambda) + F(x_s\lambda^{-1}) = 2F(x_s) \tag{2.57}$$

Using the function X in eq (2.57), we obtain the rather simple relationship

$$X(x_s\lambda) + X(x_s\lambda^{-1}) = N \tag{2.58}$$

since $X(x_s) = N/2$. It follows immediately from eq (2.58) that if X is symmetric, then in a plot of X versus $\ln x$ the values of X computed at points equidistant from the center of symmetry always add up to the total number of sites N. This situation is illustrated in Figure 2.15.

Let $\delta \equiv |\ln \lambda|$ be the distance of a point from the center of symmetry $\ln x_s$ along the $\ln x$ axis. It follows from the condition (2.58) that the area under the curve X from $\ln x_s - \delta$ to $\ln x_s$ must equal the area above the curve X, up to the asymptote $X = N$, from $\ln x_s$ to $\ln x_s + \delta$. This is because X is reproduced exactly upon a rotation of $180°$ around its center of symmetry. In mathematical terms one has

$$\int_{\ln x_s - \delta}^{\ln x_s} X \, d\ln x = \int_{\ln x_s}^{\ln x_s + \delta} (N - X) \, d\ln x \tag{2.59}$$

or

$$\int_{\ln x_s - \delta}^{\ln x_s + \delta} X \, d\ln x = N\delta \tag{2.60}$$

The area under the curve X within two points symmetrically disposed at a distance δ from the center of symmetry equals the area of a rectangle of base δ and height N (see Figure 2.15). This condition holds regardless of the value of δ in the range from zero to ∞. For a symmetric binding curve the center of symmetry necessarily coincides with the value of x at half saturation, i.e., $x_s = x_{1/2}$. We have already pointed out that the mean ligand activity x_m does not always coincide with $x_{1/2}$. This is illustrated by a

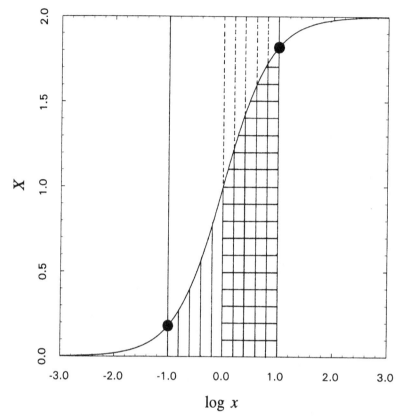

Figure 2.15 Symmetric binding curve of a two-site macromolecule. The two points shown are equidistant from the center of symmetry, $\ln x_s = 0$, and their X values sum up to two, as implied by eq (2.58). The area under the curve, from the lower point to the center of symmetry (vertical shades), is equal to the area above the curve, from the center of symmetry to the upper point (vertical dashed shades). This holds for any pair of points at a distance δ from the center of symmetry, as implied by eq (2.59). It also follows from eq (2.60) that the area under the curve between any two points symmetrically disposed at a distance δ from the center of symmetry (vertical shades and crossed shades) equals the area of a rectangle of base δ and height N (vertical dashed shades and crossed shades).

transformation of eq (2.60). Integration by parts yields

$$\int_{N/2-\alpha}^{N/2+\alpha} \ln \frac{x}{x_s}\, dX = 0 \tag{2.61}$$

Here $0 \leqslant \alpha \leqslant N/2$ represents the difference between the value of X at a distance δ from the center of symmetry and half saturation. The similarity between eq (2.61) and eq (2.50) is evident. However, eq (2.50) holds only for $\alpha = N/2$, while eq (2.61) is valid for any α in the range from zero to

$N/2$. Hence, if X is symmetric, then $x_s = x_m = x_{1/2}$ follows necessarily and the mean ligand activity coincides with the value of x at half saturation. The reverse is not necessarily true, since an asymmetric binding curve can be such that $x_m = x_{1/2}$. On the other hand, if $x_m \neq x_{1/2}$, the binding curve is necessarily asymmetric. In conclusion, the value of $x_{1/2}$ should *never* be used as an estimate of x_m, although $x_m = x_{1/2}$ holds in the case of symmetric binding curves and other special cases.

Another interesting consequence of the symmetry of X is the symmetry of the affinity function. If X is symmetric, then $\ln {}^x K$ is itself symmetric and has the same center of symmetry (Di Cera *et al.*, 1992). In fact, from eq (2.58) it follows that

$$X(x_s \lambda) = N - X(x_s \lambda^{-1}) \tag{2.62a}$$

$$X(x_s \lambda^{-1}) = N - X(x_s \lambda) \tag{2.62b}$$

Upon multiplying eqs (2.62a) and (2.62b) together, rearranging and taking the logarithm of both sides, one has

$$\ln {}^{x_s \lambda} K + \ln {}^{x_s \lambda^{-1}} K = -2 \ln x_s \tag{2.63}$$

Application of eq (2.57) and the definition of ${}^x K$ in eq (2.25) show that if X is symmetric, then $\ln {}^x K$ is also symmetric and has the same center of symmetry. The term at the right-hand side of eq (2.63) is in fact twice the value of $\ln {}^x K$ computed at its center of symmetry, which coincides with that of the function X.

Symmetry imposes constraints on the coefficients of the partition function. Some qualitative, although important, conclusions can be derived intuitively. Since a symmetric binding curve can be reproduced exactly by rotation around the center of symmetry, it follows that knowledge of half of the binding curve (say X from zero to $N/2$) completely defines the other half. Accordingly, only half of the coefficients of the partition function are truly independent. In fact, integration of eq (2.60) yields

$$\Psi(x_s \lambda) = \lambda^N \Psi(x_s \lambda^{-1}) \tag{2.64}$$

which implies that the partition function Ψ is a symmetric polynomial when x is expressed in x_s units. This is demonstrated by rewriting eq (2.64) as follows

$$\sum_{j=0}^{N} (A_j x_s^j - A_{N-j} x_s^{N-j}) \lambda^j = 0 \tag{2.65}$$

Since eq (2.65) must hold for any λ, the coefficients of each power of λ necessarily vanish. The term with $j = 0$ leads to $x_s = A_N^{-1/N}$, and hence

$x_s = x_m$ as already established. Elimination of x_s from all the other terms in eq (2.65) yields the condition of symmetry for the equilibrium constants as follows (Wyman, 1948)

$$A_j A_N^{-j/N} = A_{N-j} A_N^{-(N-j)/N} \tag{2.66}$$

which holds for any j if X is symmetric. The condition above states in more explicit form the fact that the partition function Ψ is itself symmetric when x is expressed in x_s or x_m units. The change of variable $y = x/x_m$ in Ψ yields a transformation whereby any value along the $\ln x$ axis is shifted by a constant amount $\ln x_m$. The transformation is equivalent to a 'normalization' and does not affect the shape of any of the response functions X, B or ${}^x K$. The normalized form of the partition function is

$$\Psi(y) = 1 + A_1 x_m y + A_2 x_m^2 y^2 + \ldots + A_N x_m^N y^N$$
$$= 1 + a_1 y + a_2 y^2 + \ldots + y^N \tag{2.67}$$

where

$$a_j = \frac{A_j}{A_N^{j/N}} \tag{2.68}$$

Symmetry implies that the coefficient of the jth power of y in Ψ is the same as that of the $(N - j)$th power of y, as seen for symmetric normalized polynomials. As a result, knowledge of half of the equilibrium constants completely defines the partition function. For $N = 1$ or $N = 2$ the condition (2.66) is a mere tautology. For $N = 3$ the condition is $A_3 = (A_2/A_1)^3$ or $k_1 k_3 = k_2^2$. For $N = 4$ the condition of symmetry is $A_4 = (A_3/A_1)^2$, or $k_1 k_4 = k_2 k_3$ and does not depend on A_2. In general, the condition of symmetry involves all As in the partition function for N odd and all As but $A_{N/2}$ for N even.

2.4 Properties of the binding capacity

In Section 1.3 we established that the binding capacity parallels physical response functions such as the heat capacity and the compressibility and is strictly related to the stability principles embodied by the second law of thermodynamics. The function B is always positive and an increase/decrease in the chemical potential of X always leads to an increase/decrease of X. From a mathematical point of view, the positivity of B implies that X is a monotonic function of $\ln x$ whose derivative never vanishes for finite values of x. In other words, the equation $B = 0$ has no solutions for finite x. This important property of B can be proved mathematically in a direct way (Di Cera, 1992b). Let us rewrite eq (2.39) as follows

$$B = \frac{\sum_{j=0}^{N} j^2 A_j x^j \sum_{j=0}^{N} A_j x^j - \left(\sum_{j=0}^{N} j A_j x^j\right)^2}{\left(\sum_{j=0}^{N} A_j x^j\right)^2} \tag{2.69}$$

The denominator of eq (2.69) is always positive, since the partition function has all positive coefficients and cannot vanish for any $x \geqslant 0$. The sign of B depends solely on the expression in the numerator, which can be rewritten as

$$p(x) = \sum_{i=0}^{N} \sum_{j=0}^{N} (i^2 - ij) A_i A_j x^{i+j} \tag{2.70}$$

Since it makes no difference in which order the double summation is carried out, interchanging the indices i and j yields

$$p(x) = \sum_{j=0}^{N} \sum_{i=0}^{N} (j^2 - ji) A_j A_i x^{j+i} \tag{2.71}$$

Hence,

$$p(x) = \frac{1}{2} \sum_{i=0}^{N} \sum_{j=0}^{N} (i - j)^2 A_i A_j x^{i+j} \tag{2.72}$$

and p is a polynomial with all positive coefficients and cannot vanish for any finite value of x. The only two values of x for which $B = 0$ are $x = 0$, since $p(0) = 0$, and $x = \infty$, since the denominator of (2.69) is a polynomial of higher degree in x than p itself.

Another property of B is of considerable interest. Measurements of X yield information on x_m and hence the last coefficient of the partition function A_N, through the integral equation (2.46). Information on the other coefficients of the partition function cannot be derived by numerical integration of X and requires nonlinear least-squares analysis of the data. However, there is another set of relationships such as (2.46) which allows us, at least in principle, to determine uniquely the coefficients of Ψ from numerical integration of X and B. Let us rewrite eq (2.46) as follows, using the definition of B

$$\ln x_m = \frac{1}{N} \int_{-\infty}^{\infty} \ln x B \, \mathrm{d} \ln x \tag{2.73}$$

A close examination of eq (2.73) reveals an important aspect of B, first pointed out by Sturgill and Biltonen (1976). The integral (2.73) is equivalent to the calculation of the average value of $\ln x$ over the entire range of definition of this variable. The weighting function for the integration is the binding capacity B divided by N. The normalization condition

$$\int_{-\infty}^{\infty} B \, \mathrm{d}\ln x = N \tag{2.74}$$

makes this property of B even more evident. The area under the B curve is always finite and equals the total number of sites, independent of the particular form of B. The function B/N can thus be regarded as the *probability density function* of the variable $\ln x$ (Di Cera and Chen, 1993). Calculation of $\ln x_m$ is equivalent to calculation of the first moment of B/N. Since $\Delta G_m = k_B T \ln x_m$ is the average work spent in ligating one site of the macromolecule, the ratio B/N computed at $x = \xi$ and multiplied by $\mathrm{d}\ln \xi$ gives the probability fraction, $\mathrm{d}v$, that ΔG_m has a value lying between $k_B T \ln \xi$ and $k_B T (\ln \xi + \mathrm{d}\ln \xi)$ (see Figure 2.16(*b*)). Alternatively, this probability fraction can be derived from the plot of X versus $\ln x$ by computing the slope at $x = \xi$ and dividing the result by the number of sites N, as shown in Figure 2.16(*a*). The most probable value of ΔG_m is derived from the first moment of the B/N distribution.

We now consider higher amounts of B/N, i.e.,

$$\langle \ln^k x \rangle = \frac{1}{N} \int_{-\infty}^{\infty} \ln^k x B \, \mathrm{d}\ln x \tag{2.75}$$

The connection with the coefficients of the partition function is obtained from the analytical solution of eq (2.75). Sturgill and Biltonen (1976) have provided the solution for the first two moments in a number of cases of interest. Here we give the solution in the general case, using the moment generating function (Di Cera and Chen, 1993)

$$G(\omega) = \int_0^{\infty} x^{\omega - 1} B \, \mathrm{d}x = \int_0^N x^\omega \, \mathrm{d}X \tag{2.76}$$

The function G can be cast in terms of B or X, as indicated by eq (2.76), and as such it can be constructed directly from experimental data collected either as B or X measurements. However, for the purpose of obtaining an analytical solution for G we will use the properties of B which, unlike X, is related to a probability density function. From the definition (2.76), the kth moment of the B/N distribution is derived as (Feller, 1950)

$$\langle \ln^k x \rangle = \frac{1}{N} \frac{\partial^k G(0)}{\partial \omega^k} \tag{2.77}$$

The generating function G exists for $0 < \omega < 1$, by virtue of the normalization condition (2.74).

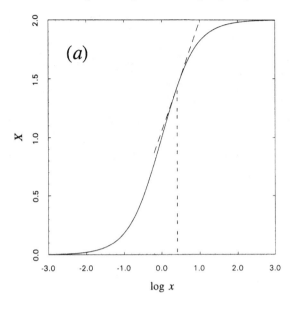

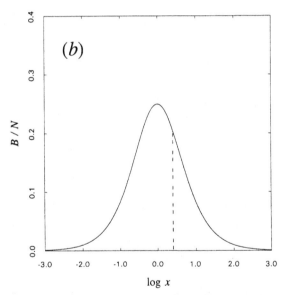

Figure 2.16 Binding capacity as a probability density function. The value of B/N at $x = \xi$, times $\mathrm{d}\ln\xi$, gives the probability fraction, $\mathrm{d}v$, that ΔG_m lies between $k_B T \ln\xi$ and $k_B T(\ln\xi + \mathrm{d}\ln\xi)$. This is equivalent to computing the probability density that the logarithm of the mean ligand activity lies between $\ln\xi$ and $\ln\xi + \mathrm{d}\ln\xi$. The value of B/N can be derived either from the curve X as the slope at $x = \xi$ divided by N (a), or directly from a plot of B/N (b). In the example shown one has $B/N = 0.2$ at $x = \xi$ $(\log\xi = 0.4)$, which implies that $\mathrm{d}v/\mathrm{d}\ln\xi = 0.2$. The probability density that $\log x_m$ lies between 0.4 and $0.4 + \Delta\log\xi$ is $0.2\Delta\log\xi$.

We now rewrite the partition function Ψ in the factorial form

$$\Psi(x) = (1 + \alpha_1 x)(1 + \alpha_2 x) \ldots (1 + \alpha_N x) \qquad (2.78)$$

where α_j is the jth coefficient of the factorization related to the jth root of Ψ. Klotz (1985, 1993) refers to these coefficients as binding constants of 'ghost' sites. Since Ψ is a polynomial of degree N with real and positive coefficients, the partition function is analytical everywhere on the positive real axis, which is the one of physical significance. The coefficients α can be real and positive, or complex conjugate. The connection between the nature of the coefficients α and cooperativity will soon appear evident. Since Ψ can be cast in a simple form such as (2.78), regardless of the nature of its roots, one has

$$X = \sum_{j=1}^{N} \frac{\alpha_j x}{1 + \alpha_j x} \qquad (2.79)$$

$$B = \sum_{j=1}^{N} \frac{\alpha_j x}{(1 + \alpha_j x)^2} \qquad (2.80)$$

Hence, the solution of eq (2.76) is itself a linear combination of N similar terms of the form

$$g(\omega) = \int_0^\infty \frac{\alpha x^\omega}{(1 + \alpha x)^2} \, dx \qquad (2.81)$$

and convergence of the integral necessarily demands $0 < \omega < 1$. If α is real, then the transformation $\theta = \alpha x / (1 + \alpha x)$ yields

$$g(\omega) = \int_0^1 \alpha^{-\omega} \theta^\omega (1 - \theta)^{-\omega} \, d\theta = \alpha^{-\omega} \frac{\pi \omega}{\sin \pi \omega} \qquad (2.82)$$

If α is complex, then g is obtained from the theorem of residues (Wilson, 1911) by properly choosing an integration path in the complex plane. Introducing the complex variable $z = \alpha x$ in eq (2.81) yields

$$g(\omega) = \int_\gamma \frac{\alpha^{-\omega} z^\omega}{(1 + z)^2} \, dz \qquad (2.83)$$

where γ denotes integration along the semi-line in the direction of α in the complex plane. Since the function $z^\omega / (1 + z)^2$ is holomorphic in the region I_r^R shown in Figure 2.17, the path integral along a closed contour C must vanish by virtue of Cauchy's theorem. Hence,

$$\int_C \frac{\alpha^{-\omega} z^\omega}{(1 + z)^2} \, dz = \int_{I_1} + \int_{I_2} + \int_{I_3} + \int_{I_4} = 0 \qquad (2.84)$$

The integral I_1 computed in the complex plane along the semi-line in the

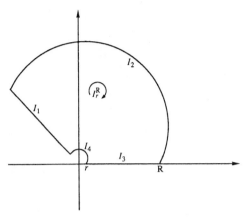

Figure 2.17 Integration path in complex plane for the integral (2.84). The argument of the integral is a holomorphic function in I_r^R and the path integral vanishes according to Cauchy's theorem.

direction of α for $R \to \infty$ and $r \to 0$ is the required solution for g. The integral I_2 is such that

$$I_2 = \left| \int_{I_2} \frac{\alpha^{-\omega} z^{\omega}}{(1 + z)^2} \, dz \right| \leq \int_{I_2} \frac{|\alpha^{-\omega}||z^{\omega}|}{|1 + z|^2} |dz| \leq \int_0^{\pi} \frac{|\alpha^{-\omega}| R^{1+\omega}}{(R - 1)^2} \, d\varphi = \frac{\pi |\alpha^{-\omega}| R^{1+\omega}}{(R - 1)^2}$$

(2.85)

and goes to zero for $R \to \infty$, since $\omega < 1$. The integral I_3 involves the positive real axis only. The integral I_4 is such that

$$I_4 = \left| \int_{I_4} \frac{\alpha^{-\omega} z^{\omega}}{(1 + z)^2} \, dz \right| \leq \int_{I_4} \frac{|\alpha^{-\omega}||z^{\omega}|}{|1 + z|^2} |dz| \leq \int_0^{\pi} \frac{|\alpha^{-\omega}| r^{1+\omega}}{(1 - r)^2} \, d\varphi = \frac{\pi |\alpha^{-\omega}| r^{1+\omega}}{(1 - r)^2}$$

(2.86)

and goes to zero for $r \to 0$. The solution for g is determined solely by the value of $-I_3$, i.e.,

$$g(\omega) = \int_0^{\infty} \frac{\alpha^{-\omega} x^{\omega}}{(1 + x)^2} \, dx = \int_0^{\infty} \frac{\alpha y^{\omega}}{(1 + \alpha y)^2} \, dy = \alpha^{-\omega} \frac{\pi \omega}{\sin \pi \omega} \quad (2.87)$$

where $y = x/\alpha$. The solution is identical to that found in the case of α real and positive. Hence, the general form of G is

$$G(\omega) = \frac{\pi \omega}{\sin \pi \omega} \sum_{j=1}^{N} \alpha_j^{-\omega} \quad (2.88)$$

To obtain the moments of eq (2.77), we take the Taylor expansion of

$G = fh$, where

$$f(\omega) = \frac{\pi\omega}{\sin \pi\omega} \tag{2.89a}$$

$$h(\omega) = \alpha_1^{-\omega} + \alpha_2^{-\omega} + \ldots + \alpha_N^{-\omega} \tag{2.89b}$$

The Taylor expansion of f is

$$f(\omega) = 1 + \sum_{k=1}^{\infty} \frac{2(2^{2k-1} - 1)|B_{2k}|(\pi\omega)^{2k}}{(2k)!} \tag{2.90}$$

where B_{2k} is the $2k$th Bernoulli number (Gradshtein and Ryzhik, 1980). The Taylor expansion of h is

$$h(\omega) = N - \sum_{j=1}^{N} \omega \ln \alpha_j + \sum_{j=1}^{N} \frac{\omega^2 \ln^2 \alpha_j}{2!} - \sum_{j=1}^{N} \frac{\omega^3 \ln^3 \alpha_j}{3!} + \ldots$$

$$= N - C_1\omega + \frac{C_2\omega^2}{2!} - \frac{C_3\omega^3}{3!} + \ldots \tag{2.91}$$

Hence,

$$\langle \ln x \rangle = -\frac{C_1}{N} \tag{2.92a}$$

$$\langle \ln^2 x \rangle = \frac{\pi^2}{3} + \frac{C_2}{N} \tag{2.92b}$$

$$\langle \ln^3 x \rangle = -\pi^2 C_1 - \frac{C_3}{N} \tag{2.92c}$$

$$\langle \ln^4 x \rangle = \frac{7\pi^4}{15} + 2\pi^2 C_2 + \frac{C_4}{N} \tag{2.92d}$$

and so on. The first moment equals $\ln x_{\mathrm{m}}$. The second moment is proportional to the sum of squares of the logarithm of each coefficient α, and in general the kth moment is expressed as the sum of all even/odd powers of each $\ln \alpha$ term up to order k, for k even/odd. The Cs are uniquely defined in terms of the moments and themselves uniquely define the sum of the various powers of $\ln \alpha_j$s. Hence (Di Cera and Chen, 1993),

Theorem: The first N moments of the binding capacity uniquely define the N independent coefficients of the partition function.

This theorem demonstrates how the statistical properties of B can be exploited for the determination of the coefficients of Ψ. The theorem is general and applies to any thermodynamic ensemble whose partition function can be cast in terms of a polynomial of finite order. Binding capacity measurements have been used to resolve the moments of the B/N distribution (Mountcastle, Freire and Biltonen, 1976; Sturgill, 1978; Sturgill, Johnson and Biltonen, 1978; Di Cera and Chen, 1993). These measurements have demonstrated that it is possible to decouple the first two moments of the B/N distribution in a satisfactory way.

The first two moments of the B/N distribution are of particular importance in practical applications. The first moment contains information on the average free energy of binding (in $k_B T$ units) per site. However, this quantity *per se* provides no information on cooperativity. Cooperativity is reflected by the steepness of the binding curve and hence by the binding capacity. This situation is illustrated in Figure 2.18 for the case of positive

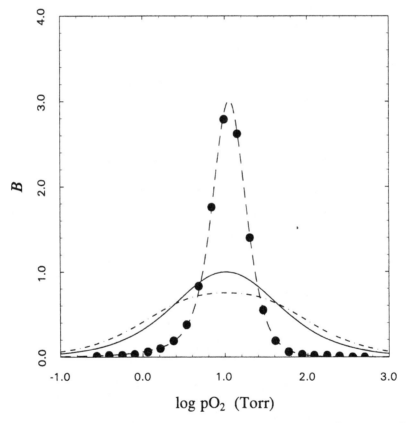

Figure 2.18 Binding capacity for $N = 4$ for various cooperative cases. Points depict the B distribution for oxygen binding to hemoglobin (Di Cera, Doyle and Gill, 1988b). The dashed curve is the best fit of the data. The continuous curve refers to the reference system, while the chain curve refers to a negatively cooperative system. The three curves have the same value of N and mean ligand activity, but differ substantially in shape.

and negative cooperativity. The three curves depict different B distributions with the same value of x_m. The difference in cooperativity stems from the dispersion of the distributions rather than the position along the $\ln x$ axis, which defines $\ln x_m$. The dispersion is related to the variance

$$\sigma^2 = \langle \ln^2 x \rangle - \langle \ln x \rangle^2 \tag{2.93}$$

The variance of B/N is related to the cooperativity of the system. This is intuitively obvious because the area under the B curve is finite and if B increases and becomes more sharply peaked, then σ^2 must decrease to keep the area constant and *vice versa*. We know from eq (2.45) that the ratio of the binding capacity of the system with respect to that of a reference system containing the same number of sites gives the Hill coefficient, i.e., a classical measure of cooperativity. The more peaked B, the higher the cooperativity. Hence, we expect the smaller the variance, the higher the cooperativity of the macromolecule. Information on the cooperative nature of the binding curve must be contained in the second moment of the binding capacity. Shifting the B curve along the $\ln x$ axis does not affect its shape and hence the variance. If the normalized form of the partition function is used, the first moment vanishes and the variance is given by the second moment only. The value of σ^2 can therefore be thought of as the variance of the normalized form of the binding capacity, shifted along the $\ln x$ axis to make its first moment equal to zero.

We now calculate σ^2 explicitly for a cooperative system. First, let us compute the first two moments for the reference system, for which the partition function is given by eq (2.9) and clearly $\alpha_1 = \alpha_2 = \ldots = \alpha_N = K$. The moment generating function is

$$G(\omega) = NK^{-\omega} \frac{\pi\omega}{\sin \pi\omega} \tag{2.94}$$

The first two moments and the variance are

$$\langle \ln x \rangle = -\ln K \tag{2.95a}$$

$$\langle \ln^2 x \rangle = \frac{\pi^2}{3} + \ln^2 K \tag{2.95b}$$

$$\sigma_0^2 = \frac{\pi^2}{3} \tag{2.95c}$$

The first two moments are independent of N and depend solely on the binding constant K. The variance is independent of N and K, as expected, and reflects an important property of the B distribution in the reference system. The value of $\sigma_0^2 = \pi^2/3$ sets the boundary between positive ($\sigma^2 < \sigma_0^2$) and negative ($\sigma^2 > \sigma_0^2$) cooperativity. In the general case where the macromolecule has a partition function such as eq (2.37), or equivalently eq (2.78), the generating function is given by (2.88) and we have

$$\langle \ln x \rangle = -\frac{\ln \alpha_1 + \ln \alpha_2 + \ldots + \ln \alpha_N}{N} \qquad (2.96a)$$

$$\langle \ln^2 x \rangle = \frac{\pi^2}{3} + \frac{\ln^2 \alpha_1 + \ln^2 \alpha_2 + \ldots + \ln^2 \alpha_N}{N} \qquad (2.96b)$$

Upon expressing each coefficient α in polar coordinates in the complex plane as $\alpha = r \exp i\varphi$, with a radius r and an angle φ, we obtain

$$\sigma^2 = \frac{\pi^2}{3} + \frac{1}{2N^2}\sum_{i=1}^{N}\sum_{j=1}^{N} \ln^2 \frac{r_i}{r_j} - \frac{1}{N}\sum_{j=1}^{N}\varphi_j^2 = \sigma_0^2 + \sigma_r^2 - \sigma_\varphi^2 \qquad (2.97)$$

which embodies the mathematical 'driving forces' for cooperativity.

There are three components that determine the value of σ^2. The first component is the variance of the reference system. The second and third components are derived from the distribution of r and φ values associated with the N coefficients α in the partition function (2.78). The second component, σ_r^2, is the variance of the r values, while the third component, σ_φ^2, is the variance of the φ values. These variances oppose each other and their net balance determines whether the variance of the B/N distribution in the general case is greater or smaller than that of the reference system. Cooperativity is determined by the balance of these two opposite 'forces'. The variance of the norm of the coefficients, σ_r^2, tends to increase the value of σ^2 with respect to σ_0^2. The variance of the angles formed by the α in the complex plane, σ_φ^2, tends to decrease the value of σ^2 with respect to σ_0^2. Therefore a necessary, although not sufficient, condition for positive cooperativity is that at least two coefficients α in the partition function are complex conjugate. No positive cooperativity is possible if Ψ can be written in factorial form with coefficients that are all real and positive. The necessary and sufficient condition for positive cooperativity is that $\sigma_\varphi^2 > \sigma_r^2$, i.e., that the dispersion of the angles formed by the coefficients α in the complex plane exceeds the dispersion of their radii. This is always guaranteed if the αs are distributed along the unit circle, since in this case σ_r^2 is at its minimum value of zero. The maximum value allowable for σ_φ^2 is exactly σ_0^2, or $\pi^2/3$, since σ^2 cannot be negative. On the other hand, negative cooperativity has as a necessary condition that $\sigma_r^2 > 0$, or that at least two coefficients α have a different norm. The necessary and sufficient condition is that $\sigma_\varphi^2 < \sigma_r^2$, i.e., that the dispersion of the angles formed by the coefficients α in the complex plane does not exceed the dispersion of their radii. This is always guaranteed if the αs are all real and distinct, since in this case σ_φ^2 is at its minimum value of zero. Unlike the case of σ_φ^2, the value of σ_r^2 can grow without bounds since σ^2 remains positive. The distribution of coefficients α in the complex plane shown in Figure 2.19 illustrates practically the connection between the properties of the B distribution and cooperativity.

The predictions derived from eq (2.97) are confirmed by the analysis of two special cases of interest. The highest degree of positive cooperativity is reached when all terms of the partition function are negligible compared to those reflecting the contribution of the unligated and fully ligated forms. In this case the partition function tends to the asymptotic form

$$\Psi(x) = 1 + A_N x^N \qquad (2.98)$$

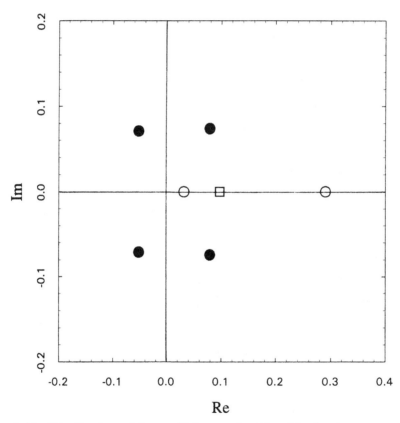

Figure 2.19 Distribution of the coefficients α for $N = 4$ in the three cases shown in Figure 2.18. Positive cooperativity is linked to complex conjugate values (filled circles), while negative cooperativity is characterized by distinct values on the positive real axis (open circles). In the case shown, the coefficients are equal in pairs. The reference system is depicted as four identical values on the positive real axis (open squares).

Although this form is physically unrealistic, it is instructive to analyze the behavior of σ^2 in such a limit. The coefficients α are given by

$$\alpha_{k+1} = A_N^{1/N} \exp\left[\frac{i\pi(N - 1 - 2k)}{N}\right] \quad (k = 0, 1, \ldots N - 1) \quad (2.99)$$

and map on the unit circle in the complex plane upon a simple conformal transformation. In this case $\sigma_r^2 = 0$ and we need only to consider the distribution of angles

$$\varphi_{k+1} = \frac{\pi(N - 1 - 2k)}{N} \quad (k = 0, 1, \ldots N - 1) \quad (2.100)$$

Application of eqs (2.96) and (2.97) yields

$$\langle \ln x \rangle = -\frac{\ln A_N}{N} \tag{2.101a}$$

$$\langle \ln^2 x \rangle = \frac{\pi^2}{3N^2} + \frac{\ln^2 A_N}{N^2} \tag{2.101b}$$

$$\sigma^2 = \frac{\pi^2}{3N^2} = \frac{\sigma_0^2}{N^2} \tag{2.101c}$$

The variance is identical to that of a reference system divided by N^2. This is the minimum possible value of σ^2 for a system containing N sites. The variance tends to zero in the limit $N \to \infty$, indicating that the binding capacity of a fully cooperative system of infinite size is a δ-function peaked at $\ln x_m = \langle \ln x \rangle$. In this limit, the αs become uniformly distributed on the unit circle with a variance $\sigma_\varphi^2 = \sigma_0^2$. Hence, the term $\sigma_0^2 = \pi^2/3$ in eq (2.97) is the variance of the distribution of the angles when the αs are uniformly distributed on the unit circle. It also follows from the foregoing argument that, if B is finite anywhere along the $\ln x$ axis, then the αs cannot be uniformly distributed on the unit circle. This result states in a form much simpler than the Lee–Yang theorem of statistical mechanics (Lee and Yang, 1952) the impossibility of obtaining a phase transition-like divergence of B ($\sigma^2 = 0$) for finite N, since in this case B is always finite and so is its variance.

The other limiting case is observed when the partition function factors out into N binomial terms with all real and positive coefficients, i.e.,

$$\Psi(x) = (1 + \alpha_1 x)(1 + \alpha_2 x) \ldots (1 + \alpha_N x) \tag{2.102}$$

This situation arises when dealing with strong negative cooperativity, or when all sites are different and independent. In this case $\sigma_\varphi^2 = 0$ and we need only to consider the distribution of the norms. The first two moments and the variance σ^2 are in this case

$$\langle \ln x \rangle = -\frac{\ln \alpha_1 + \ln \alpha_2 + \ldots + \ln \alpha_N}{N} \tag{2.103a}$$

$$\langle \ln^2 x \rangle = \frac{\pi^2}{3} + \frac{\ln^2 \alpha_1 + \ln^2 \alpha_2 + \ldots + \ln^2 \alpha_N}{N} \tag{2.103b}$$

$$\sigma^2 = \frac{\pi^2}{3} + \frac{1}{2N^2}\sum_{i=1}^{N}\sum_{j=1}^{N} \ln^2 \frac{\alpha_i}{\alpha_j} \tag{2.103c}$$

The variance grows without bounds with the heterogeneity of the binding affinity of the individual sites.

2.5 Linkage effects: The reference cycle

We now turn to the properties of a macromolecule reacting with two ligands, X and Y. This provides the simplest case of interest in the analysis of linkage effects that form the basis of our subsequent discussion of site-specific binding processes. At constant T and P, a macromolecule containing two distinct binding sites for ligands X and Y is a generalized ensemble whose properties can be derived from the definition of the partition function

$$\Psi(x, y) = Y_{00} + Y_{10}x + Y_{01}y + Y_{11}xy \tag{2.104}$$

The first index refers to ligand X, whose activity is x and the second index to ligand Y, whose activity is y. Each ligation intermediate is characterized by a mechano-thermal partition function Y_{ij}, related to the Gibbs free energy $G_{ij} = -k_B T \ln Y_{ij}$. The standard free energy change for the reaction

$$M + iX + jY \Leftrightarrow MX_iY_j \tag{2.105}$$

is given by

$$\Delta G_{ij} = G_{ij} - G_{00} = k_B T \ln \frac{Y_{00}}{Y_{ij}} = -k_B T \ln A_{ij} \tag{2.106}$$

The mechano-thermal partition function Y_{00} has been set equal to one without loss of generality and the unligated form of the macromolecule is taken as reference. The term A_{ij} gives the overall equilibrium binding constant associated with (2.105). When dealing with linkage effects, we are specifically interested in the way the two ligands affect the binding of each other. Analysis of linkage effects in terms of thermodynamic principles allows us to derive the properties of either ligand from the binding of the other, with the obvious practical advantage that information on either ligand can be derived from a set of a few carefully designed experiments.

In the presence of ligand Y, the response functions for ligand X can be derived from the partition function by differentiation, keeping the activity of Y constant. We focus our discussion on ligand X, but of course it makes no difference which ligand is chosen. The function X is

$$X = \left(\frac{\partial \ln \Psi}{\partial \ln x}\right)_y = \frac{A_{10}x + A_{11}xy}{1 + A_{10}x + A_{01}y + A_{11}xy} = \frac{{}^xKx}{1 + {}^xKx} \quad (2.107)$$

which is the familiar form involving the affinity function. The function ${}^xK$ is given by

$$^xK = \frac{A_{10} + A_{11}y}{1 + A_{01}y} \quad (2.108)$$

It is independent of x, as expected for a macromolecule containing a single site for ligand X, but is a function of y. The effect of the second ligand can be quantified by studying how the affinity function of ligand X changes with the ligand activity of Y. In the absence of Y, the affinity function of X is equal to A_{10}. In the presence of saturating concentrations of Y, the affinity function of X becomes equal to A_{11}/A_{01}. Hence, the affinity of X increases with y if $A_{11} > A_{01}A_{10}$, and decreases otherwise. The presence of Y has no effect on the binding of X if $A_{11} = A_{01}A_{10}$. Linkage is reciprocal, as can be seen from consideration of the binding properties of Y in the presence of X. The relevant expression for Y is

$$Y = \left(\frac{\partial \ln \Psi}{\partial \ln y}\right)_x = \frac{A_{01}y + A_{11}xy}{1 + A_{10}x + A_{01}y + A_{11}xy} = \frac{{}^yKy}{1 + {}^yKy} \quad (2.109)$$

and the affinity function ${}^yK$ is given by

$$^yK = \frac{A_{01} + A_{11}x}{1 + A_{10}x} \quad (2.110)$$

Hence, if increasing y increases ${}^xK$, then an increase of x must increase ${}^yK$ and *vice versa*. The condition for linkage between X and Y is one and the same for both ligands and depends solely on the value of the product $A_{01}A_{10}$ compared to A_{11}. The significance of these terms can be understood from the definition (2.106).

We now introduce the free energy of coupling (Weber, 1975, 1992; Hill, 1977)

$$\Delta G_c = \Delta G_{11} - \Delta G_{01} - \Delta G_{10} \quad (2.111)$$

If the free energy change for binding X and Y is different from the sum of the free energies of binding of X and Y separately, then binding of either ligand must affect the binding of the other. Linkage is observed whenever $\Delta G_c \neq 0$. Specifically,

$$\Delta G_c < 0 \quad \text{positive linkage} \tag{2.112a}$$

$$\Delta G_c > 0 \quad \text{negative linkage} \tag{2.112b}$$

$$\Delta G_c = 0 \quad \text{no linkage} \tag{2.112c}$$

The significance of the foregoing arguments becomes clear upon examination of the reference cycle depicted in Figure 2.20. The binding reactions of X and Y to the macromolecule are depicted as edges of a cycle with the ligation intermediates located at the vertices. Let ${}^0K_X$ and ${}^1K_X$ be the binding affinities of X to the unligated and Y-ligated forms of the macromolecule and, likewise, let ${}^0K_Y$ and ${}^1K_Y$ be the analogous quantities for Y. Then, by definition

$${}^0K_X = \frac{[MX]}{[M]x} \tag{2.113a}$$

$${}^1K_X = \frac{[MXY]}{[MY]x} \tag{2.113b}$$

$${}^0K_Y = \frac{[MY]}{[M]y} \tag{2.113c}$$

$${}^1K_Y = \frac{[MXY]}{[MX]y} \tag{2.113d}$$

Hence,

$${}^0K_X{}^1K_Y = {}^0K_Y{}^1K_X \tag{2.114}$$

Of the four reactions depicted in Figure 2.20, only three are independent and the fourth one can be derived unequivocally from the others. The connection between the parameters defined in the reference cycle and the coefficients of the partition function is derived in a straightforward way and is given by

$${}^0K_X = A_{10} \tag{2.115a}$$

$$\begin{array}{ccccc} & & {}^0K_X x & & \\ & \text{M} & \Leftrightarrow & \text{MX} & \\ {}^0K_Y y & \Updownarrow & & \Updownarrow & {}^1K_Y y \\ & \text{MY} & \Leftrightarrow & \text{MXY} & \\ & & {}^1K_X x & & \end{array}$$

Figure 2.20 The reference cycle.

$$^1K_X = \frac{A_{11}}{A_{01}} \tag{2.115b}$$

$$^0K_Y = A_{01} \tag{2.115c}$$

$$^1K_Y = \frac{A_{11}}{A_{10}} \tag{2.115d}$$

The free energy of coupling is then

$$\Delta G_c = -k_B T \ln \frac{^1K_X}{^0K_X} = -k_B T \ln \frac{^1K_Y}{^0K_Y} \tag{2.116}$$

and is given by the difference in the free energy of binding of either ligand induced by the presence of the other. If the binding affinity increases in the presence of the second ligand, then $\Delta G_c < 0$ and so forth for the other cases.

The affinity function plays an important role in the analysis of linkage effects in the particular case of the reference cycle. Consider the derivative of $\ln{}^xK$ in eq (2.108) with respect to the ligand activity y of Y

$$\frac{d\ln{}^xK}{d\ln y} = \frac{A_{11}y}{A_{10} + A_{11}y} - \frac{A_{01}y}{1 + A_{01}y} \tag{2.117}$$

This is a key thermodynamic property whose significance can be understood as follows. The four configurations in the reference cycle can be split into two manifolds, one reflecting the binding of Y to M to yield MY, and the other reflecting the binding of Y to MX to yield MXY. Each manifold can be assigned an appropriate partition function

$$^0\Psi(y) = 1 + {}^0K_Y y \tag{2.118a}$$

$$^1\Psi(y) = 1 + {}^1K_Y y \tag{2.118b}$$

The configuration M is used as reference in the manifold $\{M, MY\}$, while MX is used as reference in the manifold $\{MX, MXY\}$. The connection with the partition function of the system as a whole is provided by the relationship

$$\Psi(x, y) = {}^0\Psi(y) + {}^1\Psi(y){}^0K_X x \tag{2.119}$$

which is identical to (2.104), by virtue of eqs (2.115a)–(2.115d) and the relationship between the As and the Ys. The index 0 or 1 refers to the ligation state of the macromolecule with respect to X. The function Y in the two manifolds is

$$^1Y = \frac{d\ln{}^1\Psi}{d\ln y} = \frac{^1K_Y y}{1 + {}^1K_Y y} = \frac{A_{11}y}{A_{10} + A_{11}y} \tag{2.120a}$$

$$^0Y = \frac{\mathrm{d} \ln {}^0\Psi}{\mathrm{d} \ln y} = \frac{{}^0K_\mathrm{Y} y}{1 + {}^0K_\mathrm{Y} y} = \frac{A_{01} y}{1 + A_{01} y} \qquad (2.120\mathrm{b})$$

These expressions are identical to those at the right-hand side of eq (2.117) so that

$$\frac{\mathrm{d} \ln {}^x K}{\mathrm{d} \ln y} = {}^1Y - {}^0Y = \Delta Y \qquad (2.121)$$

The change in the function $\ln {}^x K$ gives the difference in Y-ligated sites between the two manifolds. Since the two manifolds represent the unligated and X-ligated configurations of the macromolecule, eq (2.121) expresses the change in ligation with respect to ligand Y when ligand X is bound, or the net number, ΔY, of molecules of Y exchanged upon binding of X. If $\Delta Y > 0$, binding of Y facilitates binding of X and the affinity for X increases with increasing y. On the other hand, if $\Delta Y < 0$, binding of Y opposes binding of X and the affinity for X decreases with increasing y. Finally, if $\Delta Y = 0$, binding of Y has no net effect on binding of X. The trivial case where Y does not bind to the macromolecule is included in this third possibility as a special case.

The function $\ln {}^x K$ for the interaction of thrombin with the inhibitor hirudin is plotted in Figure 2.21 as a function of the logarithm of Na^+ concentration. Hirudin represents ligand X in the reference cycle, while Na^+ plays the role of ligand Y (Ayala and Di Cera, 1994). Increasing the concentration of Na^+, at constant ionic strength, increases the binding affinity of hirudin, so that the two ligands are positively linked. The slope of $\ln {}^x K$ is given in Figure 2.22 and represents the net number of Na^+ ions taken up upon hirudin binding. Analysis of the data in Figure 2.21 according to eq (2.108) yields all the coefficients of the partition function (2.104). The free energy of coupling for the thermodynamic cycle involving hirudin and Na^+ binding to thrombin is $\Delta G_c = -1.7 \pm 0.1 \mathrm{\ kcal/mol}$, which indicates that hirudin and Na^+ enhance each other's binding free energy by 1.7 kcal/mol. This is a paradigmatic example of the power of approaching macromolecular interactions in terms of linkage effects. The binding properties of Na^+ which are difficult to measure directly are derived through the effect on the binding properties of hirudin. Hence, binding of a ligand X to the macromolecule contains information not only on the properties of X, but also on any other process linked to the binding of X. The self-consistency of the results obtained from linkage studies can be checked by direct measurements of binding, when available. In the case of Na^+ binding to thrombin, such measurements are shown in Figure 2.3 as determined by fluorescence titration (Ayala and Di Cera, 1994). A

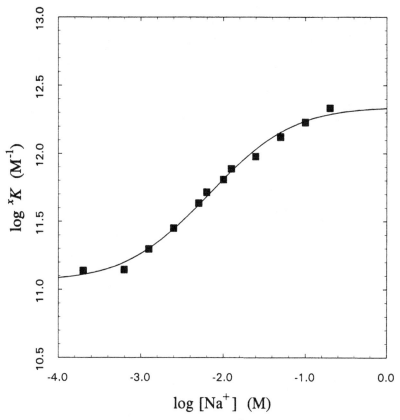

Figure 2.21 Linkage between hirudin and Na^+ binding to thrombin (Ayala and Di Cera, 1994) represented as the change in the logarithm of hirudin binding affinity as a function of the logarithm of Na^+ concentration. Experimental conditions are: 5 mM Tris, 0.1% PEG, I = 0.2 M, pH 8.0 at 25 °C. The curve was drawn according to eq (2.108), with parameter values $A_{10} = 1.1 \times 10^{11}$ M^{-1}, $A_{01} = 34$ M^{-1}, $A_{11} = 7.2 \times 10^{13}$ M^{-2}.

value of $K = 35$ M^{-1} is derived for the binding constant of Na^+ to the free enzyme. This parameter is the same as A_{01} in the reference cycle and agrees extremely well with the value derived independently through linkage studies of the effect of Na^+ on hirudin binding.

A special case of the reference cycle that is of particular interest arises when binding of either X or Y to the macromolecule excludes binding of the other ligand. This situation arises when the two ligands share the same site or portions of it, or when they bind to distinct sites that affect each other through substantial conformational transitions of the system. The difference between these cases deserves much consideration in practical applications. The ligated intermediate MXY is automatically excluded

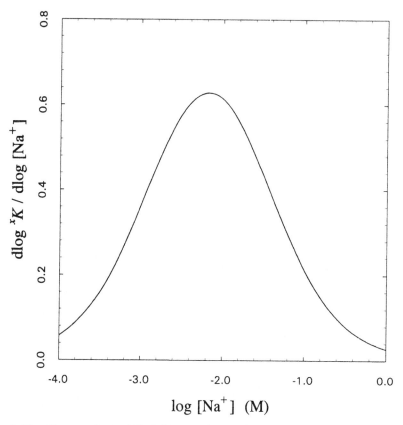

Figure 2.22   Net number of Na^+ ions exchanged upon hirudin binding to thrombin. The curve is the derivative of the data shown in Figure 2.21 and was drawn according to eq (2.121).

from both the cycle and the partition function, since $A_{11} = 0$. The two ligands are negatively linked in this case. Hence,

$$^xK = \frac{A_{10}}{1 + A_{01}y} \qquad (2.122)$$

The inverse of ^xK increases linearly with the concentration of Y. This situation is illustrated in Figure 2.23 for the competition of hirudin and Cl^- binding to thrombin. The slope in the affinity plot

$$\frac{d\ln {^xK}}{d\ln y} = -{^0Y} = \Delta Y \qquad (2.123)$$

gives the number of Y molecules exchanged upon binding of X and is a direct measure of the binding isotherm of Y. In this case, the binding curve of a second ligand (Cl^-) can be constructed directly using the linkage with the first ligand (hirudin). When direct competition between X and Y cannot be proved experimentally by independent sources, assessment of binding domains for the two ligands may be misleading. In principle, the effect reported in Figure 2.23 can be due to non-competitive inhibition induced by binding of Y to a site completely different from the X binding site. This fact has nontrivial consequences in molecular recognition studies and drug design. The neurokinin-1 receptor provides a convincing and important example. This receptor has competitive-like antagonists that bear no chemical or structural resemblance with agonists, thereby indicating that inhibition is mediated by binding to a different site (Fong, Huang and Strader, 1992; Jensen, Gerard, Schwartz and Gether, 1994; Pradier *et al.*, 1994).

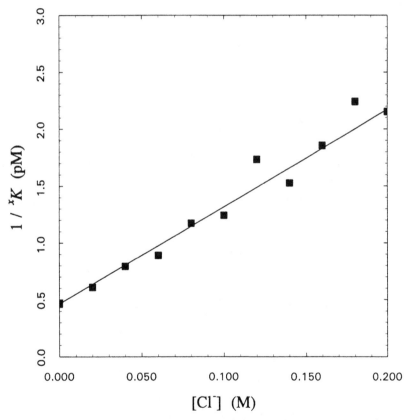

Figure 2.23 Linkage between hirudin and Cl^- binding to thrombin (Ayala and Di Cera, 1994), represented as the change in the inverse of hirudin binding affinity as a function of Cl^- concentration. Experimental conditions are: 5 mM Tris, 0.1% PEG, I = 0.2 M, pH 8.0 at 25 °C. The curve was drawn according to eq (2.122), with parameter values $A_{10} = 2.2 \times 10^{12}$ M^{-1}, $A_{01} = 12$ M^{-1}.

We will now use the reference cycle to illustrate the importance of arguments developed in Section 1.3. At constant T and P, the thermodynamic potential of interest is Π and specifically

$$d\Pi = -X \, d\mu_X - Y \, d\mu_Y = -k_B T \, d\ln \Psi(\mu_X, \mu_Y) \qquad (2.124)$$

The reciprocal nature of the linkage between X and Y is a consequence of the first law of thermodynamics and the nature of exact differential of $d\Pi$. The Euler relationship

$$\left(\frac{\partial X}{\partial \mu_Y}\right)_{\mu_X} = \left(\frac{\partial Y}{\partial \mu_X}\right)_{\mu_Y} \qquad (2.125)$$

indicates that if ligand Y facilitates binding of ligand X $(\partial X/\partial \mu_Y > 0)$, then ligand X must facilitate binding of ligand Y $(\partial Y/\partial \mu_X > 0)$. The relationship above can be rewritten in the more useful form for practical applications (Wyman, 1948, 1964)

$$\left(\frac{\partial X}{\partial \ln y}\right)_x = \left(\frac{\partial Y}{\partial \ln x}\right)_y \qquad (2.126)$$

since T and P are constant. The matrix $\mathbf{G}'$ associated with the potential $-\Pi$ is equal to

$$\mathbf{G}' = \begin{pmatrix} g'_{11} & g'_{12} \\ g'_{21} & g'_{22} \end{pmatrix} \qquad (2.127)$$

where the indices 1 and 2 refer to ligands X and Y respectively. The matrix $\mathbf{G}'$ is two-dimensional instead of four-dimensional, since T and P are held constant and make no contribution to $d\Pi$. The diagonal elements of $\mathbf{G}'$ give the binding capacities of X and Y

$$B_X = \left(\frac{\partial^2 \ln \Psi}{\partial \ln x^2}\right)_y = \frac{(1 + A_{01}y)(A_{10} + A_{11}y)x}{(1 + A_{10}x + A_{01}y + A_{11}xy)^2} = X(1 - X) \qquad (2.128)$$

$$B_Y = \left(\frac{\partial^2 \ln \Psi}{\partial \ln y^2}\right)_x = \frac{(1 + A_{10}x)(A_{01} + A_{11}x)y}{(1 + A_{10}x + A_{01}y + A_{11}xy)^2} = Y(1 - Y) \qquad (2.129)$$

These quantities are always positive, as implied by the second law of thermodynamics. The off-diagonal elements give the linkage coefficients, which are identical by virtue of the linkage relationship (2.125) and are explicitly given by

$$L_{XY} = \left(\frac{\partial^2 \ln \Psi}{\partial \ln y \partial \ln x}\right)_x = \frac{(A_{11} - A_{10}A_{01})xy}{(1 + A_{10}x + A_{01}y + A_{11}xy)^2} \qquad (2.130)$$

$$L_{YX} = \left(\frac{\partial^2 \ln \Psi}{\partial \ln x \partial \ln y}\right)_y = \frac{(A_{11} - A_{10}A_{01})xy}{(1 + A_{10}x + A_{01}y + A_{11}xy)^2} \quad (2.131)$$

The linkage coefficient can be positive, negative or zero depending upon the value of the difference in the numerator. This difference defines the cutoff for positive, negative or no linkage between the ligands, as already seen. Also, the linkage is always negative if $A_{11} = 0$, i.e., if binding of either ligand excludes binding of the other. The determinant associated with the matrix $\mathbf{G}'$ is given by the expression

$$\det \mathbf{G}' = xy$$

$$\times \frac{(1 + A_{01}y)(A_{10} + A_{11}y)(1 + A_{10}x)(A_{01} + A_{11}x) - (A_{11} - A_{10}A_{01})^2 xy}{(1 + A_{10}x + A_{01}y + A_{11}xy)^4}$$

$$(2.132)$$

The second law of thermodynamics demands that $\det \mathbf{G}'$ is always positive. Inspection of the numerator proves that this is indeed the case.

The fundamental consequences of thermodynamic stability dealt with in Section 1.3 can now be understood in terms of the elements of the matrix $\mathbf{G}'$. The binding capacity of ligand X, as defined in eq (2.128), is computed at constant chemical potential of ligand Y. The same quantity can be computed at constant Y and the results compared. The second law states that B_X computed at constant y is never smaller than B_X computed at constant Y. The same holds for B_Y computed at constant x or X. Consider the definition of Ψ for the reference cycle analogous to eq (2.119)

$$\Psi(x, y) = {}^0\Psi(x) + {}^1\Psi(x){}^0K_Y y \quad (2.133)$$

with

$${}^0\Psi(x) = 1 + {}^0K_X x \quad (2.134a)$$

$${}^1\Psi(x) = 1 + {}^1K_X x \quad (2.134b)$$

and

$${}^1X = \frac{d \ln {}^1\Psi}{d \ln x} = \frac{{}^1K_X x}{1 + {}^1K_X x} = \frac{A_{11}x}{A_{01} + A_{11}x} \quad (2.135a)$$

$${}^0X = \frac{d \ln {}^0\Psi}{d \ln x} = \frac{{}^0K_X x}{1 + {}^0K_X x} = \frac{A_{10}x}{1 + A_{10}x} \quad (2.135b)$$

Then

$$X = {}^0X(1 - Y) + {}^1XY \quad (2.136)$$

The binding capacity B'_X at constant Y is derived from differentiation as

$$B'_X = {}^0B(1 - Y) + {}^1BY \tag{2.137}$$

where ${}^0B$ and ${}^1B$ are the binding capacities of X when the site binding Y is free or bound respectively. On the other hand, the binding capacity B_X at constant y is

$$B_X = {}^0B(1 - Y) + {}^1BY + L_{XY}({}^1X - {}^0X) = B'_X + L_{XY}({}^1X - {}^0X) \tag{2.138}$$

Hence, B_X differs from B'_X by an additive term involving the linkage coefficient L_{XY} multiplied by the difference ${}^1X - {}^0X$. The sign of this difference is the same as the sign of $A_{11} - A_{10}A_{01}$. However, this expression also determines the sign of L_{XY}, as implied by eq (2.130). Hence,

$$L_{XY}({}^1X - {}^0X) \geqslant 0 \tag{2.139}$$

This result is readily understood from a physical point of view. If $L_{XY} > 0$, binding of X is facilitated by binding of Y and consequently ${}^1X > {}^0X$. On the other hand, if $L_{XY} < 0$, binding of X is opposed by binding of Y and consequently ${}^1X < {}^0X$. In either case, the nature of the linkage between X and Y leaves the sign of expression (2.139) unchanged. Hence,

$$B_X \geqslant B'_X \tag{2.140}$$

as implied by the second law of thermodynamics. The special condition $B_X = B'_X$ applies in the absence of linkage. Similar arguments hold for the binding capacity of ligand Y.

The foregoing analysis leads to the conclusion that the cooperativity of the reference system can be controlled in terms of thermodynamic stability according to eq (2.140) in a model-independent fashion. If the binding properties of ligand X are studied by keeping Y constant, then the binding capacity of X is lower than that observed at constant y. A decreased binding capacity implies the presence of apparent negative cooperativity in the reference system, which is by definition noncooperative if y is kept constant. This result is embodied by the properties of the matrix $\mathbf{G}'$ and is a consequence of Hadamard's inequality involving the metric coefficients g' (Gantmacher, 1959). The binding capacity B_X is the diagonal element g'_{11} of $\mathbf{G}'$ in (2.127), and B'_X is the value of g'_{11} 'contracted' over the manifold spanned by ligand Y, i.e.,

$$B'_X = {}^2g'_{11} - \frac{g'_{21}g'_{12}}{g'_{22}} = \frac{\det \mathbf{G}'}{g'_{22}} \tag{2.141}$$

Hence, $B_X = g'_{11} > {}^2g'_{11} = B'_X$, due to the Euler relationship $g'_{12} = g'_{21}$.

2.6 Linkage effects: General treatment involving two ligands

Some features of the foregoing treatment of the reference cycle bear on the more general case of an arbitrary number of sites for ligands X and Y. The two-dimensional array of possible ligated species in the general case is sketched in Figure 2.24. This array can be seen as an ensemble of reference cycles each composed of a 2×2 array of ligated species. It is, however, more appropriate to consider the ensemble of species in terms of separate manifolds for X and Y binding to particular Y- or X-ligated configurations. The partition function for this generalized ensemble at constant T and P is analogous to eq (2.104), except that the double summation extends to N sites for X and M sites for Y, i.e.,

$$\Psi(x, y) = \sum_{i=0}^{N}\sum_{j=0}^{M} \Upsilon_{ij} x^i y^j \tag{2.142}$$

Again the first index refers to ligand X, whose activity is x, and the second index to ligand Y, whose activity is y. Each ligation intermediate is characterized by the mechano-thermal partition function Υ_{ij} and the Gibbs free energy $G_{ij} = -k_B T \ln \Upsilon_{ij}$. The standard free energy change for the reaction

$$M + iX + jY \Leftrightarrow MX_iY_j \tag{2.143}$$

$$\begin{array}{ccccccccc}
M & \Leftrightarrow & MX & \Leftrightarrow & MX_2 & \Leftrightarrow & \ldots\ldots & \Leftrightarrow & MX_N \\
\Updownarrow & & \Updownarrow & & \Updownarrow & & & & \Updownarrow \\
MY & \Leftrightarrow & MXY & \Leftrightarrow & MX_2Y & \Leftrightarrow & \ldots\ldots & \Leftrightarrow & MX_NY \\
\Updownarrow & & \Updownarrow & & \Updownarrow & & & & \Updownarrow \\
MY_2 & \Leftrightarrow & MXY_2 & \Leftrightarrow & MX_2Y_2 & \Leftrightarrow & \ldots\ldots & \Leftrightarrow & MX_NY_2 \\
\Updownarrow & & \Updownarrow & & \Updownarrow & & & & \Updownarrow \\
\ldots\ldots & & \ldots\ldots & & \ldots\ldots & & & & \ldots\ldots \\
\Updownarrow & & \Updownarrow & & \Updownarrow & & & & \Updownarrow \\
MY_M & \Leftrightarrow & MXY_M & \Leftrightarrow & MX_2Y_M & \Leftrightarrow & \ldots\ldots & \Leftrightarrow & MX_NY_M
\end{array}$$

Figure 2.24 Linkage scheme for the binding of two ligands.

is given by

$$\Delta G_{ij} = G_{ij} - G_{00} = k_{\mathrm{B}} T \ln \frac{Y_{00}}{Y_{ij}} = -k_{\mathrm{B}} T \ln A_{ij} \qquad (2.144)$$

These equations are identical to eqs (2.105) and (2.106), except for the fact that they apply to the general case of $i = 0, 1, \ldots N$ and $j = 0, 1, \ldots M$. As usual, all equilibria can be recast in terms of the overall constants A_{ij} letting $Y_{00} = 1$ without loss of generality. This is equivalent to selecting M in Figure 2.24 as the reference configuration. The partition function (2.142) is therefore

$$\Psi(x, y) = \sum_{i=0}^{N} \sum_{j=0}^{M} A_{ij} x^i y^j \qquad (2.145)$$

The functions X and Y can be derived from Ψ by differentiation as follows

$$X = \left(\frac{\partial \ln \Psi}{\partial \ln x} \right)_y = \frac{\displaystyle\sum_{i=0}^{N} \sum_{j=0}^{M} i A_{ij} x^i y^j}{\displaystyle\sum_{i=0}^{N} \sum_{j=0}^{M} A_{ij} x^i y^j} \qquad (2.146)$$

$$Y = \left(\frac{\partial \ln \Psi}{\partial \ln y} \right)_x = \frac{\displaystyle\sum_{i=0}^{N} \sum_{j=0}^{M} j A_{ij} x^i y^j}{\displaystyle\sum_{i=0}^{N} \sum_{j=0}^{M} A_{ij} x^i y^j} \qquad (2.147)$$

The binding capacities of the two ligands are

$$B_{\mathrm{X}} = \left(\frac{\partial X}{\partial \ln x} \right)_y = \frac{\displaystyle\sum_{i=0}^{N} \sum_{j=0}^{M} i^2 A_{ij} x^i y^j}{\displaystyle\sum_{i=0}^{N} \sum_{j=0}^{M} A_{ij} x^i y^j} - \left(\frac{\displaystyle\sum_{i=0}^{N} \sum_{j=0}^{M} i A_{ij} x^i y^j}{\displaystyle\sum_{i=0}^{N} \sum_{j=0}^{M} A_{ij} x^i y^j} \right)^2 \qquad (2.148)$$

$$B_{\mathrm{Y}} = \left(\frac{\partial Y}{\partial \ln y} \right)_x = \frac{\displaystyle\sum_{i=0}^{N} \sum_{j=0}^{M} j^2 A_{ij} x^i y^j}{\displaystyle\sum_{i=0}^{N} \sum_{j=0}^{M} A_{ij} x^i y^j} - \left(\frac{\displaystyle\sum_{i=0}^{N} \sum_{j=0}^{M} j A_{ij} x^i y^j}{\displaystyle\sum_{i=0}^{N} \sum_{j=0}^{M} A_{ij} x^i y^j} \right)^2 \qquad (2.149)$$

We know from Section 2.2 that X can be written in terms of the affinity function as in eq (2.42), regardless of the form of the partition function.

The affinity function for the general case of the linkage between two ligands is given by

$$
{}^xK = \frac{\displaystyle\sum_{i=0}^{N}\sum_{j=0}^{M} i A_{ij} x^{i-1} y^{j}}{\displaystyle\sum_{i=0}^{N}\sum_{j=0}^{M} (N-i) A_{ij} x^{i} y^{j}}
\tag{2.150}
$$

$$
{}^yK = \frac{\displaystyle\sum_{i=0}^{N}\sum_{j=0}^{M} j A_{ij} x^{i} y^{j-1}}{\displaystyle\sum_{i=0}^{N}\sum_{j=0}^{M} (M-j) A_{ij} x^{i} y^{j}}
\tag{2.151}
$$

In order to clarify the significance of eqs (2.150) and (2.151), it is useful to consider the ensemble of species in Figure 2.24 in terms of separate manifolds, as done in the case of the reference cycle. Each column of the two-dimensional matrix contains the ligated intermediates for Y binding to a particular X-ligated configuration. In general, the ith column contains all possible Y-ligated configurations of the ith X-ligated species. Each column is a manifold which can be characterized in terms of a generalized partition function. The ith column has a partition function

$$
{}^i\Psi(y) = \sum_{j=0}^{M} \frac{A_{ij}}{A_{i0}} y^{j}
\tag{2.152}
$$

The form of the coefficients of eq (2.152) is readily understood in view of the fact that the ith X-ligated species, MX_i, is used as reference in the ith manifold and this species contributes a term $A_{i0} x^i$ to Ψ. If all terms of the ith manifold are expressed relative to MX_i, a polynomial expansion in the variable y alone is obtained as in eq (2.152). The partition function of the system as a whole can be rewritten as follows

$$
\Psi(x, y) = \sum_{i=0}^{N} {}^i\Psi(y) A_{i0} x^{i}
\tag{2.153}
$$

Using the manifolds relative to each X-ligated intermediate, we have essentially grouped configurations with the same number of X-ligated sites in Figure 2.24 into *ad hoc* partition functions, and expressed Ψ as a polynomial expansion in x with coefficients written in terms of these partition functions. It should be pointed out that, although $A_{00} = 1$ since the species M is used as reference in Ψ, the partition function ${}^0\Psi(y)$ is not equal to one. In fact, this term contains all intermediates of the first column in Figure 2.24. If this partition function is factored out, the resulting polynomial expansion

$$\Psi(x, y) = {}^0\Psi(y)[1 + {}^yA_1x + {}^yA_2x^2 + \ldots + {}^yA_Nx^N] \qquad (2.154)$$

differs from eq (2.37) in the prefactor, which makes no contribution to the derivation of the properties of X since it does not depend on x. Also, the ${}^yA$s are functions of y as follows

$$^yA_i = A_{i0}\frac{{}^i\Psi(y)}{{}^0\Psi(y)} \qquad (2.155)$$

The coefficient ${}^yA_i$ represents the overall equilibrium constant for the reaction $M + iX \Leftrightarrow MX_i$ in the presence of ligand Y. Had we defined manifolds of X-ligated species for given Y-ligated intermediates, we would have obtained a set of coefficients ${}^xA_j$ for the reaction $M + jY \Leftrightarrow MY_j$ in the presence of ligand X. In either case, the definition of partition functions over each manifold of interest 'contracts' the original two-dimensional array of species in the linkage scheme in Figure 2.24 into a one-dimensional array of species with coefficients depending on the activity of the second ligand.

Measurements of the coefficients ${}^yA$ as a function of y reveal important properties of the system. The derivative

$$\frac{d\ln {}^yA_i}{d\ln y} = {}^iY - {}^0Y = \Delta Y_i \qquad (2.156)$$

shows that the change in $\ln {}^yA_i$ gives the difference in Y-ligated sites between the ith and zeroth X-ligated species. The quantity ΔY_i is the net number of Y molecules exchanged when i molecules of X bind to the X-unligated form of the macromolecule. The sign of ΔY_i denotes the nature of the linkage involved in binding i molecules of X. If $\Delta Y_i > 0$, binding of i molecules of X to the X-unligated form of the macromolecule leads to an increase in the number of Y-ligated sites. In this case the two ligands are positively linked. On the other hand, if $\Delta Y_i < 0$, binding of i molecules of X to the X-unligated form gives rise to a decrease in the number of Y-ligated sites and binding of either ligand opposes binding of the other. The case $\Delta Y_i = 0$ denotes that binding of i molecules of X has no net effect on the amount of ligand Y bound to the macromolecule. Once all values of ΔY_i are known, it is possible to dissect the linkage between X and Y at the level of stepwise reaction steps. In fact, the net number of Y molecules exchanged upon adding one molecule of X to the $(i - 1)$th X-ligated intermediate is clearly

$$\Delta y_i = \Delta Y_i - \Delta Y_{i-1} \qquad (2.157)$$

If all values of Δy_i are added up from $i = 1$ to $i = N$, the total net number of Y molecules exchanged in fully ligating the macromolecule with X is derived as

$$\Delta y_1 + \Delta y_2 + \ldots + \Delta y_N = \Delta Y_N \tag{2.158}$$

This quantity is the same as the derivative $d \ln {}^y A_N / d \ln y$. The overall equilibrium constant ${}^y A_N$ is related to the mean ligand activity for X as follows

$${}^y x_m = {}^y A_N^{-1/N} \tag{2.159}$$

Hence,

$$\frac{d \ln {}^y x_m}{d \ln y} = -\frac{\Delta Y_N}{N} = -\Delta y \tag{2.160}$$

The change in the mean ligand activity of X, which can be derived from X or B_X by numerical integration, yields an important property of the system, i.e., the net number of Y molecules exchanged on the average per site ligated to X.

The linkage between two ligands may affect the cooperative properties of the macromolecule. Assume that each column in Figure 2.24 depicts a reference system. In other words, let the binding of Y to the ith X-ligated intermediate be described by the simple polynomial expansion

$${}^i\Psi(y) = (1 + K_i y)^M \tag{2.161}$$

We assume that $A_{ij} A_{i0}^{-1} = C_{j,M} K_i^j$ in eq (2.152). Then, let the first row in Figure 2.24 be itself a reference system, so that binding of X to the macromolecule in the absence of Y occurs with no cooperativity, or $A_{i0} = C_{i,N} K^i$. It follows from eq (2.155) that

$${}^y A_i = C_{i,N} K^i \left(\frac{1 + K_i y}{1 + K_0 y} \right)^M \tag{2.162}$$

Although binding of Y to each X-ligated intermediate is noncooperative, and binding of X to the Y-unligated form is noncooperative, binding of X in the presence of Y becomes cooperative if Y binds with different affinity to the different X-ligated intermediates. Cooperativity in the binding of X arises as a result of the linkage between X and Y, only requiring a different binding affinity for different manifolds. Linkage between X and Y translates into linkage between X and X. If binding of X progressively increases the affinity of Y, say $K_i > K_{i-1}$, then, after the ith X molecule binds, binding of Y is favored. This, in turn, favors the binding of other molecules of X. The resulting positive cooperativity in X binding is a consequence of the effect of X on Y, and the reciprocal effect of Y on X, which results in an effect of X on the binding of itself. This is the thermodynamic basis of allosteric transitions (Wyman and Allen, 1951; Wyman, 1967), whereby cooperativity is generated indirectly through linkage effects using either a second ligand, or multiple conformational states (Monod, Changeux and Jacob, 1963; Monod, Wyman and Changeux, 1965; Koshland, Nemethy and Filmer, 1966).

Aspects related to thermodynamic stability can be obtained in a straightforward way as was done for the reference cycle. The properties of the matrix $\mathbf{G}'$ depend only on the number of ligands involved and not on the number of sites. Hence, the inequality $B_X > B'_X$ in the presence of linkage between X and Y is a direct consequence of the Hadamard inequality involving the metric coefficients g' and holds quite generally for an arbitrary number of sites for binding of X and Y. Other important aspects arise when the two ligands compete for the same site, or share portions of it. In this case of 'identical linkage' (Di Cera, Doyle, Connelly and Gill, 1987a; Wyman and Gill, 1990), binding of either ligand excludes binding of the other. In the reference cycle, identical linkage is characterized by a negative value of the linkage coefficient L_{XY}. Intuitively, one would expect $L_{XY} < 0$ also in the general case, due to the competitive nature of the linkage. However, this is not the case and under certain conditions L_{XY} can change sign with saturation. The partition function for two identically linked ligands is given by

$$\Psi(x, y) = \sum_{i=0}^{N} \sum_{j=0}^{N-i} A_{ij} x^i y^j \tag{2.163}$$

and is identical to eq (2.145), except for the fact that all ligated intermediates $MX_i Y_j$ with $i + j > N$ are eliminated from the scheme and the double summation. The linkage coefficient L_{XY} is derived by differentiation as follows

$$L_{XY} = \frac{\partial^2 \ln \Psi(x, y)}{\partial \ln x \, \partial \ln y} = \frac{\displaystyle\sum_{i=0}^{N} \sum_{j=0}^{N-i} ij A_{ij} x^i y^j}{\displaystyle\sum_{i=0}^{N} \sum_{j=0}^{N-i} A_{ij} x^i y^j} - \frac{\displaystyle\sum_{i=0}^{N} \sum_{j=0}^{N-i} i A_{ij} x^i y^j \sum_{i=0}^{N} \sum_{j=0}^{N-i} j A_{ij} x^i y^j}{\left(\displaystyle\sum_{i=0}^{N} \sum_{j=0}^{N-i} A_{ij} x^i y^j\right)^2}$$

$$\tag{2.164}$$

The sign of L_{XY} is determined by the sign of the expression

$$L(x, y) = \sum_{i=0}^{N} \sum_{j=0}^{N-i} ij A_{ij} x^i y^j \sum_{i=0}^{N} \sum_{j=0}^{N-i} A_{ij} x^i y^j - \sum_{i=0}^{N} \sum_{j=0}^{N-i} i A_{ij} x^i y^j \sum_{i=0}^{N} \sum_{j=0}^{N-i} j A_{ij} x^i y^j$$

$$\tag{2.165}$$

For $N = 1$, the first term vanishes and the linkage is always negative, as already seen for the reference cycle. For $N = 2$, one has

$$L(x, y) = xy(A_{11} - A_{10}A_{01} - 2A_{10}A_{02}y - A_{11}A_{02}y^2 - 2A_{01}A_{20}x$$
$$- A_{11}A_{20}x^2 - 4A_{20}A_{02}xy - A_{20}A_{11}x^2) \quad (2.166)$$

The sign of $L(x, y)$ is negative, except for small values of x and y, in which case it depends on the sign of the difference $A_{11} - A_{10}A_{01}$. If $A_{11} > A_{10}A_{01}$, $L(x, y)$ is positive and so is the linkage coefficient L_{XY}, notwithstanding the fact that binding of X opposes binding of Y and *vice versa*. The case $A_{11} > A_{10}A_{01}$ corresponds to the situation where the two ligands facilitate binding of each other when they are both bound to the macromolecule. At low concentrations of X and Y the dominant species are M, MX, MY and MXY, and MXY is favored due to the condition $A_{11} > A_{10}A_{01}$. At low concentrations, both ligands can bind to the macromolecule due to the presence of multiple sites, and they can do so with positive linkage. On the other hand, increasing the concentration of one ligand leads to an increase of the amount bound and a necessary displacement of the other ligand. Hence, at high saturation the linkage between X and Y is always negative, regardless of the value of A_{11}. This result is true in general for arbitrary N. The case of oxygen and carbon monoxide binding to hemoglobin is a beautiful example of how ligands competing for the heme sites can favor each other's binding at other sites. At low oxygen saturation, an increase in carbon monoxide partial pressure first causes a 'paradoxical' increase of the amount of oxygen bound at other sites, and then leads to total oxygen desorption (Douglas, Haldane and Haldane, 1912; Di Cera *et al.*, 1989).

We conclude our analysis of linkage effects with a case of particular importance in practical applications. In the general treatment involving two ligands we have assumed that all response functions for one ligand can be derived by keeping the activity of the other ligand constant. However, in many cases a change in the chemical potential of ligand X cannot be obtained experimentally without a concomitant change in the activity of Y. This is especially true when water represents the second ligand and hydration effects are involved in structural changes of functional significance (Timasheff, 1990; Colombo, Rau and Parsegian, 1992). Consider the case of ligand X and water W (ligand Y) binding to the macromolecule under conditions where both X and W are in large excess with respect to M. This is the situation most often encountered in practice. The Pfaffian form of the potential for the generalized ensemble at constant T and P is

$$d\Pi = -k_B T(X \, d\ln x + W \, d\ln w) \quad (2.167)$$

The result above implies that the macromolecule is taken as our system, as we know from arguments discussed in Chapter 1. Since X and W are in large excess compared to M, the chemical potential of the macromolecule, $d\Pi$, is expected to make only a small contribution to the properties of the solution. The Gibbs–Duhem equation for a solution containing X and W alone is

$$0 = -k_B T(n_X \, \mathrm{d} \ln x + n_w \, \mathrm{d} \ln w) \tag{2.168}$$

where n_X and n_w are the mols of X and water present. Due to the fact that the chemical potential of the macromolecule is negligibly small compared to the terms in eq (2.168), a change of x in eq (2.167) will inevitably induce a change of w, because of eq (2.168). Hence, the number of ligands X bound to the macromolecule must be calculated as the *total* derivative $\mathrm{d} \ln \Pi / \mathrm{d} \ln x$, rather than the partial derivative $(\partial \ln \Pi / \partial \ln x)_w$, since w cannot be kept constant while changing x. Substitution of $\mathrm{d} \ln w$ in eq (2.167) from eq (2.168) yields the derivative

$$X' = \frac{\mathrm{d} \ln \Pi}{\mathrm{d} \ln w} = X - m_X W \tag{2.169}$$

This important result obtained by Tanford (1969) shows that the response function X', measured as the apparent number of ligands X bound to the macromolecule, equals the number expected if the water activity were constant, X, minus a term that includes the number of water molecules bound and the mole ratio of ligand X relative to water, $m_X = n_X / n_w$. In the case of ligands binding in the μM–nM range, as usually found in practice, the correcting term in eq (2.169) is negligible and the effect of changing the water activity can be neglected. However, in those cases where ligand X is used in the mM–M range, as in the case of salts (von Hippel and Schleich, 1969; Record, Anderson and Lohman, 1978; Anderson and Record, 1993; Record and Anderson, 1995), or denaturants (Schellman, 1978, 1987, 1990b, 1994; Tanford, 1970), the term $m_X W$ may have a significant influence on the measured value of X' relative to X. The value of X' is the *preferential binding* (Tanford, 1969) defined as the amount of ligand X bound to the macromolecule in excess of the amount 'bound' to the solution. Ligand X binds to the macromolecule in the usual sense only if X largely exceeds $m_X W$ in eq (2.169). In the case of $X = m_X W$, the number of ligands bound to the macromolecule per water molecule bound is the same as the mole ratio of ligand X relative to water in solution and there is no 'preference' for the macromolecule relative to the solution. Preferential binding can also assume negative values if the ligand is 'excluded' from the macromolecule ($X < m_X W$) relative to the solvent. The concept of preferential binding emphasizes the importance of hydration in macromolecular binding and the role of water as the ubiquitous 'second ligand'. An important generalization of the treatment of preferential interaction phenomena has been derived by Record and Anderson (1995).

3

Local binding processes

We are now in a position to deal with the main subject of this monograph, that is, the description of binding and linkage effects arising locally at the level of individual sites of a multi-site macromolecule. Consider a macromolecule containing N binding sites for a ligand X. We seek to resolve the thermodynamics of site-specific effects arising when binding of X can be monitored at each site separately. If the N sites are independent, then the description of site-specific effects becomes a trivial task, since each site behaves according to the rules outlined for a reference system in Section 2.1. When the sites are not only independent but also alike, then the macromolecule as a whole is a reference system. The case we are interested in is the one where all sites are different and linked. In this instance each separate site of the macromolecule is a subsystem open to interactions with other subsystems. Hence, the behavior of an individual site coupled to the rest of the macromolecule is expected to be rather complex, as it must reflect not only the binding properties of the site under consideration, but also the result of interactions with other sites.

3.1 The reference cycle and site-specific cooperativity

The simplest system of interest in the discussion of binding processes at the local level is that composed of a macromolecule M containing two binding sites for ligand X. Again we assume T and P constant and that the macromolecule does not change its aggregation state. In the global description the partition function of such a system is given by

$$\Psi(x) = Y_0 + Y_1 x + Y_2 x^2 = 1 + A_1 x + A_2 x^2 \qquad (3.1)$$

The significance of the Ys and As in eq (3.1) has already been dealt with at length. The global description tells us that for a system of two sites there are two independent coefficients to be determined experimentally. The cooperativity is positive or negative depending on the sign of the expression

$$\Delta = 4A_2 - A_1^2 = 4k_1(k_2 - k_1) \tag{3.2}$$

with $\Delta = 0$ providing the noncooperative case. The key question to be addressed in what follows is whether the global, macroscopic condition for cooperativity, $\Delta \neq 0$, reflects the true pattern of interaction between the sites. The partition function (3.1) embodies the overall behavior of the system and as such all quantities derived from it reflect properties of the system as a whole. When the behavior of constituent sites becomes the focus of our investigation, other properties of the system, more specific and local, are taken into account. Analysis of these properties reveals that the global description provides a gross simplification of the communication among the sites. Cooperativity patterns in the global description can underestimate, or even distort, the effective nature of the interactions. The importance of the thermodynamic treatment of site-specific effects stems entirely from the possibility of decoupling linkage and cooperativity in a direct way.

Consider the reference cycle in Figure 3.1. A comparison with the cycle in Figure 2.20 reveals an important parallel to be discussed in detail in Section 3.3. The binding reactions of the two ligands in Figure 2.20 have been replaced by the reactions of ligand X to site 1 and 2 of the macromolecule. The substitution is perfectly valid, since the properties of the reference cycle in Figure 2.20 do not depend on the nature of the ligands. Hence, X and Y in Figure 2.20 can be treated as identical to give rise to the cycle in Figure 3.1. We now write down the partition function for the reference cycle in Figure 3.1, using arguments developed in Section 2.5, as follows

$$\Psi(x) = Y_{00} + (Y_{10} + Y_{01})x + Y_{11}x^2 = 1 + (A_{10} + A_{01})x + A_{11}x^2 \tag{3.3}$$

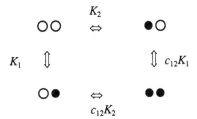

Figure 3.1 The reference cycle for site-specific binding to a macromolecule containing two sites.

The indices refer to sites 1 and 2 and the Ys and As were defined in Section 2.5. Specifically, Y_{10} is the mechano-thermal partition function of the configuration with site 1 ligated and site 2 unligated, A_{10} is the overall equilibrium constant for X binding to site 1 when site 2 is unligated and so forth. The unligated form of the macromolecule is taken as reference as usual, which makes $Y_{00} = 1$. We know from Section 2.5 that, out of the four reactions in the cycle in Figure 3.1, only three are truly independent. Consistent with this fact, the partition function (3.3) contains three independent coefficients. However, this partition function must be equivalent to eq (3.1), which describes the properties of the same system. Hence, the question arises as to how many independent parameters are to be considered in the analysis of a two-site macromolecule. The global description requires two, but the approach depicted in Figure 3.1 based on the local, site-specific description demands three. The contradiction is only apparent. In fact, the coefficients A_{10} and A_{01} in eq (3.3) are coupled to the same power of x, so that

$$A_1 = A_{10} + A_{01} \tag{3.4a}$$

$$A_2 = A_{11} \tag{3.4b}$$

What the apparent contradiction means is that it is not possible to derive uniquely the site-specific parameters A_{10}, A_{01} and A_{11} from knowledge of the global parameters A_1 and A_2. On the other hand, if the site-specific parameters are known, then the global parameters can be determined uniquely. Hence, the global description is incapable of providing information on what goes on at the level of individual sites. This fact has been recognized for a long time and the need for a local description of binding processes finds its origin in the early work on the dissociation of polyvalent substances (Wegscheider, 1895; Adams, 1916; Simms, 1926; Edsall and Blanchard, 1933; Edsall and Wyman, 1958). The description of site-specific effects demands resolution of more parameters than the global description can yield.

With this important conclusion in mind, we proceed to examine the effect on cooperativity. It is convenient to define new site-specific parameters that will henceforth be used. The equilibrium binding constants A_{10} and A_{01} reflecting the binding of X to either site when the other site is unligated will be designated K_1 and K_2 respectively. The binding constant A_{11} is defined as the product $c_{12}K_1K_2$, where c_{12} expresses an interaction constant between the two sites. It is particularly important to understand the thermodynamic significance of these parameters. The binding constants K_1 and K_2 are related to the standard free energy changes

$$\Delta G_1 = -k_B T \ln \frac{Y_{10}}{Y_{00}} = -k_B T \ln K_1 \qquad (3.5a)$$

$$\Delta G_2 = -k_B T \ln \frac{Y_{01}}{Y_{00}} = -k_B T \ln K_2 \qquad (3.5b)$$

that measure the difference in the contribution of the mechano-thermal partition functions of each ligated intermediate relative to the reference configuration. It should be stressed that the reference configuration is the unligated form of the macromolecule and therefore K_1 measures the binding affinity of site 1 in the macromolecule when site 2 is kept unligated. The interaction constant c_{12}, on the other hand, is a measure of cooperativity between the sites. In fact, it is easy to show from eq (3.3) that

$$c_{12} = \frac{Y_{00} Y_{11}}{Y_{10} Y_{01}} = \frac{A_{11}}{A_{10} A_{01}} \qquad (3.6)$$

The free energy change associated with c_{12} is therefore the free energy of coupling for the cycle in Figure 3.1, as already demonstrated in Section 2.5 with eq (2.111). In other words,

$$\Delta G_{12} = -k_B T \ln c_{12} = \Delta G_{11} - \Delta G_{10} - \Delta G_{01}$$
$$= \Delta G_{11} - \Delta G_1 - \Delta G_2 = \Delta G_c \qquad (3.7)$$

The significance of c_{12} stems exactly from its connection with a thermodynamic coupling parameter and can be considered the equilibrium constant for the 'dismutation'

$$M_{10} + M_{10} \Leftrightarrow M_{00} + M_{11} \qquad (3.8)$$

as follows directly from eq (3.6).

The partition function rewritten in terms of K_1, K_2 and c_{12} assumes the simple form

$$\Psi(x) = 1 + (K_1 + K_2)x + c_{12}K_1 K_2 x^2 \qquad (3.9)$$

and comparison with the global form (3.1) yields

$$A_1 = K_1 + K_2 \qquad (3.10a)$$

$$A_2 = c_{12}K_1 K_2 \qquad (3.10b)$$

The condition (3.2) for macroscopic cooperativity written in terms of site-specific parameters becomes

$$\Delta = 4c_{12}K_1 K_2 - (K_1 + K_2)^2 \qquad (3.11)$$

Since Δ is second-order homogeneous in the Ks, scaling K_1 and K_2 by the same constant factor leaves the sign of Δ unchanged. Hence, the sign of Δ

must depend solely on the ratio K_1/K_2, or equivalently K_2/K_1. Letting $\rho = K_1/K_2$ yields

$$\Delta = K_2^2[4c_{12}\rho - (1 + \rho)^2] \tag{3.12}$$

The sign of Δ depends solely on the expression in brackets. Symmetry allows us to study eq (3.12) in the range $0 \leqslant \rho \leqslant 1$ only. In fact, $\rho > 1$ is equivalent to swapping K_1 and K_2 and defining a new parameter $\rho' = K_2/K_1 < 1$, which leaves the expression in brackets unchanged. The cutoff between positive and negative macroscopic cooperativity is

$$4c_{12}\rho - (1 + \rho)^2 = 0 \tag{3.13}$$

This condition for the absence of macroscopic cooperativity does not coincide with the analogous microscopic condition $c_{12} = 1$, which implies that the sites bind ligand X independent of one another and the free energy of coupling is zero. The two conditions coincide only if $\rho = 1$, or if the two sites are identical and independent as when the macromolecule is a reference system. When $\rho < 1$, values of $c_{12} > 1$ may yield $\Delta = 0$ as well. The relationship between microscopic and macroscopic cooperativity is depicted in Figure 3.2. The continuous curve represents eq (3.13) and gives the condition for which macroscopic cooperativity vanishes. The region above the curve is characterized by positive macroscopic cooperativity, while the region below the curve is characterized by negative macroscopic cooperativity. The dashed line $c_{12} = 1$ gives the condition for the absence of microscopic cooperativity. The two conditions together define three regions in the $c_{12}-\rho$ plane. In region A, defined by $c_{12} < 1$, there is no ambiguity between macroscopic and microscopic cooperativity. If binding to one site opposes binding to the other site, negative cooperativity is seen at both levels. At the microscopic boundary $c_{12} = 1$, the system is always negatively cooperative in the macroscopic sense, unless $\rho = 1$. Region B, where $(1 + \rho)^2/4\rho > c_{12} > 1$, is a most interesting one. Here negative cooperativity in the macroscopic description is observed even though the two sites interact in a positive fashion. Region C, where $c_{12} > (1 + \rho)^2/4\rho$, leaves again no ambiguity between macroscopic and microscopic cooperativity. In this region binding to one site favors binding to the other site.

The conclusion to be drawn from Figure 3.2 is that macroscopic cooperativity is at most a crude approximation of the true pattern of interaction between the sites. In the case of positive macroscopic cooperativity, interactions are always underestimated if $\rho \neq 1$. In fact, for any given point in region C such that $c_{12} > 1$, there is always a value of ρ such that $\Delta = 0$, or even $\Delta < 0$. This means that, regardless of the strength of

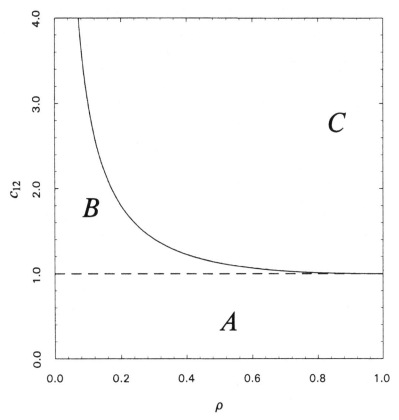

Figure 3.2 Microscopic and macroscopic cooperativity patterns for $N = 2$, plotted as the site-specific coupling constant, c_{12}, versus the parameter for binding heterogeneity, ρ. The continuous line was drawn according to eq (3.13) and represents the boundary between negative (regions A and B) and positive (region C) macroscopic cooperativity. The dashed line provides the boundary between negative (region A) and positive (regions B and C) microscopic cooperativity. In regions A and C microscopic and macroscopic cooperativity coincide at least qualitatively. In region B positive microscopic cooperativity manifests itself macroscopically as negative cooperativity.

positive coupling between the sites, macroscopic cooperativity can be made small, vanishing or even negative by changing the value of ρ. The heterogeneity of the binding affinities of the sites opposes the effect of the positive coupling between them and tends to cancel, or even reverse, the cooperativity pattern measured macroscopically. As the heterogeneity tends to increase, $\rho \to 0$ and $\Delta > 0$ only if $c_{12} > 1/4\rho$. For a system where the affinities of the two sites differ by a factor of 100 ($\rho = 0.01$), but binding to one site increases the affinity of the other by a factor of 10 ($c_{12} = 10$), one has $\Delta < 0$ and macroscopic negative cooperativity is measured notwithstanding that the microscopic picture indicates positive coupling between the sites. For this system to show macroscopic positive cooperativity the value of c_{12} must exceed 25, whilst in a system where $\rho = 0.001$ the value of c_{12} must exceed 250, and so on. The limitations of the global description of cooperative interactions are not confined to the case of positive cooperativity. When negative cooperativity is observed in the macroscopic case, the system can actually be positively or negatively cooperative in the microscopic picture. Particularly interesting is the case $c_{12} = 1$, which always leads to macroscopic negative cooperativity if $\rho \neq 1$. The heterogeneity of the binding affinities of the sites generates *per se* a misleading pattern of interaction. In summary, macroscopic cooperativity does not reflect the true pattern of interaction between the sites, unless the sites are identical. If the sites bind with different affinities, then positive cooperativity in the global description always underestimates the coupling between the sites. Negative cooperativity can be totally misleading, since it may result from distortion of positive coupling between the sites, or absence of interactions. In the case of negative coupling between the sites, on the other hand, the interaction may be overestimated in the global picture. Absence of cooperativity in the global description can be misleading as well. Positive coupling between the sites can be canceled by heterogeneity of their binding affinity. The true cooperative nature of the interactions between the sites can only be resolved from analysis of experimental data if the value of c_{12} is known. This value cannot be obtained from the global description of the facts. The example we have analyzed for the sake of simplicity is paradigmatic for the general case of $N > 2$, as is intuitively obvious. Without resolving site-specific parameters, our understanding of cooperative phenomena remains limited, if not misleading. As we realize the importance of the local picture, we must abandon the conceptual framework of the global description and enter a different, more complex scenario provided by the thermodynamics of site-specific energetics.

The partition function of the system of two interacting sites can be cast in a number of equivalent forms in the local description. The four possible ligation states in Figure 3.1 can be partitioned in two manifolds, as indicated in Figure 3.3. One manifold contains all configurations with site 1 ligated and the other contains all configurations with site 1 unligated. Hence,

$$\Psi(x) = 1 + (K_1 + K_2)x + c_{12}K_1K_2x^2 = {}^0\Psi_1(x) + {}^1\Psi_1(x)K_1x \quad (3.14)$$

Here ${}^0\Psi_1$ is the partition function of the manifold or subsystem where site

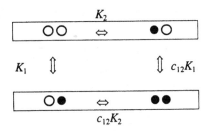

Figure 3.3 The reference cycle for site-specific binding to a macromolecule containing two sites. The four ligation states are partitioned into two manifolds where site 1 is kept unligated (top) or ligated (bottom).

1 is kept unligated and likewise $^1\Psi_1$ is the partition function of the subsystem with site 1 ligated. Had we partitioned the linkage scheme in Figure 3.1 according to site 2, we would have obtained

$$\Psi(x) = 1 + (K_1 + K_2)x + c_{12}K_1K_2x^2 = {}^0\Psi_2(x) + {}^1\Psi_2(x)K_2x \quad (3.15)$$

Hence,

$$^0\Psi_1(x) + {}^1\Psi_1(x)K_1x = {}^0\Psi_2(x) + {}^1\Psi_2(x)K_2x \quad (3.16)$$

The importance of defining partition functions for separate subsystems will be made clear in the next section. It is now rather straightforward to arrive at the key thermodynamic quantities reflecting binding to sites 1 and 2 separately. The site-specific binding curve of site 1 is given by the ratio of all configurations with site 1 ligated with respect to the total, i.e.,

$$X_1 = K_1 x \frac{{}^1\Psi_1(x)}{\Psi(x)} = 1 - \frac{{}^0\Psi_1(x)}{\Psi(x)} = \frac{{}^xK_1x}{1 + {}^xK_1x} \quad (3.17)$$

The site-specific binding isotherm X_1 can be cast in a form analogous to X in the global description by introducing the site-specific affinity function

$$^xK_1 = \frac{X_1}{(1 - X_1)x} = K_1 \frac{{}^1\Psi_1(x)}{{}^0\Psi_1(x)} = K_1 \frac{1 + c_{12}K_2x}{1 + K_2x} \quad (3.18)$$

Likewise, for site 2 one has

$$X_2 = K_2 x \frac{{}^1\Psi_2(x)}{\Psi(x)} = 1 - \frac{{}^0\Psi_2(x)}{\Psi(x)} = \frac{{}^xK_2x}{1 + {}^xK_2x} \quad (3.19)$$

$$^xK_2 = \frac{X_2}{(1 - X_2)x} = K_2 \frac{{}^1\Psi_2(x)}{{}^0\Psi_2(x)} = K_2 \frac{1 + c_{12}K_1x}{1 + K_1x} \quad (3.20)$$

It is evident from the foregoing definitions that a site-specific binding isotherm is similar to that of a cooperative macromolecule, although it reflects the binding properties of an individual site. This is the consequence of the site being a subsystem open to interaction with the rest of the macromolecule.

Many properties of site-specific processes can be derived from analysis of the affinity function using arguments already developed in the global description. For example, the limiting values of $^xK_1$ are

$$^0K_1 = K_1 \tag{3.21a}$$

$$^\infty K_1 = c_{12}K_1 \tag{3.21b}$$

Due to coupling between the sites, the affinity of site 1 increases with saturation when $c_{12} > 1$ and decreases when $c_{12} < 1$. In the absence of true interactions $c_{12} = 1$, the affinity function is constant and

$$X_1 = \frac{K_1x}{1 + K_1x} \tag{3.22}$$

The site-specific binding isotherm is identical to the binding curve of a macromolecule composed of a single site, as expected. Similar arguments hold for site 2. The reciprocity of the linkage between the sites is noteworthy. The limiting values of $^xK_2$ are

$$^0K_2 = K_2 \tag{3.23a}$$

$$^\infty K_2 = c_{12}K_2 \tag{3.23b}$$

and therefore, if binding to site 2 increases the affinity of site 1, then binding to site 1 must increase the affinity of site 2. The derivative of the logarithm of the affinity function is an important linkage quantity. This derivative is given by

$$\frac{d\ln{}^xK_1}{d\ln x} = \frac{d\ln{}^1\Psi_1(x)}{d\ln x} - \frac{d\ln{}^0\Psi_1(x)}{d\ln x} = {}^1X_1 - {}^0X_1 \tag{3.24}$$

where $^1X_1$ is the number of ligated sites of the macromolecule, other than site 1, when site 1 is ligated and $^0X_1$ is the analogous quantity when site 1 is unligated. If $^1X_1 = {}^0X_1$ for all values of x, then $^xK_1$ is a constant and site 1 behaves as an independent site. On the other hand, if $^1X_1 \neq {}^0X_1$ binding to site 1 affects the saturation of site 2 and, conversely, binding to site 2 affects the binding to site 1. This translates into a departure from a simple noncooperative binding curve and site 1 behaves 'cooperatively'. The driving force for site-specific cooperativity is provided by the difference $^1X_1 - {}^0X_1$ in eq (3.24) and leads to positive cooperativity when $^1X_1 > {}^0X_1$, or negative cooperativity when $^1X_1 < {}^0X_1$. This driving force is completely analogous to that involving two different ligands, as seen in Section 2.5. Hence, the molecule of X binding to site 2 can be seen as a 'second' ligand affecting the properties of the molecule of X binding to site

1. The change in the function $\ln {}^x K_1$ gives the difference in ligation of site 2 upon binding to site 1, or else the net number of molecules exchanged at site 2 when site 1 is ligated. If ${}^1 X_1 - {}^0 X_1$ is positive, binding to site 2 facilitates binding to site 1 and the affinity of site 1 increases with ligation. On the other hand, if ${}^1 X_1 - {}^0 X_1$ is negative, binding to site 2 opposes binding to site 1 and the affinity of site 1 decreases with ligation. Finally, if ${}^1 X_1 - {}^0 X_1 = 0$, then binding to site 2 has no effect on binding to site 1 and the two sites are independent. It should be pointed out that in this case the two sites are truly independent, since ${}^1 X_1 - {}^0 X_1 = 0$ only if $c_{12} = 1$.

The binding capacity of site 1 is computed as in the global description by differentiating X_1 or X_2 with respect to $\ln x$, i.e.,

$$B_1 = \frac{dX_1}{d \ln x} = X_1(1 + {}^1 X_1 - X) \tag{3.25}$$

By virtue of the definition of Ψ in terms of ${}^0\Psi_1$ and ${}^1\Psi_1$, one also has

$$B_1 = (1 - X_1)(X - {}^0 X_1) = X_1(1 - X_1)(1 + {}^1 X_1 - {}^0 X_1) \tag{3.26}$$

Likewise for site 2

$$B_2 = X_2(1 + {}^1 X_2 - X) = (1 - X_2)(X - {}^0 X_2)$$
$$= X_2(1 - X_2)(1 + {}^1 X_2 - {}^0 X_2) \tag{3.27}$$

The site-specific binding capacity has a peculiar feature, which makes it quite different from the analogous global quantity B. In fact, the sign of the binding capacity of an individual site, say site 1, is set by the sign of the expression $1 + {}^1 X_1 - {}^0 X_1$. We will see that, in general, this expression is not always positive. However, for the simple case of two sites, the sign of $1 + {}^1 X_1 - {}^0 X_1$ is always positive, since both ${}^1 X_1$ and ${}^0 X_1$ are bounded from zero to one. The expression $1 + {}^1 X_1 - {}^0 X_1$ is the same as the slope of the site-specific affinity plot, where $\ln {}^x K_1$ is plotted versus $\ln x$, plus one. We conclude from Section 2.2. that this slope must equal the site-specific Hill coefficient. In fact, the Hill coefficients of sites 1 and 2 are

$$n_1 = \frac{d \ln\left(\dfrac{X_1}{1 - X_1}\right)}{d \ln x} = 1 + {}^1 X_1 - {}^0 X_1 \tag{3.28a}$$

$$n_2 = \frac{d \ln\left(\dfrac{X_2}{1 - X_2}\right)}{d \ln x} = 1 + {}^1 X_2 - {}^0 X_2 \tag{3.28b}$$

The Hill coefficient in the local description is an important linkage

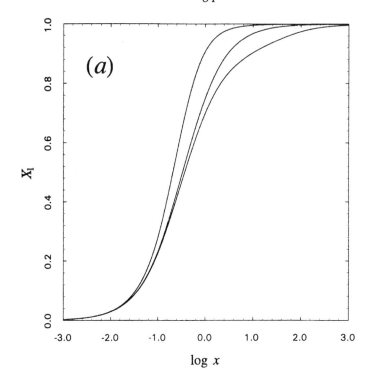

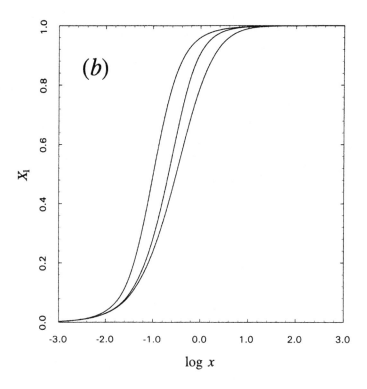

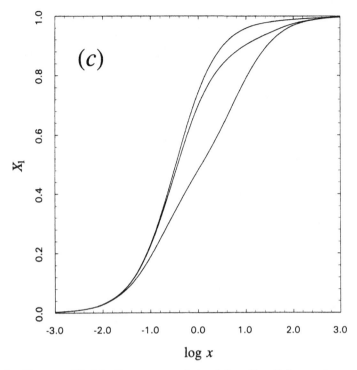

Figure 3.4 Site-specific binding curve of site 1 for $N = 2$ for various cooperative cases: (*a*) effect of changing c_{12} when $K_1 = 3$ and $K_2 = 0.33$: $c_{12} = 10$ (left), $c_{12} = 1$ (middle), $c_{12} = 0.1$ (right); (*b*) effect of changing K_2 when $K_1 = 3$ and $c_{12} = 10$: $K_2 = 3.3$ (left), $K_2 = 0.33$ (middle), $K_2 = 0.033$ (right); (*c*) effect of changing K_2 when $K_1 = 3$ and $c_{12} = 0.1$: $K_2 = 3.3$ (left), $K_2 = 0.33$ (middle), $K_2 = 0.033$ (right).

quantity, as will be shown in Sections 3.2 and 3.3. The difference $^1X_1 - {}^0X_1$, which is the slope of the affinity plot, gives the number of ligands exchanged at site 2 upon ligation of site 1. Hence, the Hill coefficient n_1 gives the net number of ligands exchanged upon ligation of site 1, including the ligand bound to this site. The response functions for site-specific effects dealt with so far are illustrated in Figures 3.4–3.7.

The local and global descriptions are linked through a number of conservation relationships (Klotz and Hunston, 1975; Ackers, Shea and Smith, 1983; Di Cera, 1989). For the simple system of two sites dealt with above one has

$$X = X_1 + X_2 \tag{3.29a}$$

$$B = B_1 + B_2 \tag{3.29b}$$

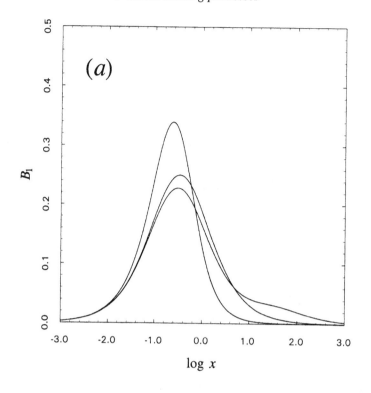

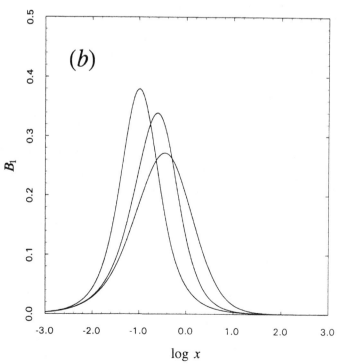

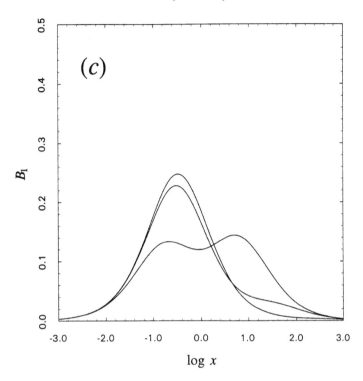

Figure 3.5 Binding capacity of site 1 for $N = 2$ for various cooperative cases: (*a*) effect of changing c_{12} when $K_1 = 3$ and $K_2 = 0.33$: $c_{12} = 10$ (left), $c_{12} = 1$ (middle), $c_{12} = 0.1$ (right); (*b*) effect of changing K_2 when $K_1 = 3$ and $c_{12} = 10$: $K_2 = 3.3$ (left), $K_2 = 0.33$ (middle), $K_2 = 0.033$ (right); (*c*) effect of changing K_2 when $K_1 = 3$ and $c_{12} = 0.1$: $K_2 = 3.3$ (left), $K_2 = 0.33$ (middle), $K_2 = 0.033$ (right).

X and B are extensive properties of the macromolecule when referred to those of the individual sites. Using the definition of Ψ in eq (3.1) one also has (Di Cera, 1989,1990)

$$X_j = \frac{\partial \ln \Psi(x)}{\partial \ln K_j} \qquad (3.30)$$

where $j = 1, 2$. The partial derivative indicates that all other variables in Ψ are kept constant. The partial derivative $\partial/\partial \ln K_j$ acts as a filter for the terms in Ψ that are linear in K_j. From the definition of $X = d \ln \Psi/d \ln x$ in the global description, the following operator representation holds

$$\frac{d}{d \ln x} = \frac{\partial}{\partial \ln K_1} + \frac{\partial}{\partial \ln K_2} = \nabla \qquad (3.31)$$

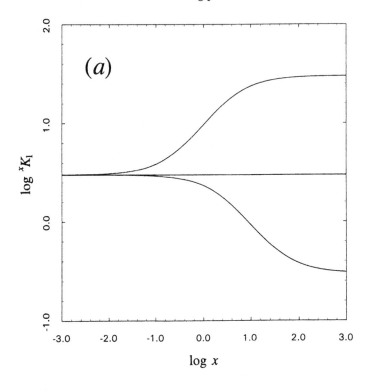

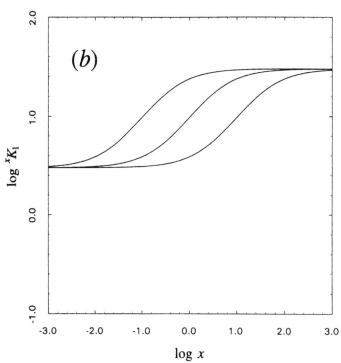

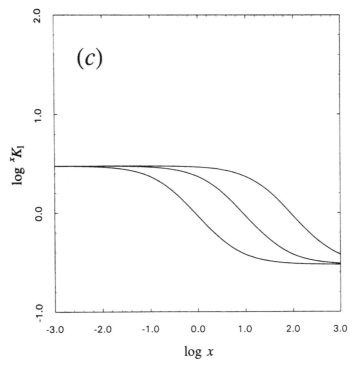

Figure 3.6 Affinity function of site 1 for $N = 2$ for various cooperative cases: (*a*) effect of changing c_{12} when $K_1 = 3$ and $K_2 = 0.33$: $c_{12} = 10$ (top), $c_{12} = 1$ (middle), $c_{12} = 0.1$ (bottom); (*b*) effect of changing K_2 when $K_1 = 3$ and $c_{12} = 10$: $K_2 = 3.3$ (left), $K_2 = 0.33$ (middle), $K_2 = 0.033$ (right); (*c*) effect of changing K_2 when $K_1 = 3$ and $c_{12} = 0.1$: $K_2 = 3.3$ (left), $K_2 = 0.33$ (middle), $K_2 = 0.033$ (right).

Here ∇ is the 'gradient' operator associated with the system. The additivity of site-specific effects giving rise to the global picture should be appreciated in those cases outlined in Figure 3.2, where local and global cooperativity patterns do not coincide. A paradigmatic case is depicted in Figure 3.8. Positive coupling between sites 1 and 2 translates into negative cooperativity in the global picture due to site heterogeneity. The 'discrepancy' between local and global effects is particularly striking when examining the affinity plot. Although the affinity function of each site increases with ligation, that of the system as a whole decreases.

3.2 Contracted partition functions

In this section we deal with the concept of the *contracted partition function* (Di Cera, 1990), which plays a central role in our discussion of site-specific

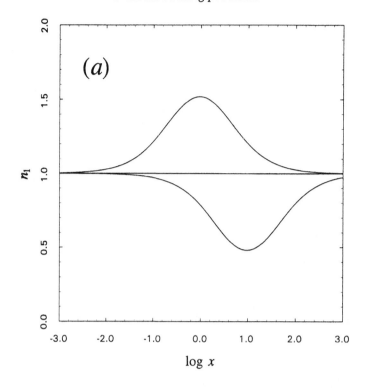

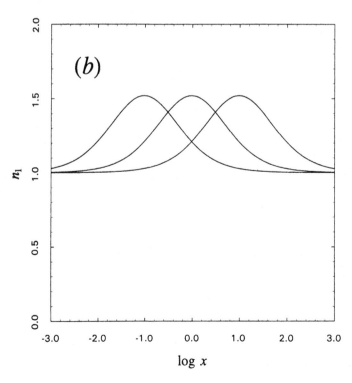

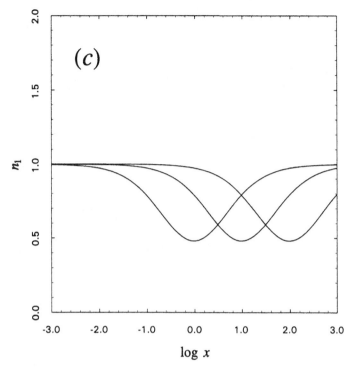

Figure 3.7 Hill coefficient of site 1 for $N = 2$ for various cooperative cases: (a) effect of changing c_{12} when $K_1 = 3$ and $K_2 = 0.33$: $c_{12} = 10$ (top), $c_{12} = 1$ (middle), $c_{12} = 0.1$ (bottom); (b) effect of changing K_2 when $K_1 = 3$ and $c_{12} = 10$: $K_2 = 3.3$ (left), $K_2 = 0.33$ (middle), $K_2 = 0.033$ (right); (c) effect of changing K_2 when $K_1 = 3$ and $c_{12} = 0.1$: $K_2 = 3.3$ (left), $K_2 = 0.33$ (middle), $K_2 = 0.033$ (right).

binding processes. We have already made use of these functions in previous sections (2.5 and 3.1), but without stressing their importance in general. Consider a macromolecule containing N binding sites. In the global description the partition function of such a system contains $N + 1$ terms, of which N are independent. In the local description, on the other hand, the ligation state of the N sites separately should be explicitly taken into account. Since each site can exist in two possible configurations, ligated or unligated, there are a total of 2^N terms to be considered in the local description. Hence, the number of independent terms in the partition function of the local description is

$$\nu = 2^N - 1 \qquad (3.32)$$

The operational complexity of the local description becomes fully evident at this point. While the number of independent terms in the partition

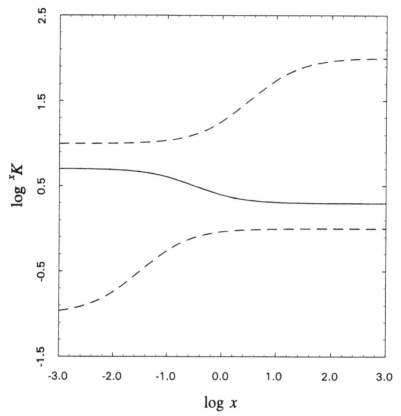

Figure 3.8   Affinity plot for $N = 2$ and site-specific parameter values $K_1 = 10$, $K_2 = 0.1$ and $c_{12} = 10$. The affinity functions for site 1 (top) and 2 (bottom) are depicted by dashed lines and show positive microscopic cooperativity. The affinity function of the macromolecule as a whole is depicted by a continuous line and shows negative macroscopic cooperativity.

function of the global description grows linearly with N, the analogous number in the local description grows exponentially with N. It is convenient to consider each configuration as an N-dimensional vector of binary digits, 0 and 1, that depict the unligated and ligated states of a site. For example, the vector

$$\xi = [0011] \tag{3.33}$$

labels one of the sixteen possible configurations of a macromolecule containing four binding sites ($N = 4$), where sites 1 and 2 are unligated and sites 3 and 4 are ligated. In the general case of N sites

$$\xi = [\alpha\beta \ldots \omega] \tag{3.34}$$

with $\alpha, \beta, \ldots \omega = 0, 1$. The first index refers to site 1, the second index to site 2 and so on, up to the ω index for site N. The partition function for this generalized ensemble at constant T and P is obtained by summing over all possible configurations, i.e.,

$$\Psi(x) = \sum_{\alpha=0}^{1} \sum_{\beta=0}^{1} \cdots \sum_{\omega=0}^{1} Y_{\alpha\beta\ldots\omega} x^{\alpha+\beta+\ldots+\omega} \tag{3.35}$$

The mechano-thermal partition function $Y_{\alpha\beta\ldots\omega}$ is linked to the configuration labeled by the vector $\xi = [\alpha\beta\ldots\omega]$ and to the Gibbs free energy $G_{\alpha\beta\ldots\omega} = -k_B T \ln Y_{\alpha\beta\ldots\omega}$. Again, without loss of generality, we can set $Y_{00\ldots0} = 1$ and take the unligated form of the macromolecule $\xi = [00 \ldots 0]$ as reference. The standard free energy change for the reaction

$$M_{00\ldots0} + (\alpha + \beta + \ldots + \omega)X \Leftrightarrow M_{\alpha\beta\ldots\omega} \tag{3.36}$$

or equivalently,

$$[00 \ldots 0] + (\alpha + \beta + \ldots + \omega)X \Leftrightarrow [\alpha\beta \ldots \omega] \tag{3.37}$$

is given by

$$\Delta G_{\alpha\beta\ldots\omega} = G_{\alpha\beta\ldots\omega} - G_{00\ldots0} = k_B T \ln \frac{Y_{00\ldots0}}{Y_{\alpha\beta\ldots\omega}} = -k_B T \ln A_{\alpha\beta\ldots\omega} \tag{3.38}$$

As usual, we have cast all equilibria in terms of the overall equilibrium constants $A_{\alpha\beta\ldots\omega}$. The partition function (3.35) written in terms of these parameters is then

$$\Psi(x) = \sum_{\alpha=0}^{1} \sum_{\beta=0}^{1} \cdots \sum_{\omega=0}^{1} A_{\alpha\beta\ldots\omega} x^{\alpha+\beta+\ldots+\omega} \tag{3.39}$$

The linkage among the N sites is uniquely defined in the site-specific description once the $v = 2^N - 1$ independent coefficients A in eq (3.39) are known.

We now cast the coefficients A in terms of site-specific binding constants K and interaction coefficients c as follows. The binding constant

$$K_j = A_{00\ldots1\ldots0} \tag{3.40}$$

reflects the binding of X to site j of the macromolecule when all other sites are unligated. The value of K_j is associated with the standard free energy change

$$\Delta G_j = -k_B T \ln \frac{Y_{00\ldots1\ldots0}}{Y_{00\ldots0\ldots0}} = -k_B T \ln K_j \tag{3.41}$$

The coefficients A for which $\alpha + \beta + \ldots + \omega = 1$ define the site-specific binding constants K. The remaining coefficients define interaction constants for site-specific cooperativity. Consider the coefficient $A_{00\ldots1\ldots1\ldots0}$

associated with the configuration with all sites unligated, except sites i and j. The interaction constant c_{ij} defines the ratio

$$c_{ij} = \frac{Y_{00\ldots0\ldots0\ldots0} Y_{00\ldots1\ldots1\ldots0}}{Y_{00\ldots1\ldots0\ldots0} Y_{00\ldots0\ldots1\ldots0}} = \frac{A_{00\ldots1\ldots1\ldots0}}{A_{00\ldots1\ldots0\ldots0} A_{00\ldots0\ldots1\ldots0}} \qquad (3.42)$$

or the equilibrium constant for the dismutation

$$M_{00\ldots1\ldots0\ldots0} + M_{00\ldots0\ldots1\ldots0} \Leftrightarrow M_{00\ldots0\ldots0\ldots0} + M_{00\ldots1\ldots1\ldots0} \qquad (3.43)$$

The interaction coefficient c_{ij} establishes the stability of the doubly-ligated configuration involving sites i and j relative to the parent singly-ligated configurations. A value of $c_{ij} > 1$ indicates that the ij-doubly-ligated intermediate is more stable than the sum of the parent singly-ligated configurations. A value of $c_{ij} < 1$ indicates reduced stability and $c_{ij} = 1$ indicates no difference. The standard free energy associated with c_{ij} is

$$\Delta G_{ij} = -k_B T \ln c_{ij} = \Delta G_{00\ldots1\ldots1\ldots0} - \Delta G_{00\ldots1\ldots0\ldots0} - \Delta G_{00\ldots0\ldots1\ldots0}$$
$$= \Delta G_{00\ldots1\ldots1\ldots0} - \Delta G_i - \Delta G_j \qquad (3.44)$$

and gives the 'excess' free energy or driving force associated with the dismutation reaction (3.43). The definition of c_{ij} allows the coefficient $A_{00\ldots1\ldots1\ldots0}$ to be written in the simple form

$$A_{00\ldots1\ldots1\ldots0} = c_{ij} K_i K_j \qquad (3.45)$$

The recipe for defining interaction coefficients involving three or more sites is a straightforward generalization of the foregoing method. Consider the triply-ligated intermediate with sites h, i and j ligated. Then

$$A_{00\ldots1\ldots1\ldots1\ldots0} = c_{hij} K_h K_i K_j \qquad (3.46)$$

and c_{hij} is the equilibrium constant for the reaction

$$M_{00\ldots1\ldots0\ldots0\ldots0} + M_{00\ldots0\ldots1\ldots0\ldots0} + M_{00\ldots0\ldots0\ldots1\ldots0}$$
$$\Leftrightarrow 2M_{00\ldots0\ldots0\ldots0\ldots0} + M_{00\ldots1\ldots1\ldots1\ldots0} \qquad (3.47)$$

In general, the coefficient $A_{\alpha\beta\ldots\omega}$ of the partition function can be written as the product of an interaction constant c and $\alpha + \beta + \ldots + \omega$ site-specific binding constants K. The index of c is assigned by listing the ligated sites of the intermediate in lexicographic order and, for each ligated site, the appropriate K is entered. For example, the product $c_{257} K_2 K_5 K_7$ refers to the triply-ligated intermediate with sites 2, 5 and 7 bound. The interaction coefficient is a measure of the coupling among sites and the number of linked sites defines the order of the interaction. The zero-order coupling constant for the reference configuration and the first-order interaction constants for all singly-ligated configurations are identically equal to one, as follows directly from their definition. First-order interaction constants are unnecessary, since they would reflect the change in binding affinity of a given site relative to a reference state other than the macromolecule under consideration. The binding constant K_j encapsulates all the molecular events, tertiary and/or quaternary, linked to the binding to site j when all other sites are unligated. This constant is a property of the macromolecule under consideration and should not be expected to be the same as that of site j isolated from the rest of the macromolecule.

The foregoing definitions of cs and Ks have the rather nontrivial consequence that all ligated intermediates are labeled in a linear fashion according to the ligated sites. Since each site is labeled as a separate entity in a unique way, it follows that each time site j is ligated the constant K_j must appear in the related coefficient of the partition function. On the other hand, if site j is unligated, then the related coefficient of the partition function does not contain K_j. Hence, although the partition function is a polynomial of degree N in the ligand activity x, it behaves as a polynomial of first degree in K_j, and this is true for all $j = 1, 2, \ldots N$. This suggests that Ψ can be manipulated using the Ks as independent variables in a rather simple way. It is at this point that contracted forms of Ψ come into the picture.

Consider a particular site, say site j, out of the total N sites. The probability of this site being ligated as a function of x is the site-specific binding curve X_j of site j. To compute X_j we proceed as follows. The whole ensemble of 2^N configurations of the system is partitioned in two subsets: one containing all configurations with site j ligated and the other containing all configurations with site j unligated. Each subset constructed in this way contains a total of 2^{N-1} terms. Each configuration belongs either to the first subset or to the second, and the information stored in each subset is complementary. Using the property that all configurations with site j ligated necessarily contain the term K_j, while those with site j unligated do not, the partition function (3.35) of the whole system can be written as

$$\Psi(x) = {}^0\Psi_j(x) + {}^1\Psi_j(x)K_jx \qquad (3.48)$$

${}^0\Psi_j$ is the partition function of the subsystem composed of $N - 1$ sites other than j with site j kept unligated and ${}^1\Psi_j$ is the partition function of the subsystem composed of $N - 1$ sites other than j with site j ligated. These functions are *contracted partition functions* in the sense that they provide information on contracted forms of the macromolecule generated by keeping a given site in a particular ligation state.[†] Like Ψ, the contracted forms are polynomial expansions in the ligand activity x with positive coefficients. The various contractions involving the 2^N configurations for $N = 3$ are shown in Figure 3.9. The function X_j is simply the ratio of all configurations with site j ligated to the total, i.e.,

$$X_j = K_j x \frac{{}^1\Psi_j(x)}{\Psi(x)} = 1 - \frac{{}^0\Psi_j(x)}{\Psi(x)} \tag{3.49}$$

If we exploit K_j as an independent variable in eq (3.48), we also have (Di Cera, 1989, 1990)

$$X_j = \frac{\partial \ln \Psi(x)}{\partial \ln K_j} \tag{3.50}$$

which is a consequence of the fact that neither ${}^0\Psi_j$ nor ${}^1\Psi_j$ are functions of K_j. The importance of eq (3.49) stems from the fact that the local quantity

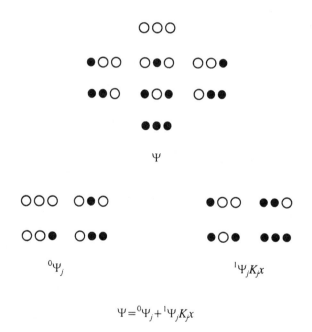

Figure 3.9   Manifold of ligated configurations for $N = 3$ (top). The ensemble can be split into two sets, one containing all intermediates with site j unligated (left), and the other containing all intermediates with site j ligated (right). The appropriate partition function is indicated for each case. The basic relationship (3.48) is readily derived from the figure.

[†] The term 'contracted' is introduced by analogy to the contraction of secular equations in quantum theory (Löwdin, 1964), which reduces the dimensionality of the space under consideration (Di Cera, 1990).

X_j can be derived from knowledge of global quantities such as Ψ and one of its contracted forms. In fact, Ψ reflects the behavior of the system as a whole, while $^0\Psi_j$ reflects the behavior of a contracted system with site j unligated. The contraction generates another macromolecule with $N-1$ sites available for binding and site j frozen in either its unligated or ligated form. The contraction has absolutely no effect on the fundamental properties of either $^0\Psi_j$ or $^1\Psi_j$ that are analogous to those of Ψ. More importantly, eq (3.49) poses no restrictions on the form of the coefficients of Ψ and $^0\Psi_j$. Information on the site-specific quantity X_j can be arrived at from Ψ and $^0\Psi_j$ even if these coefficients are known in their global form. Hence, generating contracted forms of the system as a whole is equivalent to accessing information on the site-specific components involved in the partition function. This also implies that the analysis and resolution of site-specific energetics demands knowledge of the partition function of the system as a whole, as in the global description, and also of its contracted forms. Site-specific thermodynamics provides a connection between the properties of individual sites and those of the system as a whole and its contracted forms. Intuitively, we can anticipate that many site-specific properties can be cast in terms of global properties of the system as a whole and its contracted forms. The link originates from the properties of the contracted partition functions derived from Ψ.

Each contracted partition function in eq (3.48) can be written in a form analogous to Ψ by properly choosing a reference configuration. The unligated form of the macromolecule is used as reference of Ψ, but in the case of contracted partition functions the choice may be different. Consider the case of $^1\Psi_j$. All configurations in $^1\Psi_j$ contain site j ligated and therefore it is appropriate to choose the configuration $[00\ldots 1\ldots 0]$ as reference. In the case of $^0\Psi_j$, all configurations contain site j unligated and the configuration $[00\ldots 0\ldots 0]$ may be taken as reference. Incidentally, this configuration is the same as the unligated form of the macromolecule. In either case, we note that the configuration taken as reference is the one with the maximum number of unligated sites compatible with the ensemble of ligated species. This simple rule applies quite generally even in the case of the system as a whole.

Once the reference configuration has been selected, the properties of any contracted partition function can be treated as those discussed for the partition function of the system as a whole. For example, in the case of $^1\Psi_j$, the coefficients can be treated globally as overall equilibrium constants for the reaction

$$M_{j=1} + iX \Leftrightarrow M_{j=1}X_i \qquad (3.51)$$

where $M_{j=1}$ is the same as $M_{00\ldots1\ldots0}$, or the macromolecule with site j ligated and all other sites free. If $A_{i(j=1)}$ is the equilibrium constant associated with (3.51), then

$$^1\Psi_j(x) = \sum_{i=0}^{N-1} A_{i(j=1)}x^i \tag{3.52}$$

Note that the index i runs from zero to $N-1$, since the contraction has reduced the number of sites available for binding. Likewise, for $^0\Psi_j$ one has

$$^0\Psi_j(x) = \sum_{i=0}^{N-1} A_{i(j=0)}x^i \tag{3.53}$$

where $A_{i(j=0)}$ refers to the reaction analogous to (3.51)

$$M_{j=0} + iX \Leftrightarrow M_{j=0}X_i \tag{3.54}$$

Differentiation of eqs (3.52) and (3.53) yields all the response functions of the contracted macromolecule. Specifically,

$$^1X_j = \frac{\mathrm{d}\ln {}^1\Psi_j(x)}{\mathrm{d}\ln x} = \frac{\displaystyle\sum_{i=0}^{N-1} iA_{i(j=1)}x^i}{\displaystyle\sum_{i=0}^{N-1} A_{i(j=1)}x^i} \tag{3.55}$$

$$^1B_j = \frac{\mathrm{d}^1X_j}{\mathrm{d}\ln x} = \frac{\displaystyle\sum_{i=0}^{N-1} i^2 A_{i(j=1)}x^i}{\displaystyle\sum_{i=0}^{N-1} A_{i(j=1)}x^i} - \left(\frac{\displaystyle\sum_{i=0}^{N-1} iA_{i(j=1)}x^i}{\displaystyle\sum_{i=0}^{N-1} A_{i(j=1)}x^i}\right)^2 \tag{3.56}$$

$$^{x,1}K_j = \frac{{}^1X_j}{(N-1-{}^1X_j)x} = \frac{\displaystyle\sum_{i=0}^{N-1} iA_{i(j=1)}x^{i-1}}{\displaystyle\sum_{i=0}^{N-1} (N-1-i)A_{i(j=1)}x^i} \tag{3.57}$$

are the average number of ligated sites, the binding capacity and the affinity function of the contracted macromolecule, respectively, with site j frozen in its ligated configuration. Likewise,

$$^0X_j = \frac{\mathrm{d}\ln {}^0\Psi_j(x)}{\mathrm{d}\ln x} = \frac{\displaystyle\sum_{i=0}^{N-1} iA_{i(j=0)}x^i}{\displaystyle\sum_{i=0}^{N-1} A_{i(j=0)}x^i} \tag{3.58}$$

$$
{}^0B_j = \frac{\mathrm{d}^0X_j}{\mathrm{d}\ln x} = \frac{\displaystyle\sum_{i=0}^{N-1} i^2 A_{i(j=0)}x^i}{\displaystyle\sum_{i=0}^{N-1} A_{i(j=0)}x^i} - \left(\frac{\displaystyle\sum_{i=0}^{N-1} iA_{i(j=0)}x^i}{\displaystyle\sum_{i=0}^{N-1} A_{i(j=0)}x^i}\right)^2
\tag{3.59}
$$

$$
{}^{x,0}K_j = \frac{{}^0X_j}{(N-1-{}^0X_j)x} = \frac{\displaystyle\sum_{i=0}^{N-1} iA_{i(j=0)}x^{i-1}}{\displaystyle\sum_{i=0}^{N-1} (N-1-i)A_{i(j=0)}x^i}
\tag{3.60}
$$

are the average number of ligated sites, the binding capacity and the affinity function of the contracted macromolecule, respectively, with site j frozen in its unligated configuration. All these response functions obey the properties analyzed in Section 2.2. for analogous global quantities. In particular, the quantities

$$
x_{\mathrm{m}(j=1)} = A_{N-1(j=1)}^{-1/(N-1)}
\tag{3.61a}
$$

$$
x_{\mathrm{m}(j=0)} = A_{N-1(j=0)}^{-1/(N-1)}
\tag{3.61b}
$$

give the mean ligand activities of the contracted forms of the macro-molecule and provide a measure of the average work, in $k_{\mathrm{B}}T$ units, spent in ligating one site.

Site-specific response functions become clear when cast in terms of the foregoing quantities characterizing the behavior of contracted forms of the macromolecule. The affinity function of site j leads to the familiar form

$$
X_j = \frac{{}^xK_jx}{1 + {}^xK_jx}
\tag{3.62}
$$

Hence,

$$
{}^xK_j = K_j \frac{{}^1\Psi_j(x)}{{}^0\Psi_j(x)}
\tag{3.63}
$$

In general, the affinity function changes with saturation since site j is a subsystem subject to interactions with the remaining $N-1$ sites. ${}^xK_j$ is constant for all values of x only when ${}^1\Psi_j = {}^0\Psi_j$. This case denotes the absence of linkage between site j and the rest of the macromolecule. If binding to site j does not affect the properties of the remaining $N-1$ sites, then the two contracted partition functions are identical. The limiting values of the affinity function are of particular importance. For $x \to 0$ both contracted partition functions tend to one and therefore

$$
{}^0K_j = K_j
\tag{3.64}
$$

At low saturation the affinity function approaches the value of the site-specific binding constant K_j. For $x \to \infty$, on the other hand,

$$^\infty K_j = K_j \frac{A_{N-1(j=1)}}{A_{N-1(j=0)}} = K_j \frac{x_{m(j=0)}^{N-1}}{x_{m(j=1)}^{N-1}} \tag{3.65}$$

The affinity function tends to a value that depends on the ratio of the mean ligand activities of the two contracted forms of the macromolecule. If binding to site j increases the overall affinity of the remaining $N-1$ sites, then the mean ligand activity $x_{m(j=1)}$ is smaller than $x_{m(j=0)}$. Consequently, the affinity function increases with saturation and $^\infty K_j > {}^0 K_j$. On the other hand, if binding to site j decreases the overall affinity of the remaining $N-1$ sites, then the mean ligand activity $x_{m(j=1)}$ is greater than $x_{m(j=0)}$ and $^\infty K_j < {}^0 K_j$. Hence, the behavior of the affinity function $^x K_j$ of site j provides information on the behavior of the remainder of the sites of the macromolecule. This is even more evident when considering the slope of the affinity plot

$$\frac{d \ln {}^x K_j}{d \ln x} = {}^1 X_j - {}^0 X_j = {}^j \Delta X \tag{3.66}$$

which gives a direct measure of the net number of ligands exchanged at the other $N-1$ sites upon binding to site j. Site j is positively linked to the rest of the macromolecule if $^j \Delta X > 0$, since the number of ligated sites increases upon binding to site j. It is negatively linked if $^j \Delta X < 0$, since the number of ligated sites decreases upon binding to site j. In the case of $^j \Delta X = 0$, ligation of site j has no effect on the ligation of the remaining sites. The quantity $^j \Delta X$ is the driving force for site-specific cooperativity. It is quite interesting to note that this driving force is provided by the sites of the macromolecule other than j. The response function $^x K_j$ gives a measure of how sites communicate and give rise to cooperative interactions.

The quantity $^j \Delta X$ also defines the site-specific binding capacity. Differentiation of eq (3.62) yields by virtue of (3.66)

$$B_j = X_j(1 - X_j)(1 + {}^j \Delta X) \tag{3.67}$$

The response function B_j is the product of two terms. The term $X_j(1 - X_j)$ is always positive and gives the value of B_j when $^j \Delta X = 0$. This case corresponds to the absence of linkage between site j and the rest of the macromolecule, or a site j decoupled from the other sites. In this case site j behaves as a reference system and the binding capacity is given by the familiar form

$$B_{0j} = X_j(1 - X_j) \tag{3.68}$$

where the suffix on B indicates that site j is an independent site. The second term in eq (3.67) is the slope of the affinity plot in eq (3.66) plus one. This quantity, as we have seen in the global description, defines the Hill coefficient. The Hill coefficient of site j is therefore

$$n_j = 1 + {}^j\Delta X = \frac{B_j}{B_{0j}} \qquad (3.69)$$

The definition in terms of the ratio B_j/B_{0j} is analogous to that of the global description. The Hill coefficient can be seen as the ratio of the fluctuations of the quantity X_j in the system under consideration, relative to those observed for a reference system in which site j behaves independently. The term $1 + {}^j\Delta X$, on the other hand, is peculiar to the local description and has no parallel in the global picture. This term is a linkage quantity which expresses the net number of sites ligated when site j is bound. In fact, ${}^j\Delta X$ gives the net number of ligands exchanged at the other $N - 1$ sites upon binding to site j. The unit term can be considered as the number of ligands exchanged at site j upon binding to this site. Hence (Di Cera, 1994b),

Definition: The Hill coefficient of an individual site is the net number of ligands exchanged by the macromolecule upon binding to that site.

If site j has $n_j = 2.4$, it means that binding to site j increases the number of ligated sites of the macromolecule by 2.4 units. Of this quantity, 1 refers to ligation of site j, while the remaining 1.4 units indicate the change in overall ligation of the remaining $N - 1$ sites.

Unlike the term $X_j(1 - X_j)$, the term $1 + {}^j\Delta X$ in eq (3.67) can be positive or negative. The term ${}^j\Delta X$ is the difference of two functions bounded from zero to $N - 1$ and is bounded from $-(N - 1)$ to $N - 1$. Hence, the site-specific Hill coefficient obeys the condition

$$2 - N \leqslant n_j \leqslant N \qquad (3.70)$$

For $N = 1$, this condition merely states that $n_j = 1$, as expected for a macromolecule containing a single site. For $N = 2$, the condition is analogous to that of the global case, with n_j being positive and bounded from zero to two. However, for $N \geqslant 3$, the lower bound for n_j is no longer zero, but a negative number. The upper bound, on the other hand, is always the same as that of n_H for the macromolecule as a whole. A negative value of the Hill coefficient, which would represent a violation of the second law of thermodynamics in the global picture, is perfectly legitimate in the local picture. The binding capacity, which is a positive quantity in the global picture, can be positive or negative in the local picture due to the definition (3.67). Negative values of the binding capacity have been documented experimentally (Kojima and Palmer, 1983; Hendler, Subba Reddy, Shrager and Caughey, 1986) and are shown in Figure 3.10. The physical significance of negative values of n_j or B_j should be rationalized in terms of the significance of $^j\Delta X$. The peculiar meaning of n_j in the local description reveals that this quantity is a measure of the net balance of ligation upon binding to site j. If binding to site j causes ligand X to be released at other sites due to negative linkage, then $^j\Delta X < 0$. When $^j\Delta X < -1$ the Hill coefficient of site j becomes negative and so is the site-specific binding capacity. We also note from eq (3.69) that $n_j \to 1$ for $x \to 0$ and $x \to \infty$, just as seen for the Hill coefficient

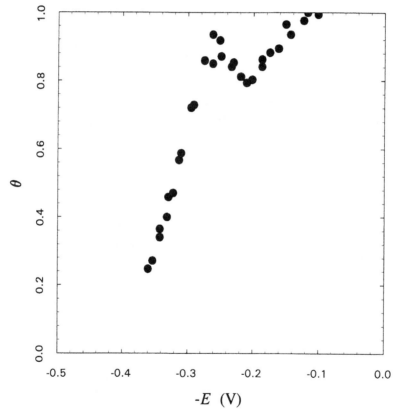

Figure 3.10   Electron binding curve of the heme center of cyt*a* in the complex of cyt*aa₃* (Kojima and Palmer, 1983). The redox potential provides the driving force for electron binding and scales linearly with the logarithm of the electron activity. The site-specific binding curve of cyt*a* shows a negative binding capacity in the potential region of 200–260 mV.

in the global description. This is because $^j\Delta X \to 0$ in those limits. A final remark should be made on the extrema of the Hill coefficient n_j, which can be derived in a straightforward way from the properties of contracted forms of the macromolecule. The function n_j has extrema for finite x whenever the derivative $dn_j/d\ln x$ vanishes, or when

$$^1B_j = {}^0B_j \tag{3.71}$$

This is also the condition for which $^j\Delta X$ has an extremum, i.e., it is a minimum or a maximum. In either case, the linkage between site j and the rest of the macromolecule, whether positive or negative, is maximized when the binding capacities of the two contracted forms with site j ligated and unligated become identical. Under this condition the cooperativity of site j is either a maximum or a minimum.

Other properties of the contracted partition functions can be derived by recognizing that the definition (3.48) does not depend on the particular value of j, so that for any two sites i and j one has

$$^0\Psi_j(x) + {}^1\Psi_j(x)K_jx = {}^0\Psi_i(x) + {}^1\Psi_i(x)K_ix \tag{3.72}$$

Clearly, since there are N sites, there must be N equivalent ways of casting Ψ in terms of contracted partition functions with one site frozen. However, each contracted partition function can in turn be split into contracted forms like Ψ. For example, $^0\Psi_j$ can be written as

$$^0\Psi_j(x) = {}^{00}\Psi_{ij}(x) + {}^{10}\Psi_{ij}(x)K_ix \tag{3.73}$$

by contracting over site i once site j is kept in the unligated configuration. The contracted partition functions $^{00}\Psi_{ij}$ and $^{10}\Psi_{ij}$ with two sites frozen depict the behavior of the remaining $N-2$ sites when site i and j are kept unligated, or when site j is unligated and site i is ligated. Likewise, for $^1\Psi_j$ one has

$$^1\Psi_j(x) = {}^{01}\Psi_{ij}(x) + {}^{11}\Psi_{ij}(x)c_{ij}K_ix \tag{3.74}$$

Hence,

$$\Psi(x) = {}^{00}\Psi_{ij}(x) + [{}^{10}\Psi_{ij}(x)K_i + {}^{01}\Psi_{ij}(x)K_j]x + {}^{11}\Psi_{ij}(x)c_{ij}K_iK_jx^2 \tag{3.75}$$

Each of the contracted partition functions in eq (3.75) contains 2^{N-2} terms and the four terms in eq (3.75) yield a total of 2^N configurations, as expected. There are $N(N-1)/2$ equivalent ways of casting Ψ in a form like eq (3.75) in terms of contracted partition functions with two sites frozen. The contraction process can be extended up to N sites, in which case the contracted partition functions contain a single term equal to one. One can see by inspection that the partition function Ψ can be written as the sum of 2^m independent terms, each containing an mth-order contracted partition function with m sites frozen, times a 'weighting factor' reflecting the probability of existence of the reference configuration in the contracted manifold relative to the unligated form of the macromolecule, which is the reference configuration for the system as a whole.

When Ψ is cast as in (3.48), the whole system is analyzed in terms of site j under the influence of the remaining $N - 1$ sites. This influence, or 'perturbation', is embodied by the contracted partition functions $^1\Psi_j$ and $^0\Psi_j$. The reciprocity of the perturbation also offers the alternative interpretation of the whole system in terms of the properties of $N - 1$ sites perturbed by site j. The perturbation in this case is provided by the weighting factors 1 and $K_j x$. Likewise, when Ψ is cast as in eq (3.75), the whole system is analyzed in terms of the interactions between sites i and j under the influence of the remaining $N - 2$ sites, as determined by $^{00}\Psi_{ij}$, $^{10}\Psi_{ij}$, $^{01}\Psi_{ij}$ and $^{11}\Psi_{ij}$, or alternatively in terms of the properties of these $N - 2$ sites under the influence of sites i and j. The weighting factors in eq (3.75) are 1, $K_i x$, $K_j x$ and $c_{ij} K_i K_j x^2$. Note that these terms are the elements of the partition function of a system containing sites i and j alone, just as the weighting factors 1 and $K_j x$ in eq (3.48) are the elements of the partition function of a system containing site j alone. Hence, we have a rather simple rule for expressing Ψ in terms of the properties of subsystems. The partition function of the system as a whole can be cast in terms of the partition function of any subsystem, provided each term of the polynomial expansion is complemented by an appropriate contracted partition function. For example, if we want to cast the properties of the system in terms of site j, we can write the partition function for a system containing site j only, i.e.,

$$^j\Psi(x) = 1 + K_j x \tag{3.76}$$

and then correct each term by *ad hoc* contracted partition functions. The choice of which function enters the definition of Ψ is determined by which configuration is represented by the terms in eq (3.76). The term 1 reflects the contribution of the unligated site j. Its complement is the contracted partition function of the remaining $N - 1$ sites with site j kept unligated. Likewise, the complement of $K_j x$ is the contracted partition function of the remaining $N - 1$ sites with site j kept ligated, hence, eq (3.48). Analogous arguments lead to eq (3.75) using the partition function for sites i and j

$$^{ij}\Psi(x) = 1 + (K_i + K_j)x + c_{ij} K_i K_j x^2 \tag{3.77}$$

and the appropriate complements for each term, and so forth. One sees that in the limit of $m = N$ the partition function contains 2^N weighting factors complemented by contracted partition functions all identically equal to one. This is the case where the expansion of Ψ in terms of contracted partition functions yields the site-specific form of Ψ. The other special case for $m = 0$ yields only one term, equal to one, complemented by a contracted partition function where none of the sites is frozen. The result is a mere tautology, since this zero-order contracted partition function is the same as Ψ.

3.3 Thermodynamic basis of site-specific effects

In the previous section we have illustrated some of the properties of site-specific response functions. In particular, we have seen that X_j cannot be derived from the partition function Ψ of the system as a whole by differentiation with respect to $\ln x$, but only from contracted forms of Ψ. We have also seen that B_j can assume negative values for $N \geqslant 3$. In this section we illustrate the thermodynamic origin of site-specific cooperativity and provide an explanation for the reason why site-specific effects are not necessarily subject to the restrictions imposed by thermodynamic stability (Di Cera, 1994b). Site-specific thermodynamics bears on fundamental features that are absent from the global description of binding phenomena dealt with in the previous chapter, as will soon be evident from consideration of the following multicomponent analogue.

Consider the N sites of the macromolecule as each binding a different ligand and let X_j be the ligand binding to site j, with $j = 1, 2, \ldots N$ (see Figure 3.1). At constant T and P the relevant potential for this generalized ensemble containing N different ligands is

$$d\Pi = -k_B T \sum_{j=1}^{N} X_j \, d\ln x_j \tag{3.78}$$

where X_j is the amount of ligand X_j bound to site j and x_j is its activity. The partition function for this system is a straightforward generalization of eq (2.142)

$$\Psi(x_1, x_2, \ldots x_N) = \sum_{\alpha=0}^{1} \sum_{\beta=0}^{1} \cdots \sum_{\omega=0}^{1} Y_{\alpha\beta\ldots\omega} x_1^\alpha x_2^\beta \ldots x_n^\omega \tag{3.79}$$

The first index refers to ligand X_1, the second index to ligand X_2 and so on. The similarity between eqs (3.79) and (3.35) is evident. The system of N ligands, each binding to a separate site, provides a multicomponent analogue of the local site-specific scenario where the same type of ligand binds to N sites of the macromolecule. Each ligation intermediate in the N-dimensional manifold spanned by the N ligands is characterized by the

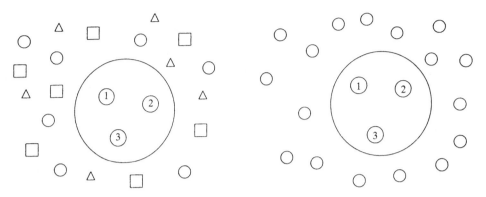

Figure 3.11 Multicomponent analogue of site-specific binding phenomena, sketched as a macromolecule containing three sites, each binding a different ligand (squares to site 1, circles to site 2, triangles to site 3). Under conditions where all ligands are present at equal concentrations under any transformation the system at left is equivalent to a macromolecule containing three sites for a single type of ligand, as shown at right.

mechano-thermal partition function $Y_{\alpha\beta\ldots\omega}$ and the Gibbs free energy $G_{\alpha\beta\ldots\omega} = -k_B T \ln Y_{\alpha\beta\ldots\omega}$. The standard free energy change for the reaction

$$M + \alpha X_1 + \beta X_2 + \ldots + \omega X_N \Leftrightarrow M(X_1)_\alpha (X_2)_\beta \ldots (X_N)_\omega \quad (3.80)$$

is given by

$$\Delta G_{\alpha\beta\ldots\omega} = G_{\alpha\beta\ldots\omega} - G_{00\ldots 0} = k_B T \ln \frac{Y_{00\ldots 0}}{Y_{\alpha\beta\ldots\omega}} = -k_B T \ln A_{\alpha\beta\ldots\omega} \quad (3.81)$$

Again, we have recast all equilibria (3.80) in terms of the overall equilibrium constants $A_{\alpha\beta\ldots\omega}$, letting $Y_{00\ldots 0} = 1$ without loss of generality. We recall that this is equivalent to selecting the unligated configuration of the macromolecule, $M_{00\ldots 0}$, as reference. The partition function (3.79) is therefore

$$\Psi(x_1, x_2, \ldots x_N) = \sum_{\alpha=0}^{1} \sum_{\beta=0}^{1} \ldots \sum_{\omega=0}^{1} A_{\alpha\beta\ldots\omega} x_1^\alpha x_2^\beta \ldots x_N^\omega \quad (3.82)$$

and contains a total of $v = 2^N - 1$ independent coefficients, since there are a total of 2^N possible configurations, of each one is used as reference. Knowledge of this set of coefficients completely defines the linkage scheme involving the N ligands. The functions X_j and B_j can be derived from Ψ by differentiation as follows

$$X_j = \left(\frac{\partial \ln \Psi}{\partial \ln x_j} \right)_{\{x_{i\neq j}\}} \quad (3.83)$$

$$B_j = \left(\frac{\partial X_j}{\partial \ln x_j} \right)_{\{x_{i \neq j}\}} \tag{3.84}$$

where $\{x_{i \neq j}\}$ denotes that all variables other than x_j are held constant.

It is easy to verify from eqs (3.83) and (3.84) that site j behaves noncooperatively as expected. The affinity function of site j can be derived by consideration of contracted forms of Ψ written as follows

$$\Psi(\{x_i\}) = {}^0\Psi(\{x_{i \neq j}\}) + {}^1\Psi(\{x_{i \neq j}\}) K_j x_j \tag{3.85}$$

The partition function of the system as a whole can be expressed as the sum of two terms, one containing all ligated configurations with site j unligated, ${}^0\Psi(\{x_{i \neq j}\})$, and the other containing all configurations with site j ligated, ${}^1\Psi(\{x_{i \neq j}\}) K_j x_j$. The coefficient $K_j = A_{00\ldots 1 \ldots 0}$ is the equilibrium constant for ligand X_j binding to site j when all other sites are in the unligated form. Note that the contracted partition functions ${}^0\Psi$ and ${}^1\Psi$ are functions of all xs but x_j. Since there are N sites, then there are N equivalent ways of casting Ψ in a form given by eq (3.85), one for each binding site. The binding isotherm X_j cast in terms of contracted partition functions is

$$X_j = K_j x_j \frac{{}^1\Psi(\{x_{i \neq j}\})}{\Psi(\{x_i\})} = 1 - \frac{{}^0\Psi(\{x_{i \neq j}\})}{\Psi(\{x_i\})} \tag{3.86}$$

It should be pointed out that the form of eq (3.86), which is of course equivalent to eq (3.83), follows directly from the definition of contracted partition functions and does not involve differentiation of Ψ. The foregoing definition of X_j allows one to write the affinity function of site j as

$$^{x_j}K = K_j \frac{{}^1\Psi(\{x_{i \neq j}\})}{{}^0\Psi(\{x_{i \neq j}\})} \tag{3.87}$$

As expected, the affinity function is a constant, since it depends on all xs but x_j. Application of arguments developed in Section 2.5, and particularly eq (2.121), yields

$$\left(\frac{\partial \ln {}^{x_j}K}{\partial \ln x_i} \right)_{\{x_{i \neq i,j}\}} = {}^1X_{i(j)} - {}^0X_{i(j)} = {}^j\Delta X_i \tag{3.88}$$

The change of the logarithm of the affinity function of ligand X_j, due to a change of the logarithm of the activity of ligand X_i, gives the net amount of ligands X_i exchanged upon binding of ligand X_j. The quantity ${}^j\Delta X_i$ is specific of X_i and X_j, and its sign defines the nature of the linkage, positive or negative, between ligands X_i and X_j.

The basic properties of the foregoing linkage scheme can be defined in terms of response functions computed when a given set of activities are kept constant. However, thermodynamics puts no restriction on what particular values this set of activities must satisfy. Consider, therefore, the case in which all xs are identical under any transformation. This is equivalent to a system of N ligands, which are present in solution at the same activity and bind to N distinct sites of the macromolecule. Any transformation that changes the activity of a given ligand also affects the activity of all others to the same extent, so that

$$x_1 = x_2 = \ldots = x_N = x \tag{3.89}$$

Under this constraint, our original linkage scheme involving N different ligands binding to N distinct sites of the macromolecule becomes equivalent to a system composed of a macromolecule containing N sites for a ligand X, whose activity is x. In fact, the partition function (3.82) under the constraint (3.89) becomes

$$\Psi(x) = \sum_{\alpha=0}^{1}\sum_{\beta=0}^{1} \ldots \sum_{\omega=0}^{1} A_{\alpha\beta\ldots\omega} x^{\alpha+\beta+\ldots+\omega} \tag{3.90}$$

The indices α, β, ... ω labeling different ligands in the linkage scheme, now label different sites: site 1 for ligand 1, site 2 for ligand 2, and so forth. The partition function (3.90) is mathematically identical to eq (3.39). The sum

$$j = \alpha + \beta + \ldots + \omega \tag{3.91}$$

runs from zero to N, as expected for the partition function of a macromolecule containing N sites for a ligand X. Also, for each value of j, there are as many as $C_{j,N}$ coefficients $A_{\alpha\beta\ldots\omega}$ in (3.90) that contribute to the overall coefficient of x^j in the polynomial expansion.

The analogy with the linkage scheme involving different ligands makes it possible to recognize the site-specific contribution to the global description. The conceptual basis of this analogy can be summarized with the equivalence

$$\text{ligands (global description)} = \text{sites (local description)} \tag{3.92}$$

The properties of individual sites of biological macromolecules can thus be derived from the properties of a multicomponent analogue composed of different ligands, each binding to a distinct site of the macromolecule, when the activities of all ligands are constrained to be the same under any transformation.

Consider again the linkage scheme involving N ligands, under the constraint (3.89). When eq (3.89) holds, it is not possible to derive the amount of ligand X_j bound to the macromolecule by differentiation of the partition function Ψ with respect to $\ln x_j$. The partial differentiation in eq (3.83) implies that all xs other than x_j are held constant. This condition, however, is at variance with the constraint (3.89), which establishes that all xs are the same under any transformation. A change in x_j necessarily leads to a change in all other xs. Under this condition, only the total derivative $d \ln \Psi / d \ln x_j$ can be computed. The differential operator for a function of N variables is given by

$$\frac{d}{d \ln x_j} = \sum_{i=1}^{N} \left(\frac{\partial}{\partial \ln x_i} \right)_{\{x_{l \neq i}\}} \frac{d \ln x_i}{d \ln x_j} \tag{3.93}$$

Under the constraint (3.89) all derivatives at the right-hand side become identically equal to one, while $d/d \ln x_j$ is the same as the derivative $d/d \ln x$. Hence,

$$\frac{d}{d \ln x} = \sum_{i=1}^{N} \left(\frac{\partial}{\partial \ln x_i} \right)_{\{x_{l \neq i}\}} = \nabla \tag{3.94}$$

The action of the gradient operator ∇ is equivalent to a total differentiation with respect to $\ln x$, or to the sum of partial differentiations with respect to all variables x in the usual way. Hence, the properties of the linkage scheme under the constraint (3.89) can be derived from those of the same system in the absence of constraint.

Partial differentiation of the partition function (3.85) with respect to $\ln x_j$, under the constraint (3.89), is equivalent to application of the ∇ operator to $\ln \Psi$, which yields

$$\nabla \ln \Psi(\{x_j\}) = \sum_{j=1}^{N} X_j \tag{3.95}$$

The presence of the constraint yields a quite different result. Instead of the function X_j, reflecting the amount of ligand X_j bound to the macromolecule, the differentiation yields the sum of the Xs for all ligands present in solution, or the total number of ligands bound to the macromolecule. This provides a simple explanation for the impossibility of deriving site-specific binding isotherms from differentiation of the partition function Ψ with respect to $\ln x$, when dealing with the binding of ligand X to N sites. The quantity X_j in the multicomponent analogue, in the presence of the constraint (3.89), acquires a site-specific character and needs to be derived in a different way. Eq (3.86) proves extremely useful in this regard. Knowledge of the partition function and its contracted forms in (3.85) provides a definition for X_j regardless of the particular conditions. Once X_j is obtained, other important response functions can be derived from it. The binding capacity of ligand X_j is defined as the derivative of X_j with respect to $\ln x_j$. Again, if eq (3.89) holds, only the total derivative $dX_j/d\ln x_j$ can be computed. Application of the operator relationship (3.94) yields.

$$\nabla X_j = B_j + L_{j1} + L_{j2} + \ldots + L_{jN} \qquad (3.96)$$

The change of X_j with $\ln x_j$ under the constraint (3.89) does not give the binding capacity in the absence of the constraint, B_j, but the sum of B_j and all other linkage coefficients L_{ji} $(i \neq j)$. Switching to the case of individual sites using the equivalence (3.92) and the significance of the gradient operator ∇, it results that the binding capacity of an individual site, $dX_j/d\ln x$, contains the contribution of the linkage between that site and the rest of the macromolecule. The important consequence of eq (3.96) is that the second law of thermodynamics, which applies to B_j, has no control on ∇X_j in the case of different ligands under the constraint (3.89). Hence, the constraints imposed by the second law do not apply in the case of the binding capacity of an individual site and eq (3.96) provides the thermodynamic basis of site-specific effects. The sign of the expression (3.96) can be positive or negative, depending upon the relative contribution of the linkage coefficients that can themselves be positive or negative.

Another important property of the multicomponent analogue under the constraint (3.89) is demonstrated by the affinity function (3.87). When the logarithm of the affinity function is changed with respect to $\ln x_j$ in the presence of the constraint, the gradient operator applies and yields the result

$$\nabla \ln {}^{x_j}\!K = {}^j\Delta X_1 + {}^j\Delta X_2 + \ldots + {}^j\Delta X_N \qquad (3.97)$$

where the various terms on the right-hand side have the same significance as in eq (3.88), and represent the amounts of ligands X_1, X_2, ... X_N exchanged upon binding of ligand j. In the absence of the constraint, the change in the affinity function with x_j is equal to zero. No such result is guaranteed in the presence of the constraint. The Hill coefficient of ligand X_j is obtained from eq (2.44) by adding one to eq (3.97), i.e.,

$$n_j = 1 + {}^j\Delta X_1 + {}^j\Delta X_2 + \ldots + {}^j\Delta X_N \qquad (3.98)$$

The value of n_j, which is always unity in the absence of constraint, can be positive or negative depending on the value of the individual ${}^j\Delta X_i$s. The upper and lower bounds for eqs (3.97) and (3.98) can easily be calculated by recalling the definition of ${}^j\Delta X_i$ in eq (3.88). For the linkage scheme under consideration, ${}^j\Delta X_i$ is bounded from -1 to 1, since this quantity expresses the difference between two binding curves that change from zero to one. Since there are $N - 1$ such terms in the sum in eq (3.97), then the right-hand side must be bound from $-(N - 1)$ to $N - 1$. Hence, the Hill coefficient of ligand X_j, in the presence of the constraint, is bounded from $2 - N$ to N. This is essentially the result derived for the Hill coefficient of an individual site in eq (3.70), and eq (3.98) provides the thermodynamic basis for site-specific cooperativity. The linkage with other sites generates cooperativity in site-specific binding curves. Eqs (3.70) and (3.98) give the quantitative expressions for this effect.

The constraint (3.89) also reduces the dimensionality of the system. In a system containing N different ligands, each binding to a distinct site of the macromolecule, the potential (3.78) is associated with an N-dimensional matrix $\mathbf{G}'$ with coefficients

$$g'_{ij} = \left(\frac{\partial X_i}{\partial \ln x_j} \right)_{\{x_{l\neq j}\}} \tag{3.99}$$

The matrix is positive-definite, as already demonstrated in Section 1.3. In the presence of the constraint (3.89), the potential (3.78) becomes

$$d\Pi = -k_B T \sum_{j=1}^{N} X_j \, d\ln x = -k_B TX \, d\ln x \tag{3.100}$$

where X denotes the total number of ligands bound to the macromolecule. The constraint has acted upon the system by contracting the dimensionality from N to one. Although N ligands are present, their activities do not change independently and are always equal, regardless of the transformation. As a result, the response function of the potential to a change in the independent variable $\ln x$ gives information on the overall quantity X, and not on the individual quantities X_j. Consequently, the derivative $dX/d\ln x$ is always positive due to thermodynamic stability. This derivative is equivalent to

$$\nabla X = \nabla \sum_{j=1}^{N} X_j > 0 \tag{3.101}$$

by virtue of (3.94). Consideration of eqs (3.96) and (3.99) yields

$$\nabla X_j = \sum_{i=1}^{N} g'_{ji} \tag{3.102}$$

The action of ∇ on X_j is equivalent to summing all metric coefficients of the jth row of $\mathbf{G}'$. We recall that the symmetric nature of $\mathbf{G}'$ makes this result equivalent to the sum of all metric coefficients of the jth column. Hence, $\nabla X = dX/d\ln x$ is the sum of all coefficients of $\mathbf{G}'$ and is always positive, since $\mathbf{G}'$ is positive-definite.

3.4 Properties of the binding curve in the local description

The properties of site-specific binding isotherms are somewhat analogous
to those discussed in the case of the global description. We start by noting
the conservation relationship

$$X = \sum_{j=1}^{N} X_j \tag{3.103}$$

The sum of the probabilities of binding to each site separately gives the
amount of ligand bound to the macromolecule. The global property X is
derived as the sum of all analogous local quantities. Since the quantity X
is the derivative of $\ln \Psi$ with respect to $\ln x$, and each X_j can be derived as
in eq (3.50), it follows that

$$\frac{d}{d \ln x} = \sum_{i=1}^{N} \frac{\partial}{\partial \ln K_i} = \nabla \tag{3.104}$$

Differentiation with respect to $\ln x$ is equivalent to the sum of partial
differentials with respect to the variables K. This is a consequence of the
fact that Ψ is a linear function of K_i and that each K_i labels the
corresponding site unambiguously. The conservation relationship (3.103)
has some important consequences in the analysis of site-specific effects.
The mean ligand activity of the macromolecule in the global description is
given by the integral equation (2.46)

$$\ln x_{\mathrm{m}} = \frac{1}{N} \int_0^N \ln x \, dX \tag{3.105}$$

We have already pointed out that this quantity expresses the mean value
of the ligand chemical potential in $k_B T$ units, or the average work spent
on ligating one site of the system. Upon substituting X in eq (3.105) with
eq (3.103), we obtain a sum of N terms of the form

$$\ln x_{\mathrm{m}j} = \int_0^1 \ln x \, dX_j \tag{3.106}$$

The quantity $\ln x_{\mathrm{m}j}$ is the logarithm of the mean ligand activity of site j, or
the work spent to ligate site j in $k_B T$ units. Hence (Ackers *et al.*, 1983; Di
Cera, 1989),

$$\ln x_{\mathrm{m}} = \frac{1}{N} \sum_{j=1}^{N} \ln x_{\mathrm{m}j} \tag{3.107}$$

which yields another conservation relationship. The mean ligand activity
of the macromolecule is the sum of the work spent to ligate each site
separately, divided by the number of sites.

The value of $\ln x_{mj}$ is the value of $\ln x$ where the area under the X_j curve, $Q_u(x)$, equals the area above the X_j curve up to the asymptote $X_j = 1$, $Q_a(x)$, this property being equivalent to that of $\ln x_m$ in the global description. Likewise, the value of $\ln x_{mj}$ can be derived from numerical analysis of the binding isotherm X_j from any point along the $\ln x$ axis using the same method as described in Section 2.3 for the function X. Specifically, one has

$$Q_u(x) = \int_{-\infty}^{\ln x} X_j \, d \ln x' \tag{3.108a}$$

$$Q_a(x) = \int_{\ln x}^{\infty} (1 - X_j) \, d \ln x' \tag{3.108b}$$

Hence,

$$\ln x_{mj} = \ln x + [Q_a(x) - Q_u(x)] \tag{3.109}$$

for any value of x. Given the arbitrary point $\ln x_1$, the value of $Q_u(x_1)$ is given by the area under the curve X_j from $-\infty$ to $\ln x_1$ and likewise the value of $Q_a(x_1)$ is given by the area above X_j up to the asymptote $X_j = 1$, from $\ln x_1$ to ∞. The difference $Q_a(x_1) - Q_u(x_1)$ added to $\ln x_1$ gives $\ln x_{mj}$, regardless of the particular point x_1.

The analytical solution of the integral in eq (3.106) for the mean ligand activity of an individual site is not as simple as in the case of the global description, since the function X_j cannot be obtained from differentiation of the partition function Ψ with respect to $\ln x$. The difficulty is immediately evident if Ψ is cast as in eq (2.78), i.e.,

$$\Psi(x) = \prod_{i=1}^{N}(1 + \alpha_i x) \tag{3.110}$$

where α_i is the ith coefficient of the factorization related to the ith root of Ψ. From (3.50) one has

$$X_j = \sum_{i=1}^{N} \frac{v_{ij}\alpha_i x}{1 + \alpha_i x} \tag{3.111a}$$

$$v_{ij} = \frac{\partial \ln \alpha_i}{\partial \ln K_j} \tag{3.111b}$$

Except for the coefficients v, X_j has a form identical to X in the global description (see eq (2.79)). These coefficients can be real or complex conjugate, depending on the nature of the αs with which they are associated. They can be thought of as 'filters' that distort the shape of the overall binding curve X to yield X_j. Some properties of these coefficients can be derived in a straightforward manner from consideration of the conservation relationship (3.103). We know from eq (2.79) that

$$X = \sum_{i=1}^{N} \frac{\alpha_i x}{1 + \alpha_i x} \tag{3.112}$$

The sum of all terms as in eq (3.111a) must give eq (3.112), because of eq (3.103). Hence,

$$\sum_{j=1}^{N} v_{ij} = 1 \tag{3.113}$$

It also follows from eq (3.111a) that

$$\sum_{i=1}^{N} v_{ij} = 1 \tag{3.114}$$

since $X_j \rightarrow 1$ when $x \rightarrow \infty$. The vs belong to an $N \times N$ matrix for which the elements of each row and column add up to one. The exact form of these coefficients depends on the form of the αs written in terms of the site-specific variables K and c. Without an explicit form for these coefficients, solution of eq (3.106) is not straightforward. An intriguing consequence of eq (3.111a) is that the site-specific binding curve X_j can be expressed as the derivative of an *ad hoc* function with respect to $\ln x$ as follows

$$X_j = \frac{\mathrm{d} \ln \Psi_j(x)}{\mathrm{d} \ln x} \tag{3.115}$$

with

$$\Psi_j(x) = \prod_{i=1}^{N} (1 + \alpha_i x)^{v_{ij}} \tag{3.116}$$

It follows from the properties of the vs that

$$\Psi(x) = \Psi_1(x)\Psi_2(x) \ldots \Psi_N(x) \tag{3.117}$$

The Ψ_js are *pseudo* partition functions for the individual sites. Each Ψ_j resembles Ψ of the system as a whole, except for the filtering effect of the vs.

Analysis of the shape of site-specific binding isotherms reveals other interesting aspects of the local description. When all sites are independent of one another, all coefficients c in the partition function Ψ are equal to one and Ψ assumes the rather simple form

$$\Psi(x) = \prod_{j=1}^{N} (1 + K_j x) \tag{3.118}$$

In this case the coefficients α are real and coincide with the binding affinities of the individual sites. Also, the matrix of the coefficients v degenerates into the identity matrix, since $v_{ij} = 1$ if $i = j$ and $v_{ij} = 0$ otherwise. Under these conditions, the site-specific binding curve is

$$X_j = \frac{K_j x}{1 + K_j x} \qquad (3.119)$$

and is always symmetric around the center of symmetry $x_{sj} = K_j^{-1}$. If all sites are not only independent but also identical, then $K_1 = K_2 = \ldots = K_N = K$ and the system as a whole becomes a reference system. The site-specific binding curve of any site is merely X/N and is therefore symmetric like the global binding curve around $x_s = x_{sj} = K^{-1}$. In general, the symmetry properties of X have no bearing on those of X_j and *vice versa* (Di Cera *et al.*, 1992). The sum of symmetric functions is not necessarily symmetric and conversely the sum of asymmetric functions is not necessarily asymmetric. Consideration of the symmetry properties of X_j is, however, of great practical utility due to the difficulty of deriving analytical expressions for x_{mj}. It should be recalled that for a symmetric binding curve the mean ligand activity is the same as the center of symmetry, or the value of x at half saturation.

In order to investigate the symmetry properties of X_j, we recall a result derived in Section 2.3 which states that if the binding curve X is symmetric, then the logarithm of the affinity function, $\ln{}^x K$, is also symmetric with the same center of symmetry. This result applies to X_j as well, since the affinity function ${}^x K$ is a simple transformation of X. From the site-specific affinity function in eq (3.63), and the condition of symmetry for $\ln{}^x K$ in eq (2.63), we obtain the condition for the site-specific binding curve X_j to be symmetric as follows

$$\frac{{}^1\Psi_j(x_{sj}\lambda){}^1\Psi_j(x_{sj}\lambda^{-1})}{{}^0\Psi_j(x_{sj}\lambda){}^0\Psi_j(x_{sj}\lambda^{-1})} = K_j^{-2} x_{sj}^{-2} = \omega^2 \qquad (3.120)$$

where x_{sj} is the center of symmetry at half saturation, or else the mean ligand activity x_{mj} of site j, and the condition must hold for any $\lambda \geqslant 0$. The condition (3.120) involves the contracted partition functions ${}^0\Psi_j$ and ${}^1\Psi_j$, but imposes no constraints on the symmetry properties of the contracted forms of the macromolecule. In fact, symmetry of X_j does not require symmetry of ${}^0\Psi_j$ and ${}^1\Psi_j$. The explicit analytical condition for symmetry of X_j involves the coefficients of ${}^0\Psi_j$ and ${}^1\Psi_j$, along with K_j. We now rewrite ${}^0\Psi_j$ and ${}^1\Psi_j$ in eqs (3.52) and (3.53) in the more convenient form

$$^1\Psi_j(x) = \sum_{i=0}^{N-1} \beta_i x^i \qquad (3.121a)$$

$$^{0}\Psi_{j}(x) = \sum_{i=0}^{N-1} \gamma_{i}x^{i} \tag{3.121b}$$

with $\beta_{0} = \gamma_{0} = 1$. Substitution of eqs (3.121a) and (3.121b) into (3.120) yields

$$\sum_{h=0}^{N-1}\sum_{i=0}^{N-1} (\beta_{i}\beta_{h} - \gamma_{i}\gamma_{h}\omega^{2})x_{sj}^{i+h}\lambda^{i-h} = 0 \tag{3.122}$$

Since eq (3.122) must hold for any $\lambda \geqslant 0$, one necessarily has

$$\sum_{h=0}^{N-1}\sum_{i=0}^{N-1} (\beta_{i}\beta_{h} - \gamma_{i}\gamma_{h}\omega^{2})x_{sj}^{i+h} = 0 \tag{3.123}$$

for any i and h such that $i - h$ is constant. The relationships above are invariant upon the substitution $i \leftrightarrow h$ and therefore only half of them are truly independent. Letting $i - h = n$ in eq (3.123), and considering only the independent relationships yields

$$\sum_{h=0}^{N-1-n} (\beta_{h+n}\beta_{h} - \gamma_{h+n}\gamma_{h}\omega^{2})x_{sj}^{2h+n} = 0 \tag{3.124}$$

There are a total of N such conditions, corresponding to the different values of $n = 0, 1, 2, \ldots N - 1$. Each condition for a given value of n is only necessary, but becomes also sufficient when the other conditions hold. On the other hand, if X_{j} is symmetric, then all the conditions (3.124) necessarily apply. One is particularly important and is obtained for $n = N - 1$. This gives $\beta_{N-1} - \omega^{2}\gamma_{N-1} = 0$, or

$$x_{sj} = \frac{1}{K_{j}}\sqrt{\frac{\gamma_{N-1}}{\beta_{N-1}}} \tag{3.125}$$

which provides a simple analytical expression for the center of symmetry of the curve X_{j}. This value coincides with the mean ligand activity of site j. It is worth pointing out that eq (3.125) is a direct consequence of the symmetry of $\ln {}^{x}K_{j}$. In fact, eq (2.63) implies

$$\ln {}^{x_{sj}\lambda}K_{j} + \ln {}^{x_{sj}\lambda^{-1}}K_{j} = -2\ln x_{sj} \tag{3.126}$$

or equivalently,

$${}^{x_{sj}\lambda}K_{j}{}^{x_{sj}\lambda^{-1}}K_{j} = x_{sj}^{-2} \tag{3.127}$$

Since eq (3.127) must hold for any $\lambda \geqslant 0$, letting $\lambda = 0$ yields

$${}^{0}K_{j}{}^{\infty}K_{j} = x_{sj}^{-2} \tag{3.128}$$

We know from eq (3.64) that ${}^{0}K_{j} = K_{j}$ and from eq (3.65) that ${}^{\infty}K_{j} = K_{j}\beta_{N-1}/\gamma_{N-1}$, which substituted into eq (3.128) yield eq (3.125).

We can now explicitly solve the problem of symmetry for local binding processes using two simple examples. In Section 2.3 we have demonstrated that X is always symmetric in the global description for $N = 2$. However, this condition does not apply in general to X_1 and X_2. Consider the case of X_1 and the contracted partition functions

$$^0\Psi_1(x) = 1 + K_2 x \qquad (3.129a)$$

$$^1\Psi_1(x) = 1 + c_{12} K_2 x \qquad (3.129b)$$

Here $\gamma_1 = K_2$ and $\beta_1 = c_{12} K_2$, and the conditions (3.124) are given by

$$\beta_1 - \gamma_1 \omega^2 = 0 \quad (n = 1) \qquad (3.130a)$$

$$1 - \omega^2 + (\beta_1^2 - \gamma_1^2 \omega^2) x_{s1}^2 = 0 \quad (n = 0) \qquad (3.130b)$$

These conditions are satisfied only if $\gamma_1 = \beta_1$, which implies $c_{12} = 1$, or if $\gamma_1 \beta_1 x_{s1}^2 = 1$, which implies $K_1 = K_2$. This means that X_1 is symmetric only if the two sites are identical ($K_1 = K_2$) or independent ($c_{12} = 1$). The former case corresponds to $X_1 = X_2 = X/2$, and the latter to independent sites. Hence, for $N = 2$ a site-specific binding curve is always asymmetric, unless the two sites are identical or independent.

For $N = 3$, the global description yields a binding curve that is symmetric if $A_3 = (A_2/A_1)^3$. The partition function of the macromolecule as a whole is

$$\Psi(x) = 1 + (K_1 + K_2 + K_3)x$$
$$+ (c_{12}K_1 K_2 + c_{13}K_1 K_3 + c_{23}K_2 K_3)x^2 + c_{123}K_1 K_2 K_3 x^3 \quad (3.131)$$

and the contracted partition functions for X_1 are

$$^0\Psi_1(x) = 1 + (K_2 + K_3)x + c_{23}K_2 K_3 x^2 \qquad (3.132a)$$

$$^1\Psi_1(x) = 1 + (c_{12}K_2 + c_{13}K_3)x + c_{123}K_2 K_3 x^2 \qquad (3.132b)$$

Here $\gamma_1 = K_2 + K_3$, $\gamma_2 = c_{23}K_2 K_3$, $\beta_1 = c_{12}K_2 + c_{13}K_3$, $\beta_2 = c_{123}K_2 K_3$ and the conditions (3.124) are given by

$$\beta_2 - \gamma_2 \omega^2 = 0 \quad (n = 2) \qquad (3.133a)$$

$$\beta_1 - \gamma_1 \omega^2 + (\beta_1 \beta_2 - \gamma_1 \gamma_2 \omega^2) x_{s1}^2 = 0 \quad (n = 1) \qquad (3.133b)$$

$$1 - \omega^2 + (\beta_1^2 - \lambda_1^2 \omega^2) x_{s1}^2 + (\beta_2^2 - \gamma_2^2 \omega^2) x_{s1}^4 = 0 \quad (n = 0) \qquad (3.133c)$$

Symmetry of X_1 demands $\beta_2 \gamma_1^2 = \gamma_2 \beta_1^2$ and $K_1^2 \beta_1 = \gamma_1 \gamma_2$, or

$$c_{123}(K_2 + K_3)^2 = c_{23}(c_{12}K_2 + c_{13}K_3)^2 \qquad (3.134)$$

and

$$K_1^2(c_{12}K_2 + c_{13}K_3) = c_{23}K_2 K_3(K_2 + K_3) \qquad (3.135)$$

It is possible for a site-specific binding curve to be symmetric for $N = 3$. The trivial case where all cs are equal to one, which again corresponds to independent sites, is embodied by the conditions (3.134) and (3.135) as a special case. However, symmetry in the global description does not necessarily imply symmetry in the local description. We have seen that for $N = 2$ the global binding isotherm is always symmetric, while the site-specific isotherms are always asymmetric, except in a few special cases. For $N = 3$, the conditions for symmetry in the global description are completely decoupled from those applying in the local description. Of particular interest is the simple case where all interaction constants are equal to one. In this case, X is the sum of three symmetric site-specific curves, X_1, X_2 and X_3, but is itself symmetric only if one of the Ks is the geometric mean of the other two, e.g., if $K_1^2 = K_2 K_3$. When all Ks are identical, X is symmetric when $27c_{123} = (c_{12} + c_{13} + c_{23})^3$, and X_1 is symmetric when $c_{123} = c_{23}^3$, provided $2c_{23} = c_{12} + c_{13}$.

3.5 Properties of the binding capacity in the local description

In Sections 3.2 and 3.3 we have demonstrated that the binding capacity of an individual site behaves quite differently from the analogous quantity reflecting the properties of the macromolecule as a whole. For a given site, j, B_j can be positive or negative if $N \geqslant 3$. This is a consequence of the significance of the Hill coefficient and the affinity function in the local picture. Using the definition of X_j in eq (3.49), the binding capacity of site j can be written in a number of equivalent forms as follows

$$B_j = X_j(1 - X_j)(1 + {}^1X_j - {}^0X_j) = X_j(1 + {}^1X_j - X) = (1 - X_j)(X - {}^0X_j)$$
$$(3.136)$$

In all these forms, B_j is expressed as the product of two terms, one always positive and the other of variable sign if $N \geqslant 3$. We also have the conservation relationship analogous to eq (3.103)

$$B = \sum_{j=1}^{N} B_j \qquad (3.137)$$

Since B is always positive, as a consequence of the second law of thermodynamics, negative values of the binding capacity of one site must be compensated by positive values at other sites to keep the sum (3.137) positive. Hence, from eq (3.137) we derive the rather obvious result that it is not possible for a macromolecule to have all sites showing negative values of the binding capacity. Given the fact that B_j can in general be positive or negative for $N \geqslant 3$, the question arises as to whether we can use B_j as a probability density function for the variable $\ln x$ in a way analogous to B.

By definition, a probability density function must be positive everywhere in the domain of definition of the independent variable (Feller, 1950). We note, however, that $B_j = 0$ for $x \to 0$ and $x \to \infty$ and that the integral

$$\int_{-\infty}^{\infty} B_j \, d \ln x = 1 \qquad (3.138)$$

exists and is finite, regardless of the properties of B_j. This fact allows us to define a moment generating function for site j as follows

$$G_j(\omega) = \int_0^\infty x^{\omega-1} B_j \, dx = \int_0^1 x^\omega \, dX_j \qquad (3.139)$$

The function exists because of eq (3.138). We note from (3.139) and (3.137), that the sum of the moment generating functions of the individual sites gives the moment generating function G of the macromolecule as a whole, i.e.,

$$G(\omega) = \sum_{j=1}^N G_j(\omega) \qquad (3.140)$$

Substitution of X_j in eq (3.139) with the expression given in eq (3.111a) gives

$$G_j(\omega) = \int_0^\infty \sum_{i=1}^N \frac{v_{ij}\alpha_i x^\omega}{(1+\alpha_i x)^2} \, dx \qquad (3.141)$$

and again convergence of the integral necessarily demands $0 < \omega < 1$. Solution of the integral (3.141) is straightforward using the treatment given in Section 2.4 and yields

$$G_j(\omega) = \frac{\pi\omega}{\sin \pi\omega} \sum_{i=1}^N v_{ij}\alpha_i^{-\omega} \qquad (3.142)$$

which is analogous to G, except for the presence of the coefficients v. The conservation relationship (3.140) follows immediately from (3.142) because of the properties of the vs embodied by (3.113). The first two moments of the B_j distribution are derived from the Taylor expansion of G_j. As in the case of G, we let $G_j = fh$ with

$$f(\omega) = \frac{\pi\omega}{\sin \pi\omega} \qquad (3.143a)$$

$$h(\omega) = \sum_{i=1}^N v_{ij}\alpha_i^{-\omega} \qquad (3.143b)$$

The Taylor expansion of f up to second order is derived from eq (2.90) as

$$f(\omega) = 1 + \frac{\pi^2 \omega^2}{6}$$

(3.144)

The Taylor expansion of h up to second order is

$$h(\omega) = 1 - \left(\sum_{i=1}^{N} v_{ij} \ln \alpha_i\right)\omega + \frac{1}{2}\left(\sum_{i=1}^{N} v_{ij} \ln^2 \alpha_i\right)\omega^2$$

(3.145)

Hence, the first two moments of B_j are

$$\langle \ln x_j \rangle = -\sum_{i=1}^{N} v_{ij} \ln \alpha_i$$

(3.146a)

$$\langle \ln^2 x_j \rangle = \frac{\pi^2}{3} + \sum_{i=1}^{N} v_{ij} \ln^2 \alpha_i$$

(3.146b)

The first moment equals $\ln x_{mj}$. The second moment is proportional to the sum of squares of the logarithm of each coefficient α, multiplied by the appropriate coefficient v.

The site-specific cooperativity associated with site j is again measured by the dispersion of the B_j distribution along the $\ln x$ axis, or the variance

$$\sigma_j^2 = \langle \ln^2 x_j \rangle - \langle \ln x_j \rangle^2$$

(3.147)

Elementary transformations of eqs (3.146a) and (3.146b), using the property (3.114), lead to

$$\sigma_j^2 = \frac{\pi^2}{3} + \frac{1}{2}\sum_{l=1}^{N}\sum_{h=1}^{N} v_{lj} v_{hj} \ln^2 \frac{\alpha_l}{\alpha_h}$$

(3.148)

Upon expressing each coefficient α in polar coordinates in the complex plane as $\alpha = r \exp i\varphi$, we obtain the final result

$$\sigma_j^2 = \frac{\pi^2}{3} + \frac{1}{2}\sum_{l=1}^{N}\sum_{h=1}^{N} v_{lj} v_{hj} \ln^2 \frac{r_l}{r_h} + i\sum_{l=1}^{N}\sum_{h=1}^{N} v_{lj} v_{hj}(\varphi_l - \varphi_h)\ln\frac{r_l}{r_h}$$
$$- \frac{1}{2}\sum_{l=1}^{N}\sum_{h=1}^{N} v_{lj} v_{hj}(\varphi_l - \varphi_h)^2$$

(3.149)

As in the case of the global description, the mathematical 'driving forces' for site-specific cooperativity can be understood from the expression above involving the variance of B_j. There are four components that determine the value of σ_j^2. The first component is the variance of the reference system, σ_0^2. The second component, $\sigma_{<r>}^2$, is the variance of the r values, filtered by a proper convolution of the vs. In fact,

$$\frac{1}{2}\sum_{l=1}^{N}\sum_{h=1}^{N}v_{lj}v_{hj}\ln^2\frac{r_l}{r_h} = \sum_{h=1}^{N}v_{hj}\ln^2 r_h - \left(\sum_{h=1}^{N}v_{hj}\ln r_h\right)^2 = \sigma^2_{\langle r\rangle} \quad (3.150)$$

Likewise, the third component, $\sigma^2_{\langle\varphi\rangle}$, is the variance of the φ values weighted by the corresponding v coefficient, as shown by the following expression analogous to eq (3.150)

$$\frac{1}{2}\sum_{l=1}^{N}\sum_{h=1}^{N}v_{lj}v_{hj}(\varphi_l - \varphi_h)^2 = \sum_{h=1}^{N}v_{hj}\varphi_h^2 - \left(\sum_{h=1}^{N}v_{hj}\varphi_h\right)^2 = \sigma^2_{\langle\varphi\rangle} \quad (3.151)$$

The fourth term is most interesting, as it is peculiar of the local description and does not appear in the definition of σ^2 in the global description. This term represents the covariance, $\sigma^2_{\langle r\varphi\rangle}$, of the joint distribution of the r and φ values, convoluted with the appropriate coefficients v. In fact,

$$i\sum_{l=1}^{N}\sum_{h=1}^{N}v_{lj}v_{hj}(\varphi_l - \varphi_h)\ln\frac{r_l}{r_h} = 2i\sum_{h=1}^{N}v_{hj}\varphi_h\ln r_h -$$

$$2i\sum_{h=1}^{N}v_{hj}\varphi_h\sum_{h=1}^{N}v_{hj}\ln r_h = 2i\sigma^2_{\langle r\varphi\rangle} \quad (3.152)$$

The covariance so defined is a complex number, since it is multiplied by the imaginary unit. Hence,

$$\sigma_j^2 = \sigma_0^2 + \sigma^2_{\langle r\rangle} - \sigma^2_{\langle\varphi\rangle} + 2i\sigma^2_{\langle r\varphi\rangle} \quad (3.153)$$

It is straightforward to prove that the covariance is the result of the differences between the values of the vs. If v_{ij} is the same for all $i = 1, 2,$ ... N, then $\sigma^2_{\langle r\varphi\rangle} = 0$ because

$$\sum_{h=1}^{N}\varphi_h = 0 \quad (3.154a)$$

$$\sum_{h=1}^{N}\varphi_h\ln r_h = 0 \quad (3.154b)$$

The other terms in eq (3.153) remain finite. The various terms in eq (3.149) oppose each other in defining the value of σ_j^2. Although the analysis of the contribution of each term is complicated by the presence of the coefficients v, qualitatively the variance of the norm of the coefficients α, $\sigma^2_{\langle r\rangle}$, tends to increase the value of σ_j^2 with respect to σ_0^2, while the variance of the angles formed by the αs in the complex plane, $\sigma^2_{\langle\varphi\rangle}$, tends to decrease the value of σ_j^2 with respect to σ_0^2. In addition, the covariance term $\sigma^2_{\langle r\varphi\rangle}$ makes a contribution to the sum whose sign cannot be easily predicted.

A final point should be made on the connection between the statistical moments in the global and local descriptions. The first two moments of the B/N distribution of the macromolecule as a whole are

$$\langle \ln x \rangle = -\frac{1}{N}\sum_{j=1}^{N} \ln \alpha_j \tag{3.155a}$$

$$\langle \ln^2 x \rangle = \frac{\pi^2}{3} + \frac{1}{N}\sum_{j=1}^{N} \ln^2 \alpha_j \tag{3.155b}$$

as shown in Section 2.4. The definitions of v_{hj} and α_h are such that

$$v_{hj} = \frac{\partial \ln \alpha_h}{\partial \ln K_j} = \frac{\partial \ln r_h}{\partial \ln K_j} + \mathrm{i}\frac{\partial \varphi_h}{\partial \ln K_j} \tag{3.156}$$

Hence, the first moment in eq (3.146a) can be rewritten as

$$\langle \ln x_j \rangle = -\frac{1}{2}\sum_{h=1}^{N} \frac{\partial \ln^2 \alpha_h}{\partial \ln K_j} \tag{3.157}$$

The term in the summation is related to the second moment of the B/N distribution in eq (3.155b), and specifically

$$\langle \ln x_j \rangle = -\frac{N}{2}\frac{\partial \langle \ln^2 x \rangle}{\partial \ln K_j} \tag{3.158}$$

The derivative of the second moment of the B/N distribution with respect to $\ln K_j$ gives the first moment of the B_j distribution, times a factor that depends on the number of sites. Since $\langle \ln x_j \rangle$ is the same as the mean ligand activity of site j, $\ln x_{mj}$, and the sum of such terms is related to the mean ligand activity of the macromolecule by virtue of eq (3.107), summing eq (3.158) over all values of j yields the interesting result

$$\ln x_m = \langle \ln x \rangle = -\frac{1}{2}\nabla\langle \ln^2 x \rangle \tag{3.159}$$

The mean ligand activity of the macromolecule as a whole can be derived from the second moment of the B/N distribution through the gradient operator. This operator acts as a filter on the second moment. The importance of eq (3.159) stems from the fact that the first two moments of the B/N distribution in the global description are not truly independent. The connection, however, can only be seen from consideration of the local properties of the macromolecule and cannot be drawn in terms of the global properties. The gradient operator provides the filtering function through which the first two moments of B/N can be obtained from one another. It also follows from eq (3.155a) that

$$\nabla \ln x_m = -1 \tag{3.160}$$

due to the properties of the coefficients v. Hence, from the definition of the variance in the global description and eq (3.159) it follows that

$$\nabla \sigma^2 = 0 \qquad (3.161)$$

a rather intriguing result. The variance of the global binding capacity is independent of the gradient defined by the site-specific binding constants.

4

Specific cases and applications

In this chapter we present a systematic treatment of site-specific effects in systems composed of a small number of sites. We also deal with applications of the concepts illustrated in previous chapters to the analysis of cooperative systems of biological relevance.

4.1 The case $N = 2$

The simplest cooperative system is provided by a macromolecule containing two sites. Although some of the properties of this system have already been discussed in Section 3.1, here we give a systematic treatment for the sake of clarity and completeness. The partition function for the case $N = 2$ can be written in either of the following forms

$$\Psi(x) = 1 + A_1 x + A_2 x^2 \tag{4.1a}$$

$$\Psi(x) = 1 + (K_1 + K_2)x + c_{12} K_1 K_2 x^2 \tag{4.1b}$$

The first is the global form of Ψ, while the second is the local form. The coefficients A_1 and A_2 are the overall equilibrium constants in the global description, while K_1, K_2 and c_{12} are the relevant site-specific parameters. From the definition of Ψ it follows that

$$A_1 = K_1 + K_2 \tag{4.2a}$$

$$A_2 = c_{12} K_1 K_2 \tag{4.2b}$$

and therefore the site-specific parameters cannot be assigned uniquely from knowledge of the global coefficients A. There are a total of four distinct configurations in the local description, corresponding to the four terms of the partition function (4.1b). These configurations can be associated with the two-dimensional vectors

$$[00]$$

$$[10] \qquad [01] \qquad\qquad (4.3)$$

$$[11]$$

The unligated form of the macromolecule, M, is the same as the configuration M_{00} labeled in a site-specific fashion. This form is used as reference in the definition of the partition functions (4.1a) and (4.1b). The singly-ligated form of the macromolecule, MX, is the sum of the concentrations of the site-specific configurations M_{10} and M_{01}, corresponding to the intermediates with site 1 and site 2 ligated respectively. The doubly-ligated form, MX_2, is the same as M_{11}.

Since there are four configurations in eq (4.1b), of which one is used as reference, then there must be three independent reactions among the possible intermediates leading to a total of three independent equilibrium constants. These are the three site-specific parameters K_1, K_2 and c_{12} that represent the equilibrium constants for the reactions

$$M_{00} + X \Leftrightarrow M_{10} \qquad\qquad (4.4a)$$

$$M_{00} + X \Leftrightarrow M_{01} \qquad\qquad (4.4b)$$

$$M_{10} + M_{01} \Leftrightarrow M_{00} + M_{11} \qquad\qquad (4.4c)$$

In the first reaction site 1 is ligated while site 2 is kept unligated, and the equilibrium constant is K_1. Likewise, in the second reaction site 2 is ligated while site 1 is kept unligated, and the equilibrium constant is K_2. The third reaction is a dismutation where the unligated and doubly-ligated intermediates are generated from the two singly-ligated configurations. The equilibrium constant for this reaction is c_{12} and is a measure of the linkage between the sites. The parameters A_1 and A_2 of the global description refer to the reactions

$$M + X \Leftrightarrow MX \qquad\qquad (4.5a)$$

$$M + 2X \Leftrightarrow MX_2 \qquad\qquad (4.5b)$$

We also have, from arguments discussed in Section 2.2, two stepwise binding constants k_1 and k_2 for the reactions

$$M + X \Leftrightarrow MX \qquad\qquad (4.6a)$$

$$MX + X \Leftrightarrow MX_2 \qquad\qquad (4.6b)$$

with

$$A_1 = 2k_1 \qquad\qquad (4.7a)$$

$$A_2 = k_1 k_2 \qquad\qquad (4.7b)$$

and conversely

$$k_1 = \frac{A_1}{2} \tag{4.8a}$$

$$k_2 = \frac{2A_2}{A_1} \tag{4.8b}$$

In the absence of interactions between the sites, $c_{12} = 1$ and the partition function assumes the rather simple form

$$\Psi(x) = (1 + K_1 x)(1 + K_2 x) \tag{4.9}$$

Each site behaves as an independent unit which provides a binomial factor to Ψ.

We have seen in Section 3.1 that the case $c_{12} = 1$ does not correspond to the absence of cooperativity in the global picture, which demands that the expression

$$\Delta = 4c_{12}K_1 K_2 - (K_1 + K_2)^2 = 4A_2 - A_1^2 = 4k_1(k_2 - k_1) \tag{4.10}$$

should vanish. This condition reflects an important property of the partition function. Let us rewrite Ψ in terms of the coefficients α as follows

$$\Psi(x) = (1 + \alpha_1 x)(1 + \alpha_2 x) \tag{4.11}$$

The coefficients can be real and positive, or complex conjugate. The separation between these two cases is found by equating terms between (4.11) and (4.1b), i.e.

$$\alpha_1 + \alpha_2 = K_1 + K_2 \tag{4.12a}$$

$$\alpha_1 \alpha_2 = c_{12} K_1 K_2 \tag{4.12b}$$

The αs are therefore obtained as the solution of the quadratic expression

$$\alpha^2 - (K_1 + K_2)\alpha + c_{12}K_1 K_2 = 0 \tag{4.13}$$

as

$$\alpha_{1,2} = \frac{K_1 + K_2 \pm \sqrt{-\Delta}}{2} = \frac{A_1 \pm \sqrt{-\Delta}}{2} \tag{4.14}$$

The term Δ in eq (4.14) is the same as Δ in eq (4.10). Hence, the condition giving rise to positive cooperativity in the global description ($\Delta > 0$) coincides with the condition for the αs to be complex conjugate (Briggs, 1983, 1984, 1985). If the coefficients α of the partition function are real and positive, the system cannot be positively cooperative. This conclusion is a consequence of arguments discussed in Section 2.4 about the variance of the B/N distribution. It should also be noted that the αs provide information on the sum $K_1 + K_2 = A_1$ and the product

$c_{12}K_1K_2 = A_2$ only. As expected, these coefficients are sufficient to resolve global, but not site-specific parameters.

The αs written in terms of polar coordinates assume a particularly interesting form. When they are complex conjugate, then $\Delta > 0$ and eq (4.11) can be rewritten as follows

$$\Psi(x) = [1 + r \exp{(i\varphi)}x][1 + r \exp{(-i\varphi)}x] \tag{4.15}$$

Hence,

$$r = \sqrt{c_{12}K_1K_2} = \sqrt{A_2} \tag{4.16a}$$

$$\varphi = \arccos \frac{K_1 + K_2}{2\sqrt{c_{12}K_1K_2}} = \arccos \frac{A_1}{2\sqrt{A_2}} \tag{4.16b}$$

The angle φ is bounded from $\pi/2$, obtained for $c_{12} \to \infty$, to zero, corresponding to $\Delta = 0$. In the case of negative macroscopic cooperativity, $\Delta < 0$ and the two coefficients α are real and positive and defined by eq (4.14). In this case, the αs correspond to the norm of their polar coordinates in the complex plane since the angles vanish. When the αs are real, the partition function factors out in terms of two binomials with positive coefficients, α_1 and α_2. This is also the case when $c_{12} = 1$. Hence, analysis of the global properties cannot distinguish the case of two independent sites that bind with different affinities, K_1 and K_2, from the case where the two coefficients, α_1 and α_2, are real and positive, with $\Delta < 0$. As discussed in Section 3.1, $\Delta < 0$ does not necessarily demand $c_{12} = 1$. For $\rho < 1$, any value of $c_{12} < (1 + \rho)^2/4\rho$ yields $\Delta < 0$. The case of negative macroscopic cooperativity is therefore most ambiguous and misleading when it comes to unraveling the pattern of microscopic cooperativity even in qualitative terms. If the partition function factors out as in eq (4.11), with α_1 and α_2 real, interactions between the sites cannot be distinguished by a situation where the two sites are independent and bind the ligand with affinities given by eq (4.14).

Before we analyze the site-specific properties of the system, it is important to discuss the global quantities X, B and ^xK in some detail. These quantities are given by

$$X = \frac{d\ln\Psi}{d\ln x} = \frac{A_1x + 2A_2x^2}{1 + A_1x + A_2x^2} \tag{4.17}$$

$$B = \frac{dX}{d\ln x} = \frac{A_1x + 4A_2x^2 + A_1A_2x^3}{(1 + A_1x + A_2x^2)^2} \tag{4.18}$$

$$^xK = \frac{A_1 + 2A_2x}{2 + A_1x} \tag{4.19}$$

and are depicted in Figures 4.1–4.3 for the case of Ca^{2+} binding to calbindin (Linse *et al.*, 1988, 1991; Martin *et al.*, 1990). The mean ligand activity of the macromolecule is $x_m = 1/\sqrt{A_2}$. This is also the value of x that yields $X = 1$, or the half saturation of the macromolecule. We recall that the curve X is always symmetric for $N = 2$. The binding capacity at half saturation reaches an extremum value of

$$B_{1/2} = \frac{1}{1 + A_1/2\sqrt{A_2}} = \frac{1}{1 + \sqrt{k_1/k_2}} \qquad (4.20)$$

The Hill coefficient derived from eq (2.45) is given by

$$n_H = 1 + \frac{2A_2x}{A_1 + 2A_2x} - \frac{A_1x}{2 + A_1x} = 1 + \frac{k_2x}{1 + k_2x} - \frac{k_1x}{1 + k_1x} \qquad (4.21)$$

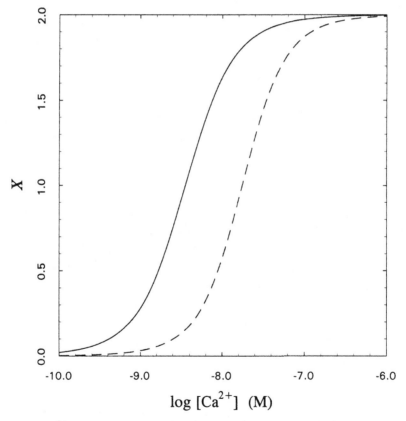

Figure 4.1 Ca^{2+} binding curve of calbindin (continuous line) and its mutant E17Q (dashed line). Curves were drawn according to eq (4.17) with parameter values $A_1 = 2.0 \times 10^8 \, M^{-1}$ and $A_2 = 7.9 \times 10^{16} \, M^{-2}$ for the wild-type, and $A_1 = 2.5 \times 10^7 \, M^{-1}$ and $A_2 = 3.2 \times 10^{15} \, M^{-2}$ for the mutant (Linse *et al.*, 1988, 1991).

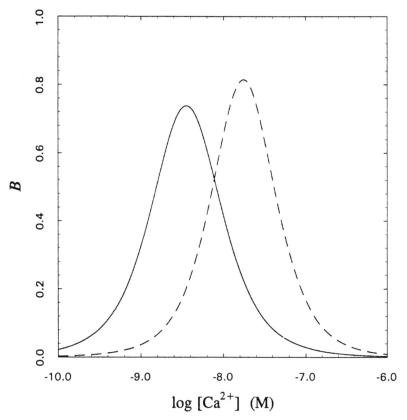

Figure 4.2 Ca^{2+} binding capacity of calbindin (continuous line) and its mutant E17Q (dashed line). Curves were drawn according to eq (4.18) with parameter values as in Figure 4.1.

As expected, the Hill coefficient tends to one for $x \to 0$ and $x \to \infty$, is positive everywhere and is bounded from zero to two. The extremum value of n_H reached at half saturation

$$n_{1/2} = \frac{2}{1 + A_1/2\sqrt{A_2}} = \frac{2}{1 + \sqrt{k_1/k_2}} \qquad (4.22)$$

confirms that the asymptotic values of $n_{1/2}$ are zero and two, reached for extreme negative and positive macroscopic cooperativity respectively. The variance of the B/N distribution is also noteworthy. If α_1 and α_2 are real, the term σ_φ^2 in eq (2.97) vanishes and the variance is

$$\sigma^2 = \sigma_0^2 + \frac{1}{4}\ln^2 \frac{A_1 + \sqrt{-\Delta}}{A_1 - \sqrt{-\Delta}} \qquad (4.23)$$

where we recall that $\sigma_0^2 = \pi^2/3$ and $\Delta < 0$. If α_1 and α_2 are complex

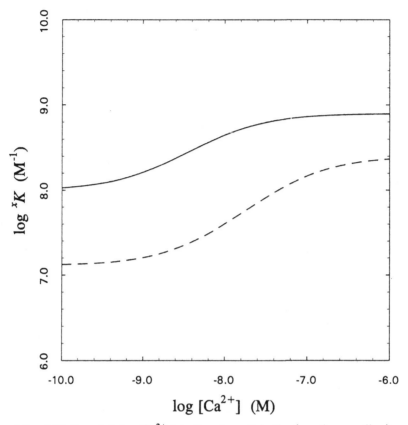

Figure 4.3   Affinity plot for Ca^{2+} binding to calbindin (continuous line) and its mutant E17Q (dashed line). Curves were drawn according to eq (4.19) with parameter values as in Figure 4.1.

conjugate, the term σ_r^2 in eq (2.97) vanishes, the angle φ is given by (4.16b) and the variance is

$$\sigma^2 = \sigma_0^2 - \arccos^2 \frac{A_1}{2\sqrt{A_2}} \tag{4.24}$$

The value of σ^2 is always greater than σ_0^2 if $\Delta < 0$ and can grow without bounds. In the case $\Delta > 0$, the variance is always less than σ_0^2 and reaches the minimum value of $\pi^2/12$ when cooperativity is a maximum.

The global parameters A_1 and A_2, or k_1 and k_2, can be derived from analysis of X, B or ^xK. Resolution of the site-specific parameters K_1, K_2 and c_{12}, on the other hand, demands measurements of at least one site-specific property of the system. The problem of resolving these parameters can be understood using contracted partition functions. Since any response function X, B or ^xK is derived by differentiation or transformation of Ψ, all the information for the resolution of binding

Table 4.1. *Contracted partition functions and weighting factors for N = 2.*

Order	Contracted partition function	Weighting factor
0	$\Psi(x) = 1 + (K_1 + K_2)x + c_{12}K_1K_2x^2$	1
1	$^0\Psi_1(x) = 1 + K_2x$	1
1	$^0\Psi_2(x) = 1 + K_1x$	1
1	$^1\Psi_1(x) = 1 + c_{12}K_2x$	K_1x
1	$^1\Psi_2(x) = 1 + c_{12}K_1x$	K_2x

parameters is stored in the partition function. The function Ψ can provide two such parameters, since it is a polynomial of second order. If three parameters are to be resolved, an additional constraint is needed. This constraint is provided by a contracted partition function. The possible contracted partition functions for $N = 2$, up to order $N - 1$, and their proper weighting terms are listed in Table 4.1. We recall from Section 3.2 that the Nth-order contracted partition functions are all equal to one and are weighted by the 2^N terms defining the partition function Ψ. The Nth-order contracted partition functions are therefore omitted from the table. The table shows that Ψ can be written in a series of alternative forms as follows

$$\Psi(x) = {}^0\Psi_1(x) + {}^1\Psi_1(x)K_1x = {}^0\Psi_2(x) + {}^1\Psi_2(x)K_2x \qquad (4.25)$$

Each first-order contracted partition function can itself be expressed in terms of second-order contracted partition functions as follows

$$^0\Psi_1(x) = {}^{00}\Psi_{12}(x) + {}^{01}\Psi_{12}(x)K_2x = 1 + K_2x \qquad (4.26a)$$

$$^0\Psi_2(x) = {}^{00}\Psi_{12}(x) + {}^{10}\Psi_{12}(x)K_1x = 1 + K_1x \qquad (4.26b)$$

$$^1\Psi_1(x) = {}^{10}\Psi_{12}(x) + {}^{11}\Psi_{12}(x)c_{12}K_2x = 1 + c_{12}K_2x \qquad (4.26c)$$

$$^1\Psi_2(x) = {}^{01}\Psi_{12}(x) + {}^{11}\Psi_{12}(x)c_{12}K_1x = 1 + c_{12}K_1x \qquad (4.26d)$$

Knowledge of any of the first-order contracted partition functions in Table 4.1, in connection with Ψ, is sufficient to resolve the three independent site-specific parameters K_1, K_2 and c_{12}. This can be done by measuring X_1 or X_2 given by the expressions

$$X_1 = K_1x\frac{^1\Psi_1(x)}{\Psi(x)} = 1 - \frac{^0\Psi_1(x)}{\Psi(x)} = \frac{K_1x(1 + c_{12}K_2x)}{1 + (K_1 + K_2)x + c_{12}K_1K_2x^2}$$

$$(4.27a)$$

$$X_2 = K_2x\frac{^1\Psi_2(x)}{\Psi(x)} = 1 - \frac{^0\Psi_2(x)}{\Psi(x)} = \frac{K_2x(1 + c_{12}K_1x)}{1 + (K_1 + K_2)x + c_{12}K_1K_2x^2}$$

$$(4.27b)$$

with $X_1 + X_2 = X$. These functions are illustrated in Figure 4.4 for Ca^{2+} binding to the two sites of calbindin. The numerator of X_1 contains a term which is a function of K_1 only, and the same holds for X_2 relative to the term K_2. In the limit $x \to 0$, X_1 grows linearly as $K_1 x$ and the site-specific parameter K_1 can be determined in a direct way. Once K_1 is known, K_2 and c_{12} can be decoupled from knowledge of A_1 and A_2. We note from the definitions (4.27a) and (4.27b) that X_1 and X_2 can be derived from the partition function Ψ as implied by eq (3.50), i.e.,

$$X_j = \frac{\partial \ln \Psi(x)}{\partial \ln K_j} \tag{4.28}$$

where $j = 1, 2$. The coefficients $v_{ij} = \partial \ln \alpha_i / \partial \ln K_j$ are derived from the definition (4.14) and form the matrix

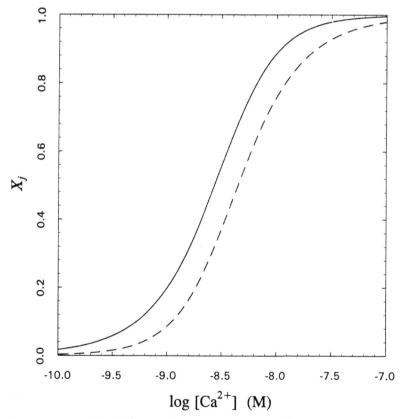

Figure 4.4.   Site-specific Ca^{2+} binding curves of site 1 (continuous line) and site 2 (dashed line) of calbindin. Curves were drawn using parameter values obtained from the analysis of structural perturbations, as discussed in Section 4.5. The site-specific   parameters   are: $K_1 = 1.7 \times 10^8 \, M^{-1}$, $K_2 = 2.7 \times 10^7 \, M^{-1}$ and $c_{12} = 18$ (see also Table 4.16 and Figure 4.22).

$$v = \begin{pmatrix} \dfrac{K_1 - \dfrac{\Delta_1'}{\sqrt{-\Delta}}}{K_1 + K_2 + \sqrt{-\Delta}} & \dfrac{K_2 - \dfrac{\Delta_2'}{\sqrt{-\Delta}}}{K_1 + K_2 + \sqrt{-\Delta}} \\[3em] \dfrac{K_1 + \dfrac{\Delta_1'}{\sqrt{-\Delta}}}{K_1 + K_2 - \sqrt{-\Delta}} & \dfrac{K_2 + \dfrac{\Delta_2'}{\sqrt{-\Delta}}}{K_1 + K_2 - \sqrt{-\Delta}} \end{pmatrix} \qquad (4.29)$$

with

$$\Delta_1' = 2c_{12} K_1 K_2 - K_1(K_1 + K_2) \qquad (4.30a)$$
$$\Delta_2' = 2c_{12} K_1 K_2 - K_2(K_1 + K_2) \qquad (4.30b)$$

It is easy to verify from the relationships above that the matrix of the coefficients v has columns and rows whose elements add up to one.

The definition of the coefficients v yields the analytical expressions for the first two moments of the B_1 and B_2 distributions. The relevant expressions for the first moment are

$$\langle \ln x_1 \rangle = \ln x_m - \left(\frac{1}{2} - v_{21} \right) \ln \frac{\alpha_1}{\alpha_2}$$
$$= \ln x_m + \frac{K_2 - K_1}{2\sqrt{-\Delta}} \ln \frac{K_1 + K_2 + \sqrt{-\Delta}}{K_1 + K_2 - \sqrt{-\Delta}} \qquad (4.31a)$$

$$\langle \ln x_2 \rangle = \ln x_m - \left(\frac{1}{2} - v_{22} \right) \ln \frac{\alpha_1}{\alpha_2}$$
$$= \ln x_m + \frac{K_1 - K_2}{2\sqrt{-\Delta}} \ln \frac{K_1 + K_2 + \sqrt{-\Delta}}{K_1 + K_2 - \sqrt{-\Delta}} \qquad (4.31b)$$

for α_1 and α_2 real and positive ($\Delta < 0$). These are the expressions for the mean ligand activity of the site-specific binding curves X_1 and X_2. For α_1 and α_2 complex conjugate ($\Delta > 0$), one has the somewhat simpler relationships

$$\langle \ln x_1 \rangle = \ln x_m - \left(\frac{1}{2} - v_{21} \right) \ln \frac{\alpha_1}{\alpha_2} = \ln x_m + \frac{K_2 - K_1}{\sqrt{\Delta}} \arccos \frac{K_1 + K_2}{2\sqrt{c_{12} K_1 K_2}}$$
$$(4.32a)$$

$$\langle \ln x_2 \rangle = \ln x_m - \left(\frac{1}{2} - v_{22} \right) \ln \frac{\alpha_1}{\alpha_2} = \ln x_m + \frac{K_1 - K_2}{\sqrt{\Delta}} \arccos \frac{K_1 + K_2}{2\sqrt{c_{12} K_1 K_2}}$$
$$(4.32b)$$

In either case, the values of $\ln x_{m1} = \langle \ln x_1 \rangle$ and $\ln x_{m2} = \langle \ln x_2 \rangle$ are symmetrically disposed around the mean ligand activity of the macromolecule as a whole, since

$$\ln x_{m1} + \ln x_{m2} = 2 \ln x_m \tag{4.33}$$

The variance of the B_1 and B_2 distributions can be found from application of (3.149) and is given by

$$\sigma_1^2 = \sigma_0^2 + v_{11} v_{21} \ln^2 \frac{\alpha_1}{\alpha_2} = \frac{\pi^2}{3} + \frac{(c_{12} - 1) K_1 K_2}{\Delta} \ln^2 \frac{K_1 + K_2 + \sqrt{-\Delta}}{K_1 + K_2 - \sqrt{-\Delta}} \tag{4.34a}$$

$$\sigma_1^2 = \sigma_0^2 + v_{12} v_{22} \ln^2 \frac{\alpha_1}{\alpha_2} = \frac{\pi^2}{3} + \frac{(c_{12} - 1) K_1 K_2}{\Delta} \ln^2 \frac{K_1 + K_2 + \sqrt{-\Delta}}{K_1 + K_2 - \sqrt{-\Delta}} \tag{4.34b}$$

for α_1 and α_2 real and positive ($\Delta < 0$). For α_1 and α_2 complex conjugate ($\Delta > 0$) one has

$$\sigma_1^2 = \sigma_0^2 - \varphi^2 = \frac{\pi^2}{3} - \arccos^2 \frac{K_1 + K_2}{2\sqrt{c_{12} K_1 K_2}} \tag{4.35a}$$

$$\sigma_2^2 = \sigma_0^2 - \varphi^2 = \frac{\pi^2}{3} - \arccos^2 \frac{K_1 + K_2}{2\sqrt{c_{12} K_1 K_2}} \tag{4.35b}$$

The interesting result is that $\sigma_1 = \sigma_2$ for $N = 2$, regardless of the sign of Δ. The binding capacities of the individual sites always have the same variance, regardless of the cooperativity pattern. In the case of macroscopic positive cooperativity, the variance of the B_1 and B_2 distributions also coincides with the variance of the B/N distribution for the macromolecule as a whole, given by eq (4.24). This shows quite simply the nonadditive character of the variances, and hence cooperativity, in the site-specific scenario. For example, in the case $c_{12} = 1$, the site-specific variances are equal to $\pi^2/3$, but the variance of the B/N distribution can grow without bounds with the heterogeneity of the sites.

For practical purposes the site-specific parameters K_1, K_2 and c_{12} can be resolved from analysis of X_1 or X_2 using the properties of the site-specific affinity functions

$$^x K_1 = \frac{X_1}{(1 - X_1) x} = K_1 \frac{^1 \Psi_1(x)}{^0 \Psi_1(x)} = K_1 \frac{1 + c_{12} K_2 x}{1 + K_2 x} \tag{4.36a}$$

$$^x K_2 = \frac{X_2}{(1 - X_2) x} = K_2 \frac{^1 \Psi_2(x)}{^0 \Psi_2(x)} = K_2 \frac{1 + c_{12} K_1 x}{1 + K_1 x} \tag{4.36b}$$

The values of K_1 and c_{12} can be determined from ${}^x K_1$ in the limits $x \to 0$ and $x \to \infty$, since ${}^0 K_1 = K_1$ and ${}^\infty K_1 = c_{12} K_1$. Alternatively, the values of K_2 and c_{12} can be determined from ${}^x K_2$, since ${}^0 K_2 = K_2$ and ${}^\infty K_2 = c_{12} K_2$. In practice, however, if K_1 and K_2 differ by a factor of ten or more, measurements of only one site-specific binding curve may be insufficient to resolve all three parameters. In fact, assume that $K_1 \gg K_2$. If X_1 is measured, the term linear in x in the denominator of (4.27a) is insensitive to K_2 and only K_1 and the product $c_{12} K_2$ can be resolved from analysis of experimental data. Since X_1 saturates in a range of x values such that $K_2 x \ll 1$, the affinity function plot of $\ln {}^x K_1$ versus $\ln x$ is characterized by an asymptote $\ln {}^0 K_1 = \ln K_1$ which allows for the determination of K_1, but no asymptote for x large. This prevents measurements of c_{12} and hence K_2 separately. In the case where $K_1 \gg K_2$ it is X_2, and not X_1, that should be measured experimentally. Although the denominator of (4.27b) may still be insensitive to K_2, the numerator grows linearly in K_2. Also, the affinity function plot yields the value of K_2 directly from the asymptote for $x \to 0$. The value of K_1 is derived from the term linear in x in (4.26b). Once K_1 and K_2 are known, c_{12} is derived from the term quadratic in x in (4.27b) without difficulty.

4.2 The case $N = 3$

The case $N = 2$ has the unique advantage of being mathematically tractable. However, it by no means represents a paradigm for the complexity arising from cooperative interactions in the global and local descriptions. In fact, the global binding isotherm for $N = 2$ is always symmetric, while the site-specific isotherms are always asymmetric. Hence, the effect of interactions and cooperativity on the shape of the response functions X cannot be studied for $N = 2$. Another important feature of site-specific binding processes, regarding the nonapplicability of thermodynamic stability to the properties of the binding capacity, does not apply to the case $N = 2$ either. The simplest cooperative system that encapsulates paradigmatically all the important aspects of global and site-specific binding processes is provided by a macromolecule containing three sites. The partition function for the case $N = 3$ can be written in the global and local descriptions as

$$\Psi(x) = 1 + A_1 x + A_2 x^2 + A_3 x^3 \tag{4.37a}$$

$$\Psi(x) = 1 + (K_1 + K_2 + K_3)x + (c_{12} K_1 K_2 + c_{13} K_1 K_3 + c_{23} K_2 K_3)x^2$$
$$+ c_{123} K_1 K_2 K_3 x^3 \tag{4.37b}$$

Again, from the uniqueness of Ψ it follows that

$$A_1 = K_1 + K_2 + K_3 \tag{4.38a}$$

$$A_2 = c_{12} K_1 K_2 + c_{13} K_1 K_3 + c_{23} K_2 K_3 \tag{4.38b}$$

$$A_3 = c_{123} K_1 K_2 K_3 \tag{4.38c}$$

There are a total of eight distinct configurations in the local description, corresponding to the eight terms of the partition function (4.37b). These configurations can be associated with the vectors

$$
\begin{gathered}
[000] \\
[100] \quad [010] \quad [001] \\
[110] \quad [101] \quad [011] \\
[111]
\end{gathered}
\tag{4.39}
$$

As usual, the unligated form of the macromolecule, M, is the same as the configuration M_{000} labeled in a site-specific fashion and is used as reference in the definition of the partition functions (4.37a) and (4.37b). The singly-ligated form of the macromolecule in the global description, MX, is the sum of the site-specific configurations M_{100}, M_{010} and M_{001}, corresponding to the intermediates with site 1, site 2 or site 3 ligated. The doubly-ligated form, MX_2, is the sum of M_{110}, M_{101} and M_{011}, while the triply- or fully-ligated form, MX_3, is the same as M_{111}.

Since there are eight configurations in eq (4.37b) and one is used as reference, then there must be seven independent reactions among the possible intermediates leading to a total of seven independent equilibrium constants. These are three site-specific binding affinities K_1, K_2 and K_3, three interaction constants c_{12}, c_{13} and c_{23} for second-order coupling and one constant c_{123} for third-order coupling. These constants are associated with the reactions

$$M_{000} + X \Leftrightarrow M_{100} \tag{4.40a}$$

$$M_{000} + X \Leftrightarrow M_{010} \tag{4.40b}$$

$$M_{000} + X \Leftrightarrow M_{001} \tag{4.40c}$$

$$M_{100} + M_{010} \Leftrightarrow M_{000} + M_{110} \tag{4.40d}$$

$$M_{100} + M_{001} \Leftrightarrow M_{000} + M_{101} \tag{4.40e}$$

$$M_{010} + M_{001} \Leftrightarrow M_{000} + M_{011} \tag{4.40f}$$

$$M_{100} + M_{010} + M_{001} \Leftrightarrow 2M_{000} + M_{111} \tag{4.40g}$$

In the first reaction site 1 is ligated, while sites 2 and 3 are kept unligated, and the equilibrium constant is K_1. Likewise, in the second reaction site 2 is ligated, while sites 1 and 3 are kept unligated, and the equilibrium constant is K_2. In the third reaction site 3 is ligated, while sites 1 and 2 are kept unligated, and the equilibrium constant is K_3. In the fourth reaction a dismutation is considered whereby the unligated and doubly-ligated intermediate with sites 1 and 2 ligated are generated from the two parent singly-ligated configurations. The equilibrium constant for this reaction is c_{12} and is a measure of the linkage between sites 1 and 2. In the fifth and sixth reactions analogous dismutations are considered for the site pairs 1 and 3, and 2 and 3, with equilibrium constants c_{13} and c_{23}. The seventh and last reaction is also a dismutation, but it involves the generation of the reference unligated form and the fully-ligated intermediate from the three parent singly-ligated species, and the equilibrium constant is the third-order interaction constant c_{123}.

The global parameters A_1, A_2 and A_3 refer to the reactions

$$M + X \Leftrightarrow MX \tag{4.41a}$$

$$M + 2X \Leftrightarrow MX_2 \tag{4.41b}$$

$$M + 3X \Leftrightarrow MX_3 \tag{4.41c}$$

The three stepwise binding constants k_1, k_2 and k_3 refer to the reactions

$$M + X \Leftrightarrow MX \tag{4.42a}$$

$$MX + X \Leftrightarrow MX_2 \tag{4.42b}$$

$$MX_2 + X \Leftrightarrow MX_3 \tag{4.42c}$$

with

$$A_1 = 3k_1 \tag{4.43a}$$

$$A_2 = 3k_1 k_2 \tag{4.43b}$$

$$A_3 = k_1 k_2 k_3 \tag{4.43c}$$

and conversely

$$k_1 = A_1/3 \tag{4.44a}$$

$$k_2 = A_2/A_1 \tag{4.44b}$$

$$k_3 = 3A_3/A_2 \tag{4.44c}$$

In the absence of interactions among the sites, all site-specific interaction constants are equal to one and the partition function simplifies into the factorial form

$$\Psi(x) = (1 + K_1 x)(1 + K_2 x)(1 + K_3 x) \tag{4.45}$$

In general, the partition function can be written as follows

$$\Psi(x) = (1 + \alpha_1 x)(1 + \alpha_2 x)(1 + \alpha_3 x) \tag{4.46}$$

where the αs can be all real and positive, or one real and positive and two complex conjugate. The separation between these two cases is found by equating terms between (4.46) and (4.37b), i.e.,

$$\alpha_1 + \alpha_2 + \alpha_3 = K_1 + K_2 + K_3 \tag{4.47a}$$

$$\alpha_1\alpha_2 + \alpha_1\alpha_3 + \alpha_2\alpha_3 = c_{12}K_1K_2 + c_{13}K_1K_3 + c_{23}K_2K_3 \tag{4.47b}$$

$$\alpha_1\alpha_2\alpha_3 = c_{123}K_1K_2K_3 \tag{4.47c}$$

In general, the αs are very complicated expressions of the site-specific parameters.

The global quantities X, B and ^xK are

$$X = \frac{d\ln\Psi}{d\ln x} = \frac{A_1x + 2A_2x^2 + 3A_3x^3}{1 + A_1x + A_2x^2 + A_3x^3} \tag{4.48}$$

$$B = \frac{dX}{d\ln x} = \frac{A_1x + 4A_2x^2 + (A_1A_2 + 9A_3)x^3 + 4A_1A_3x^4 + A_2A_3x^5}{(1 + A_1x + A_2x^2 + A_3x^3)^2} \tag{4.49}$$

$$^xK = \frac{A_1 + 2A_2x + 3A_3x^2}{3 + 2A_1x + A_2x^2} \tag{4.50}$$

and are illustrated in Figures 4.5–4.7 for the case of the λ-repressor binding to its operator sites O_R1, O_R2 and O_R3 (Ptashne, 1986; Senear and Ackers, 1990). The mean ligand activity of the macromolecule is $x_m = A_3^{-1/3}$ and, in general, does not coincide with the value of x at half saturation. This is seen only when X is symmetric, or else when $A_3 = (A_2/A_1)^3$, or equivalently $k_1k_3 = k_2^2$, as demonstrated in Section 2.3.

The patterns of macroscopic cooperativity can be understood from the properties of the affinity function. It is convenient to rewrite ^xK in terms of the stepwise binding constants as follows

$$^xK = k_1\frac{1 + 2k_2x + k_2k_3x^2}{1 + 2k_1x + k_1k_2x^2} \tag{4.51}$$

The asymptotic values of ^xK are $^0K = k_1$ and $^\infty K = k_3$. The ratio k_3/k_1 defines positive ($k_3/k_1 > 1$) or negative ($k_3/k_1 < 1$) macroscopic cooperativity. The case $k_1 = k_2 = k_3$ corresponds to the absence of macroscopic cooperativity. Unlike the case $N = 2$, the conditions for macroscopic cooperativity do not provide unambiguous answers to the cooperativity pattern of the global description. Macroscopic positive cooperativity implies $k_3 > k_1$, regardless of the value of k_2. The macroscopic pattern of cooperativity remains unchanged when $k_2 \leqslant k_1$, or $k_1 < k_2 < k_3$, or else $k_2 \geqslant k_3$, although different values of k_2 bring about significant differences in the properties of the macromolecule. Comparison of consecutive

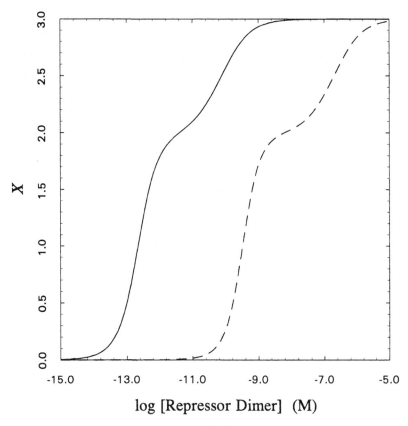

Figure 4.5 Binding curves of the λ cI repressor to its operator at pH 5.0 (continuous line) and pH 8.0 (dashed line). Curves were drawn according to eq (4.48), with parameter values $A_1 = 3.9 \times 10^{12}$ M^{-1}, $A_2 = 2.1 \times 10^{25}$ M^{-2}, $A_3 = 2.7 \times 10^{35}$ M^{-3} for pH 5.0, and $A_1 = 1.95 \times 10^9$ M^{-1}, $A_2 = 9.7 \times 10^{18}$ M^{-2}, $A_3 = 4.8 \times 10^{25}$ M^{-3} for pH 8.0 (Senear and Ackers, 1990).

stepwise constants reveals the stepwise pattern of cooperativity. When $k_3 > k_2 > k_1$, the binding affinity increases uniformly with ligation. On the other hand, when $k_3 > k_1 > k_2$ the binding affinity first decreases and then increases. Although the overall result is that the third ligation step takes place with higher affinity than the first, the second step occurs with a lower affinity than the first. When $k_2 > k_3 > k_1$, the binding affinity first increases and then decreases. Again, the overall result is that the third ligation step takes place with higher affinity than the first, but the second step now occurs with an affinity larger than the third. In the first case, the stepwise pattern of cooperativity is uniform. In the other two cases, we speak of mixed cooperativity in the global description. The possible

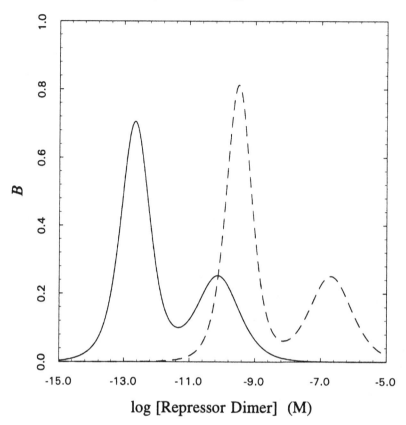

log [Repressor Dimer] (M)

Figure 4.6 Binding capacity of the λ cI repressor operator as a function of the logarithm of repressor concentration. Curves were drawn according to eq (4.49) with parameter values as in Figure 4.5.

patterns of cooperativity for $N = 3$ are listed in Table 4.2. A + sign indicates the presence of positive cooperativity in the stepwise sense, a − sign refers to negative cooperativity and 0 indicates the absence of cooperativity. Interestingly, the same pattern of stepwise cooperativity can give rise to two different macroscopic patterns. The case of λ cI repressor binding to its operator shown in Figure 4.7 is a clear example of mixed cooperativity, with the affinity first increasing and then decreasing to give a macroscopic pattern of negative cooperativity.

When there are three stepwise binding constants, the sign of stepwise cooperativity can change only once (Whitehead, 1980b, 1981). This is confirmed by the properties of the affinity function. A change in the stepwise pattern implies the existence of an extremum in the affinity plot.

Table 4.2. *Patterns of stepwise and macroscopic cooperativity for N = 3.*

Pattern	Cooperativity	Stepwise	Macroscopic
$k_1 = k_2 = k_3$	Uniform	00	None
$k_1 > k_2 > k_3$	Uniform	$--$	Negative
$k_1 > k_3 > k_2$	Mixed	$-+$	Negative
$k_3 > k_1 > k_2$	Mixed	$-+$	Positive
$k_1 < k_3 < k_2$	Mixed	$+-$	Positive
$k_3 < k_1 < k_2$	Mixed	$+-$	Negative
$k_1 < k_2 < k_3$	Uniform	$++$	Positive

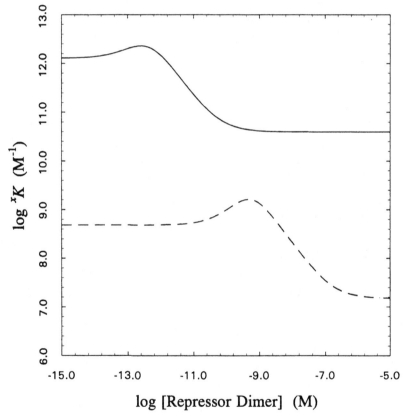

Figure 4.7 Affinity plot for λ cI repressor binding to its operator. Curves were drawn according to eq (4.50) with parameter values as in Figure 4.5. Note the macroscopic pattern of negative cooperativity, with mixed $(+-)$ stepwise cooperativity due to $k_3 < k_1 < k_2$ (see also Table 4.2).

The slope in the affinity plot, $\Delta X = d \ln {}^x K / d \ln x$ (see Section 2.2) is

$$\Delta X = \frac{2k_2 x(1 + k_3 x)}{1 + 2k_2 x + k_2 k_3 x^2} - \frac{2k_1 x(1 + k_2 x)}{1 + 2k_1 x + k_1 k_2 x^2} \tag{4.52}$$

The expression giving the extrema of $\ln {}^x K$ is obtained for $\Delta X = 0$ and is given by

$$k_1 k_2 (k_3 - k_2) x^2 + k_2 (k_3 - k_1) x + k_2 - k_1 = 0 \tag{4.53}$$

Never does this equation have two positive real roots, indicating that the slope of the affinity function can change sign only once for $N = 3$. The Hill coefficient derived from eq (2.44) is given by

$$n_{\mathrm{H}} = 1 + \Delta X = 1 + \frac{2k_2 x(1 + k_3 x)}{1 + 2k_2 x + k_2 k_3 x^2} - \frac{2k_1 x(1 + k_2 x)}{1 + 2k_1 x + k_1 k_2 x^2} \tag{4.54}$$

The Hill coefficient tends to one for $x \to 0$ and $x \to \infty$, is positive everywhere and is bounded from zero to three. When the binding curve is symmetric, n_{H} has an extremum value at $x = x_{\mathrm{m}}$ given by

$$n_{1/2} = 1 + 2\frac{k_2 + k_3}{3k_2 + k_3} - 4\frac{k_1}{k_2 + 3k_1} \tag{4.55}$$

The upper bound of n_{H} is reached in the limit $k_3 \gg k_2 \gg k_1$ and is equal to three.

Resolution of the site-specific parameters K_1, K_2, K_3, c_{12}, c_{13}, c_{23} and c_{123} demands measurements of at least two site-specific properties of the system. This is because the global partition function provides one constraint for these parameters through eqs (4.38a)–(4.38c), but A_1 and A_2 contain three independent terms, thereby demanding two additional constraints. These constraints are again provided by contracted partition functions. The possible contracted partition functions for $N = 3$, up to order $N - 1$, and their proper weighting terms are listed in Table 4.3. The partition function Ψ can be written in a series of alternative forms as follows

$$\begin{aligned}
\Psi(x) = {}^0\Psi_1(x) &+ {}^1\Psi_1(x) K_1 x = {}^0\Psi_2(x) + {}^1\Psi_2(x) K_2 x \\
&= {}^0\Psi_3(x) + {}^1\Psi_3(x) K_3 x = {}^{00}\Psi_{12}(x) \\
&\quad + [{}^{10}\Psi_{12}(x) K_1 + {}^{01}\Psi_{12}(x) K_2] x + {}^{11}\Psi_{12}(x) c_{12} K_1 K_2 x^2 \\
&= {}^{00}\Psi_{13}(x) + [{}^{10}\Psi_{13}(x) K_1 + {}^{01}\Psi_{13}(x) K_3] x + {}^{11}\Psi_{13}(x) c_{13} K_1 K_3 x^2 \\
&= {}^{00}\Psi_{23}(x) + [{}^{10}\Psi_{23}(x) K_2 + {}^{01}\Psi_{23}(x) K_3] x + {}^{11}\Psi_{23}(x) c_{23} K_2 K_3 x^2
\end{aligned} \tag{4.56}$$

In addition, each contracted partition function up to second order can

Table 4.3. *Contracted partition functions and weighting factors for N = 3.*

Order	Contracted partition function	Weighting factor
0	$\Psi(x) = 1 + (K_1 + K_2 + K_3)x + (c_{12}K_1K_2 + c_{13}K_1K_3$ $+ c_{23}K_2K_3)x^2 + c_{123}K_1K_2K_3x^3$	1
1	$^0\Psi_1(x) = 1 + (K_2 + K_3)x + c_{23}K_2K_3x^2$	1
1	$^0\Psi_2(x) = 1 + (K_1 + K_3)x + c_{13}K_1K_3x^2$	1
1	$^0\Psi_3(x) = 1 + (K_1 + K_2)x + c_{12}K_1K_2x^2$	1
1	$^1\Psi_1(x) = 1 + (c_{12}K_2 + c_{13}K_3)x + c_{123}K_2K_3x^2$	K_1x
1	$^1\Psi_2(x) = 1 + (c_{12}K_1 + c_{23}K_3)x + c_{123}K_1K_3x^2$	K_2x
1	$^1\Psi_3(x) = 1 + (c_{13}K_1 + c_{23}K_2)x + c_{123}K_1K_2x^2$	K_3x
2	$^{00}\Psi_{12}(x) = 1 + K_3x$	1
2	$^{00}\Psi_{13}(x) = 1 + K_2x$	1
2	$^{00}\Psi_{23}(x) = 1 + K_1x$	1
2	$^{01}\Psi_{12}(x) = 1 + c_{23}K_3x$	K_2x
2	$^{01}\Psi_{13}(x) = 1 + c_{23}K_2x$	K_3x
2	$^{01}\Psi_{23}(x) = 1 + c_{13}K_1x$	K_3x
2	$^{10}\Psi_{12}(x) = 1 + c_{13}K_3x$	K_1x
2	$^{10}\Psi_{13}(x) = 1 + c_{12}K_2x$	K_1x
2	$^{10}\Psi_{23}(x) = 1 + c_{12}K_1x$	K_2x
2	$^{11}\Psi_{12}(x) = 1 + c_{123}c_{12}^{-1}K_3x$	$c_{12}K_1K_2x^2$
2	$^{11}\Psi_{13}(x) = 1 + c_{123}c_{13}^{-1}K_2x$	$c_{13}K_1K_3x^2$
2	$^{11}\Psi_{23}(x) = 1 + c_{123}c_{23}^{-1}K_1x$	$c_{23}K_2K_3x^2$

itself be expressed in terms of higher-order contracted partition functions. For example,

$$^0\Psi_1(x) = {}^{00}\Psi_{12}(x) + {}^{01}\Psi_{12}(x)K_2x = {}^{00}\Psi_{13}(x) + {}^{01}\Psi_{13}(x)K_3x \quad (4.57a)$$

$$^0\Psi_2(x) = {}^{00}\Psi_{12}(x) + {}^{10}\Psi_{12}(x)K_1x = {}^{00}\Psi_{23}(x) + {}^{01}\Psi_{23}(x)K_3x \quad (4.57b)$$

$$^0\Psi_3(x) = {}^{00}\Psi_{13}(x) + {}^{10}\Psi_{13}(x)K_1x = {}^{00}\Psi_{23}(x) + {}^{10}\Psi_{23}(x)K_2x \quad (4.57c)$$

$$^{00}\Psi_{12}(x) = {}^{000}\Psi_{123}(x) + {}^{001}\Psi_{123}(x)K_3x = 1 + K_3x \quad (4.57d)$$

$$^{00}\Psi_{13}(x) = {}^{000}\Psi_{123}(x) + {}^{010}\Psi_{123}(x)K_2x = 1 + K_2x \quad (4.57e)$$

$$^{00}\Psi_{23}(x) = {}^{000}\Psi_{123}(x) + {}^{100}\Psi_{123}(x)K_1x = 1 + K_1x \quad (4.57f)$$

Knowledge of any two of the first-order contracted partition functions in Table 4.3, along with Ψ, is necessary and sufficient to resolve the seven independent site-specific parameters. These partition functions can be accessed from measurements of X_1, X_2 or X_3 given by the expressions

$$X_1 = \frac{K_1x[1 + (c_{12}K_2 + c_{13}K_3)x + c_{123}K_2K_3x^2]}{1 + (K_1 + K_2 + K_3)x + (c_{12}K_1K_2 + c_{13}K_1K_3 + c_{23}K_2K_3)x^2 + c_{123}K_1K_2K_3x^3}$$

$$(4.58a)$$

$X_2 =$

$$\frac{K_2 x[1 + (c_{12} K_1 + c_{23} K_3) x + c_{123} K_1 K_3 x^2]}{1 + (K_1 + K_2 + K_3) x + (c_{12} K_1 K_2 + c_{13} K_1 K_3 + c_{23} K_2 K_3) x^2 + c_{123} K_1 K_2 K_3 x^3}$$

(4.58b)

$X_3 =$

$$\frac{K_3 x[1 + (c_{13} K_1 + c_{23} K_2) x + c_{123} K_1 K_2 x^2]}{1 + (K_1 + K_2 + K_3) x + (c_{12} K_1 K_2 + c_{13} K_1 K_3 + c_{23} K_2 K_3) x^2 + c_{123} K_1 K_2 K_3 x^3}$$

(4.58c)

with $X_1 + X_2 + X_3 = X$. These functions are sketched in Figure 4.8 for the case of λ cI repressor binding to its operator sites. Again, the

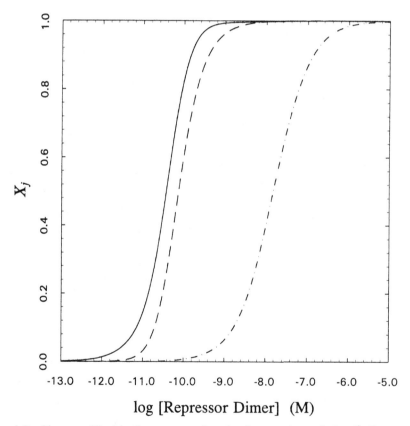

Figure 4.8 Site-specific binding curves for the interaction of the λ cI repressor with its operator sites $O_R 1$ (continuous line), $O_R 2$ (dashed line) and $O_R 3$ (chain line) at pH 7.0 (Senear and Ackers, 1990). Curves were drawn using eqs (4.58a)–(4.58c) with site-specific parameter values: $K_1 = 1.3 \times 10^{10} \, \text{M}^{-1}$, $K_2 = 1.1 \times 10^8 \, \text{M}^{-1}$, $K_3 = 1.7 \times 10^7 \, \text{M}^{-1}$, $c_{12} = 241$, $c_{13} = 1$, $c_{23} = 801$, $c_{123} = 950$.

site-specific binding curves can be derived from Ψ by differentiation as follows

$$X_j = \frac{\partial \ln \Psi(x)}{\partial \ln K_j} \tag{4.59}$$

where $j = 1, 2, 3$.

The most direct way to derive the site-specific parameters is from the properties of the affinity function. Given A_1, A_2 and A_3 from analysis of the global properties of the macromolecule, the site-specific affinity functions

$$^xK_1 = \frac{X_1}{(1 - X_1)x}$$

$$= K_1\frac{^1\Psi_1(x)}{^0\Psi_1(x)} = K_1\frac{1 + (c_{12}K_2 + c_{13}K_3)x + c_{123}K_2K_3x^2}{1 + (K_2 + K_3)x + c_{23}K_2K_3x^2} \tag{4.60a}$$

$$^xK_2 = \frac{X_2}{(1 - X_2)x}$$

$$= K_2\frac{^1\Psi_2(x)}{^0\Psi_2(x)} = K_2\frac{1 + (c_{12}K_1 + c_{23}K_3)x + c_{123}K_1K_3x^2}{1 + (K_1 + K_3)x + c_{13}K_1K_3x^2} \tag{4.60b}$$

$$^xK_3 = \frac{X_3}{(1 - X_3)x}$$

$$= K_3\frac{^1\Psi_3(x)}{^0\Psi_3(x)} = K_3\frac{1 + (c_{13}K_1 + c_{23}K_2)x + c_{123}K_1K_2x^2}{1 + (K_1 + K_2)x + c_{12}K_1K_2x^2} \tag{4.60c}$$

yield a complete solution of the problem. The values of K_1 and c_{123}/c_{23} can be determined from $^xK_1$ in the limits $x \rightarrow 0$ and $x \rightarrow \infty$. The values of K_2 and c_{123}/c_{13} can be determined from $^xK_2$. This information is sufficient to resolve all parameters since the sum $K_1 + K_2 + K_3 = A_1$ and the product $c_{123}K_1K_2K_3 = A_3$ can be derived from the global properties of the macromolecule.

We now address the problem of thermodynamic stability at the site-specific level recalling the results discussed in Sections 3.2 and 3.3 that the binding capacity of an individual site can assume negative values for $N \geqslant 3$. We address this problem in a purely mathematical fashion by noting that if B_j ($j = 1$, 2 or 3) can be negative, then there must be a condition for which the critical value $B_j = 0$ is observed, as shown in Figure 4.9. When $B_j < 0$, the binding curve X_j shown in Figure 4.9(a) necessarily cuts a straight line $X_j = \eta$ parallel to the $\ln x$ axis at three

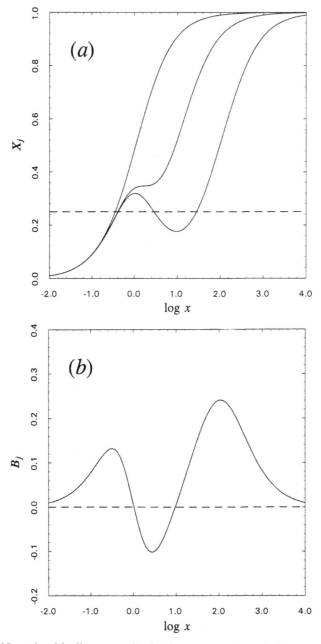

Figure 4.9   Negative binding capacity in the site-specific binding curve for $N = 3$ for a hypothetical example where $K_1 = 1$, $K_2 = K_3 = 0.1$, $c_{12} = c_{13} = 0.1$, $c_{12} = 100$ and $c_{123} = 100$ ((a), left), $c_{123} = 10$ ((a), middle) or $c_{123} = 1$ ((a), right and (b)). Curves refer to site 1. Changing the value of c_{123} affects the shape of the site-specific binding curve (a). For $c_{123} = 1$, site 1 shows negative binding capacity values (b) and a straight line parallel to the $\ln x$ axis cuts the binding curve at three points (a).

points. This is exactly what has been found experimentally in the case of electron binding to cytochrome aa_3 (Kojima and Palmer, 1983), as discussed in Section 3.2 (see Figure 3.10). Hence, for $B_j < 0$ the equation $X_j - \eta = 0$ has three real and positive roots. The critical value of $B_j = 0$ is observed when all roots are identical. We shall define the value of η corresponding to this case as X_c, to denote the critical value of X_j for which $B_j = 0$. Using X_1 as an example, the relevant equation giving three identical and positive roots is

$$^1\Psi_1(x)K_1 x - X_c\Psi(x) = 0 \tag{4.61}$$

Expanding the terms in eq (4.61) yields

$$-X_c + (X_+ - X_c)A_1 x - (X_c - X_-)A_2 x^2 + (1 - X_c)A_3 x^3 = 0 \tag{4.62}$$

where $X_+ = K_1/A_1$ and $X_- = K_1(c_{12}K_2 + c_{13}K_3)/A_2$ represent the upper and lower bounds for the value of X_c compatible with the existence of three identical and positive roots in eq (4.62). Let x_c be the value of x at the critical point. Then

$$-X_c + (X_+ - X_c)A_1 x - (X_c - X_-)A_2 x^2 + (1 - X_c)A_3 x^3$$
$$= A_3(1 - X_c)(x - x_c)^3 = 0 \tag{4.63}$$

Hence,

$$x_c^3 = \frac{X_c}{(1 - X_c)A_3} \tag{4.64a}$$

$$3x_c^2 = \frac{(X_+ - X_c)A_1}{(1 - X_c)A_3} \tag{4.64b}$$

$$3x_c = \frac{(X_c - X_-)A_2}{(1 - X_c)A_3} \tag{4.64c}$$

From eq (4.64a) follows the rather interesting result that the value of X_c is always located on a binding curve defined as

$$X_c = \frac{A_3 x_c^3}{1 + A_3 x_c^3} \tag{4.65}$$

This is the binding curve of an individual site in the limit where only the unligated and fully-ligated species are significantly populated, i.e., in the fully cooperative case. Elimination of x_c from eqs (4.64a)–(4.64c) yields the quadratic expression

$$(A_1 A_2 - 9A_3)X_c^2 - [A_1 A_2(X_+ + X_-) - 9A_3]X_c + A_1 A_2 X_- X_+ = 0 \tag{4.66}$$

which must have only one positive solution, due to the definition of X_c.

This requires as a necessary condition

$$[A_1A_2(X_+ + X_-) - 9A_3]^2 - 4A_1A_2X_-X_+(A_1A_2 - 9A_3)$$
$$= [k_1(X_+ + X_-) - k_3]^2 - 4k_1X_-X_+(k_1 - k_3) = 0 \quad (4.67)$$

which is only satisfied provided $A_1A_2 > 9A_3$, or $k_1 > k_3$. Quite interestingly, the binding capacity of site 1 can vanish if, and only if, the macroscopic cooperativity of the macromolecule is negative. This condition applies to any of the three sites, since it is independent of site-specific properties and is a necessary constraint to which the stepwise equilibrium constants of the system are subject. The necessary and sufficient condition for $B_1 = 0$ is given by eq (4.67) and yields the critical value of X_1 as follows

$$X_c = \frac{A_1A_2(X_+ + X_-) - 9A_3}{2(A_1A_2 - 9A_3)} = \frac{k_1(X_+ + X_-) - k_3}{2(k_1 - k_3)} \quad (4.68)$$

The corresponding value of x_c is

$$x_c = \sqrt[3]{\frac{K_1(X_+ + X_-) - K_3}{K_1(2 - X_+ - X_-) - K_3} \frac{1}{K_1K_2K_3}} \quad (4.69)$$

as derived from eq (4.64a).

4.3 The case $N = 4$

This is the simplest case that shows the role of second-order contracted partition functions in the resolution of site-specific parameters. The partition function for the case $N = 4$ in the global and local descriptions is

$$\Psi(x) = 1 + A_1x + A_2x^2 + A_3x^3 + A_4x^4 \quad (4.70a)$$

$$\Psi(x) = 1 + (K_1 + K_2 + K_3 + K_4)x$$
$$+ (c_{12}K_1K_2 + c_{13}K_1K_3 + c_{14}K_1K_4 + c_{23}K_2K_3 + c_{24}K_2K_4 + c_{34}K_3K_4)x^2$$
$$+ (c_{123}K_1K_2K_3 + c_{124}K_1K_2K_4 + c_{134}K_1K_3K_4 + c_{234}K_2K_3K_4)x^3$$
$$+ c_{1234}K_1K_2K_3K_4x^4 \quad (4.70b)$$

Hence,

$$A_1 = K_1 + K_2 + K_3 + K_4 \quad (4.71a)$$

$$A_2 = c_{12}K_1K_2 + c_{13}K_1K_3 + c_{14}K_1K_4 + c_{23}K_2K_3 + c_{24}K_2K_4 + c_{34}K_3K_4$$
$$\quad (4.71b)$$

$$A_3 = c_{123}K_1K_2K_3 + c_{124}K_1K_2K_4 + c_{134}K_1K_3K_4 + c_{234}K_2K_3K_4$$
$$\quad (4.71c)$$

$$A_4 = c_{1234}K_1K_2K_3K_4 \quad (4.71d)$$

The total number of distinct configurations in the local description is sixteen, corresponding to an equivalent number of terms in the partition function (4.70b). These configurations can be associated with the four-dimensional vectors

$$[0000]$$

$$[1000] \quad [0100] \quad [0010] \quad [0001]$$

$$[1100] \quad [1010] \quad [1001] \quad [0110] \quad [0101] \quad [0011] \qquad (4.72)$$

$$[1110] \quad [1101] \quad [1011] \quad [0111]$$

$$[1111]$$

The unligated form of the macromolecule, M, is the same as the configuration M_{0000} labeled in a site-specific fashion and is used as reference in the definition of the partition functions (4.70a) and (4.70b). The singly-ligated form of the macromolecule in the global description, MX, is the sum of four site-specific configurations M_{1000}, M_{0100}, M_{0010} and M_{0001}, corresponding to the intermediates with site 1, site 2, site 3 or site 4 ligated respectively. The doubly-ligated form, MX_2, is the sum of six site-specific configurations (M_{1100}, M_{1010}, M_{1001}, M_{0110}, M_{0101}, M_{0011}), the triply-ligated form, MX_3, is the sum of four configurations (M_{1110}, M_{1101}, M_{1011}, M_{0111}), while the quadruply- or fully-ligated form, MX_4, is the same as M_{1111}.

There are fifteen independent reactions among the possible intermediates leading to a total of fifteen independent equilibrium constants. These constants include four site-specific binding affinities K_1, K_2, K_3 and K_4 for the reactions

$$M_{0000} + X \Leftrightarrow M_{1000} \qquad (4.73a)$$

$$M_{0000} + X \Leftrightarrow M_{0100} \qquad (4.73b)$$

$$M_{0000} + X \Leftrightarrow M_{0010} \qquad (4.73c)$$

$$M_{0000} + X \Leftrightarrow M_{0001} \qquad (4.73d)$$

where a given site is ligated while the other sites are kept unligated. The second-order interaction constants c_{12}, c_{13}, c_{14}, c_{23}, c_{24} and c_{34}, are associated with the dismutations

$$M_{1000} + M_{0100} \Leftrightarrow M_{0000} + M_{1100} \qquad (4.74a)$$

$$M_{1000} + M_{0010} \Leftrightarrow M_{0000} + M_{1010} \qquad (4.74b)$$

$$M_{1000} + M_{0001} \Leftrightarrow M_{0000} + M_{1001} \qquad (4.74c)$$

$$M_{0100} + M_{0010} \Leftrightarrow M_{0000} + M_{0110} \qquad (4.74d)$$

$$M_{0100} + M_{0001} \Leftrightarrow M_{0000} + M_{0101} \qquad (4.74e)$$

$$M_{0010} + M_{0001} \Leftrightarrow M_{0000} + M_{0011} \qquad (4.74f)$$

where the reference species and a doubly-ligated intermediate are generated from the parent singly-ligated intermediates. The third-order interaction constants c_{123}, c_{124}, c_{134} and c_{234}, are associated with the dismutations

$$M_{1000} + M_{0100} + M_{0010} \Leftrightarrow 2M_{0000} + M_{1110} \qquad (4.75a)$$

$$M_{1000} + M_{0100} + M_{0001} \Leftrightarrow 2M_{0000} + M_{1101} \qquad (4.75b)$$

$$M_{1000} + M_{0010} + M_{0001} \Leftrightarrow 2M_{0000} + M_{1011} \qquad (4.75c)$$

$$M_{0100} + M_{0010} + M_{0001} \Leftrightarrow 2M_{0000} + M_{0111} \qquad (4.75d)$$

where the reference species and a triply-ligated intermediate are generated from the parent singly-ligated intermediates. Finally, the fourth-order interaction constant c_{1234} is associated with the dismutation

$$M_{1000} + M_{0100} + M_{0010} + M_{0001} \Leftrightarrow 3M_{0000} + M_{1111} \qquad (4.76)$$

where the reference species and the fully-ligated intermediate are generated from the parent singly-ligated intermediates. The global parameters A_1, A_2, A_3 and A_4 refer to the reactions

$$M + X \Leftrightarrow MX \qquad (4.77a)$$

$$M + 2X \Leftrightarrow MX_2 \qquad (4.77b)$$

$$M + 3X \Leftrightarrow MX_3 \qquad (4.77c)$$

$$M + 4X \Leftrightarrow MX_4 \qquad (4.77d)$$

The four stepwise binding constants k_1, k_2, k_3 and k_4 apply to the reactions

$$M + X \Leftrightarrow MX \qquad (4.78a)$$

$$MX + X \Leftrightarrow MX_2 \qquad (4.78b)$$

$$MX_2 + X \Leftrightarrow MX_3 \qquad (4.78c)$$

$$MX_3 + X \Leftrightarrow MX_4 \qquad (4.78d)$$

with

$$A_1 = 4k_1 \qquad (4.79a)$$

$$A_2 = 6k_1 k_2 \qquad (4.79b)$$

$$A_3 = 4k_1 k_2 k_3 \qquad (4.79c)$$

$$A_4 = k_1 k_2 k_3 k_4 \qquad (4.79d)$$

and conversely

$$k_1 = A_1/4 \qquad (4.80a)$$

$$k_2 = 2A_2/3A_1 \qquad (4.80b)$$

$$k_3 = 3A_3/2A_2 \qquad (4.80c)$$

$$k_4 = 4A_4/A_3 \qquad (4.80d)$$

In the absence of interactions among the sites, all cs equal one and

$$\Psi(x) = (1 + K_1 x)(1 + K_2 x)(1 + K_3 x)(1 + K_4 x) \qquad (4.81)$$

The partition function written in terms of the αs is

$$\Psi(x) = (1 + \alpha_1 x)(1 + \alpha_2 x)(1 + \alpha_3 x)(1 + \alpha_4 x) \qquad (4.82)$$

Three cases should be considered, the first where all αs are real and positive, the second where two αs are real and positive and two are complex conjugate, and the third where the αs are two pairs of complex conjugates. In general, the expression of the αs in terms of the As is very complicated and makes any algebraic manipulation hopeless.

The global quantities X, B and ^xK are

$$X = \frac{d \ln \Psi}{d \ln x} = \frac{A_1 x + 2A_2 x^2 + 3A_3 x^3 + 4A_4 x^4}{1 + A_1 x + A_2 x^2 + A_3 x^3 + A_4 x^4} \qquad (4.83)$$

$$B = \frac{A_1 x + 4A_2 x^2 + (A_1 A_2 + 9A_3)x^3 + 4(A_1 A_3 + 4A_4)x^4}{(1 + A_1 x + A_2 x^2 + A_3 x^3 + A_4 x^4)^2}$$
$$+ \frac{(A_2 A_3 + 9A_1 A_4)x^5 + 4A_2 A_4 x^6 + A_3 A_4 x^7}{(1 + A_1 x + A_2 x^2 + A_3 x^3 + A_4 x^4)^2} \qquad (4.84)$$

$$^xK = \frac{A_1 + 2A_2 x + 3A_3 x^2 + 4A_4 x^3}{4 + 3A_1 x + 2A_2 x^2 + A_3 x^3} \qquad (4.85)$$

The mean ligand activity of the macromolecule is $x_m = A_4^{-1/4}$ and does not coincide with the value of x at half saturation, unless X is symmetric. Patterns of macroscopic and stepwise cooperativity can be understood from the properties of the affinity function

$$^xK = k_1 \frac{1 + 3k_2 x + 3k_2 k_3 x^2 + k_2 k_3 k_4 x^3}{1 + 3k_1 x + 3k_1 k_2 x^2 + k_1 k_2 k_3 x^3} \qquad (4.86)$$

The asymptotic values of ^xK are $^0K = k_1$ and $^\infty K = k_4$. The ratio k_4/k_1 defines positive ($k_4/k_1 > 1$) or negative ($k_4/k_1 < 1$) macroscopic cooperativity. The case $k_1 = k_2 = k_3 = k_4$ corresponds to the absence of cooperativity. Like the case of $N = 3$, the conditions for macroscopic cooperativity provide no information on the stepwise patterns that involve the constants k_2 and k_3. Stepwise cooperativity can change sign twice for $N = 4$, as indicated by the patterns listed in Table 4.4. This is a consequence of the fact that the slope of the affinity function is

$$\Delta X = \frac{3k_2 x(1 + 2k_3 x + k_3 k_4 x^2)}{1 + 3k_2 x + 3k_2 k_3 x^2 + k_2 k_3 k_4 x^3}$$
$$- \frac{3k_1 x(1 + 2k_2 x + k_2 k_3 x^2)}{1 + 3k_1 x + 3k_1 k_2 x^2 + k_1 k_2 k_3 x^3} \qquad (4.87)$$

Table 4.4. *Patterns of stepwise and macroscopic cooperativity for N = 4.*

Pattern	Cooperativity	Stepwise	Macroscopic
$k_1 = k_2 = k_3 = k_4$	Uniform	000	None
$k_1 > k_2 > k_3 > k_4$	Uniform	$---$	Negative
$k_1 > k_2 > k_3 < k_4$	Mixed	$--+$	Negative ($k_1 > k_4$)
$k_1 > k_2 > k_3 < k_4$	Mixed	$--+$	Positive ($k_1 < k_4$)
$k_1 > k_2 < k_3 > k_4$	Mixed	$-+-$	Negative ($k_1 > k_4$)
$k_1 > k_2 < k_3 > k_4$	Mixed	$-+-$	Positive ($k_1 < k_4$)
$k_1 < k_2 > k_3 > k_4$	Mixed	$+--$	Negative ($k_1 > k_4$)
$k_1 < k_2 > k_3 > k_4$	Mixed	$+--$	Positive ($k_1 < k_4$)
$k_1 > k_2 < k_3 < k_4$	Mixed	$-++$	Negative ($k_1 > k_4$)
$k_1 > k_2 < k_3 < k_4$	Mixed	$-++$	Positive ($k_1 < k_4$)
$k_1 < k_2 > k_3 < k_4$	Mixed	$+-+$	Negative ($k_1 > k_4$)
$k_1 < k_2 > k_3 < k_4$	Mixed	$+-+$	Positive ($k_1 < k_4$)
$k_1 < k_2 < k_3 > k_4$	Mixed	$++-$	Negative ($k_1 > k_4$)
$k_1 < k_2 < k_3 > k_4$	Mixed	$++-$	Positive ($k_1 < k_4$)
$k_1 < k_2 < k_3 < k_4$	Uniform	$+++$	Positive

and can vanish for two distinct values of x. The Hill coefficient derived from eq (2.44) is given by $1 + \Delta X$, or

$$n_{\mathrm{H}} = 1 + \frac{3k_2x(1 + 2k_3x + k_3k_4x^2)}{1 + 3k_2x + 3k_2k_3x^2 + k_2k_3k_4x^3}$$
$$- \frac{3k_1x(1 + 2k_2x + k_2k_3x^2)}{1 + 3k_1x + 3k_1k_2x^2 + k_1k_2k_3x^3} \tag{4.88}$$

from which it follows that $n_{\mathrm{H}} \to 1$ for $x \to 0$ and $x \to \infty$. The Hill coefficient is also positive everywhere and is bounded from zero to four.

Resolution of the fifteen site-specific parameters demands measurements of at least five site-specific properties of the system. The global partition function provides one constraint through eqs (4.71a)–(4.71d), but A_2 contains six independent terms thereby demanding five additional constraints. Unlike the cases of $N = 2$ and $N = 3$, measurement of all site-specific binding curves X_1, X_2, X_3 and X_4 is not sufficient to resolve all site-specific parameters. This demonstrates that for $N \geqslant 4$ site-specific binding isotherms cannot provide a complete solution of the site-specific problem. The number of independent parameters grows as $C_{[N/2],N}$ (Di Cera, 1994a), where $[N/2]$ is $N/2$ for N even and $(N + 1)/2$ for N odd, while the number of constraints provided by the X_js grows as N. Clearly $C_{[N/2],N} > N$ for $N \geqslant 4$. Since measurements of the X_js provide information on contracted partition functions of first order, contracted forms of higher order must be accessed experimentally to resolve all site-specific parameters. The possible contracted partition functions for $N = 4$, up to order $N - 1$, and their proper weighting terms are listed in Table 4.5. The partition function Ψ can be written in a series of alternative forms using first-order contracted partition functions as follows

$$\Psi(x) = {}^0\Psi_1(x) + {}^1\Psi_1(x)K_1x = {}^0\Psi_2(x) + {}^1\Psi_2(x)K_2x$$
$$= {}^0\Psi_3(x) + {}^1\Psi_3(x)K_3x = {}^0\Psi_4(x) + {}^1\Psi_4(x)K_4x \quad (4.89)$$

Each first-order contracted partition function can in turn be written in terms of second-order contracted partition functions. For example,

$$^0\Psi_1(x) = {}^{00}\Psi_{12}(x) + {}^{01}\Psi_{12}(x)K_2x = {}^{00}\Psi_{13}(x) + {}^{01}\Psi_{13}(x)K_3x$$
$$= {}^{00}\Psi_{14}(x) + {}^{01}\Psi_{14}(x)K_4x \quad (4.90)$$

For higher-order contracted partition functions one has

$$^{00}\Psi_{12}(x) = {}^{000}\Psi_{123}(x) + {}^{001}\Psi_{123}(x)K_3x = {}^{000}\Psi_{124}(x) + {}^{001}\Psi_{124}(x)K_4x$$
$$(4.91)$$

$$^{000}\Psi_{123}(x) = {}^{0000}\Psi_{1234}(x) + {}^{0001}\Psi_{1234}(x)K_4x = 1 + K_4x \quad (4.92)$$

The first-order contracted partition functions in Table 4.5, along with Ψ, can be accessed from measurements of X_1, X_2, X_3 or X_4 given by the expressions

$$X_1 = \frac{K_1x}{\Psi(x)}[1 + (c_{12}K_2 + c_{13}K_3 + c_{14}K_4)x$$
$$+ (c_{123}K_2K_3 + c_{124}K_2K_4 + c_{134}K_3K_4)x^2 + c_{1234}K_2K_3K_4x^3] \quad (4.93a)$$

$$X_2 = \frac{K_2x}{\Psi(x)}[1 + (c_{12}K_1 + c_{23}K_3 + c_{24}K_4)x$$
$$+ (c_{123}K_1K_3 + c_{124}K_1K_4 + c_{234}K_3K_4)x^2 + c_{1234}K_1K_3K_4x^3] \quad (4.93b)$$

$$X_3 = \frac{K_3x}{\Psi(x)}[1 + (c_{13}K_1 + c_{23}K_2 + c_{34}K_4)x$$
$$+ (c_{123}K_1K_2 + c_{134}K_1K_4 + c_{234}K_2K_4)x^2 + c_{1234}K_1K_2K_4x^3] \quad (4.93c)$$

$$X_4 = \frac{K_4x}{\Psi(x)}[1 + (c_{14}K_1 + c_{24}K_2 + c_{34}K_3)x$$
$$+ (c_{124}K_1K_2 + c_{134}K_1K_3 + c_{234}K_2K_3)x^2 + c_{1234}K_1K_2K_3x^3] \quad (4.93d)$$

with $X_1 + X_2 + X_3 + X_4 = X$. The site-specific binding curves can also be derived by differentiation as $X_j = \partial \ln \Psi / \partial \ln K_j$, with $j = 1, 2, 3, 4$.

Table 4.5. *Contracted partition functions and weighting factors for N = 4.*

Order	Contracted partition function	Weighting factor
0	$\Psi(x) = 1 + (K_1 + K_2 + K_3 + K_4)x + (c_{12}K_1K_2 + c_{13}K_1K_3$ $+ c_{14}K_1K_4 + c_{23}K_2K_3 + c_{24}K_2K_4 + c_{34}K_3K_4)x^2$ $+ (c_{123}K_1K_2K_3 + c_{124}K_1K_2K_4 + c_{134}K_1K_3K_4$ $+ c_{234}K_2K_3K_4)x^3 + c_{1234}K_1K_2K_3K_4x^4$	1
1	${}^0\Psi_1(x) = 1 + (K_2 + K_3 + K_4)x + (c_{23}K_2K_3 + c_{24}K_2K_4$ $+ c_{34}K_3K_4)x^2 + c_{234}K_2K_3K_4x^3$	1
1	${}^0\Psi_2(x) = 1 + (K_1 + K_3 + K_4)x + (c_{13}K_1K_3 + c_{14}K_1K_4$ $+ c_{34}K_3K_4)x^2 + c_{134}K_1K_3K_4x^3$	1
1	${}^0\Psi_3(x) = 1 + (K_1 + K_2 + K_4)x + (c_{12}K_1K_2 + c_{14}K_1K_4$ $+ c_{24}K_2K_4)x^2 + c_{124}K_1K_2K_4x^3$	1
1	${}^0\Psi_4(x) = 1 + (K_1 + K_2 + K_3)x + (c_{12}K_1K_2 + c_{13}K_1K_3$ $+ c_{23}K_2K_3)x^2 + c_{123}K_1K_2K_3x^3$	1
1	${}^1\Psi_1(x) = 1 + (c_{12}K_2 + c_{13}K_3 + c_{14}K_4)x + (c_{123}K_2K_3$ $+ c_{124}K_2K_4 + c_{134}K_3K_4)x^2 + c_{1234}K_2K_3K_4x^3$	K_1x
1	${}^1\Psi_2(x) = 1 + (c_{12}K_1 + c_{23}K_3 + c_{24}K_4)x + (c_{123}K_1K_3$ $+ c_{124}K_1K_4 + c_{234}K_3K_4)x^2 + c_{1234}K_1K_3K_4x^3$	K_2x
1	${}^1\Psi_3(x) = 1 + (c_{13}K_1 + c_{23}K_2 + c_{34}K_4)x + (c_{123}K_1K_2$ $+ c_{134}K_1K_4 + c_{234}K_2K_4)x^2 + c_{1234}K_1K_2K_4x^3$	K_3x
1	${}^1\Psi_4(x) = 1 + (c_{14}K_1 + c_{24}K_2 + c_{34}K_3)x + (c_{124}K_1K_2$ $+ c_{134}K_1K_3 + c_{234}K_2K_3)x^2 + c_{1234}K_1K_2K_3x^3$	K_4x
2	${}^{00}\Psi_{12}(x) = 1 + (K_3 + K_4)x + c_{34}K_3K_4x^2$	1
2	${}^{00}\Psi_{13}(x) = 1 + (K_2 + K_4)x + c_{24}K_2K_4x^2$	1
2	${}^{00}\Psi_{14}(x) = 1 + (K_2 + K_3)x + c_{23}K_2K_3x^2$	1
2	${}^{00}\Psi_{23}(x) = 1 + (K_1 + K_4)x + c_{14}K_1K_4x^2$	1
2	${}^{00}\Psi_{24}(x) = 1 + (K_1 + K_3)x + c_{13}K_1K_3x^2$	1
2	${}^{00}\Psi_{34}(x) = 1 + (K_1 + K_2)x + c_{12}K_1K_2x^2$	1
2	${}^{01}\Psi_{12}(x) = 1 + (c_{23}K_3 + c_{24}K_4)x + c_{234}K_3K_4x^2$	K_2x
2	${}^{01}\Psi_{13}(x) = 1 + (c_{23}K_2 + c_{34}K_4)x + c_{234}K_2K_4x^2$	K_3x
2	${}^{01}\Psi_{14}(x) = 1 + (c_{24}K_2 + c_{34}K_3)x + c_{234}K_2K_3x^2$	K_4x
2	${}^{01}\Psi_{23}(x) = 1 + (c_{13}K_1 + c_{34}K_4)x + c_{134}K_1K_4x^2$	K_3x
2	${}^{01}\Psi_{24}(x) = 1 + (c_{14}K_1 + c_{34}K_3)x + c_{134}K_1K_3x^2$	K_4x
2	${}^{01}\Psi_{34}(x) = 1 + (c_{14}K_1 + c_{24}K_2)x + c_{124}K_1K_2x^2$	K_4x
2	${}^{10}\Psi_{12}(x) = 1 + (c_{13}K_3 + c_{14}K_4)x + c_{134}K_3K_4x^2$	K_1x
2	${}^{10}\Psi_{13}(x) = 1 + (c_{12}K_2 + c_{14}K_4)x + c_{124}K_2K_4x^2$	K_1x
2	${}^{10}\Psi_{14}(x) = 1 + (c_{12}K_2 + c_{13}K_3)x + c_{123}K_2K_3x^2$	K_1x
2	${}^{10}\Psi_{23}(x) = 1 + (c_{12}K_1 + c_{24}K_4)x + c_{124}K_1K_4x^2$	K_2x
2	${}^{10}\Psi_{24}(x) = 1 + (c_{12}K_1 + c_{23}K_3)x + c_{123}K_1K_3x^2$	K_2x
2	${}^{10}\Psi_{34}(x) = 1 + (c_{13}K_1 + c_{23}K_2)x + c_{123}K_1K_2x^2$	K_3x
2	${}^{11}\Psi_{12}(x) = 1 + (c_{123}K_3 + c_{124}K_4)c_{12}^{-1}x + c_{1234}c_{12}^{-1}K_3K_4x^2$	$c_{12}K_1K_2x^2$
2	${}^{11}\Psi_{13}(x) = 1 + (c_{123}K_2 + c_{134}K_4)c_{13}^{-1}x + c_{1234}c_{13}^{-1}K_2K_4x^2$	$c_{13}K_1K_3x^2$
2	${}^{11}\Psi_{14}(x) = 1 + (c_{124}K_2 + c_{134}K_3)c_{14}^{-1}x + c_{1234}c_{14}^{-1}K_2K_3x^2$	$c_{14}K_1K_4x^2$
2	${}^{11}\Psi_{23}(x) = 1 + (c_{123}K_1 + c_{234}K_4)c_{23}^{-1}x + c_{1234}c_{23}^{-1}K_1K_4x^2$	$c_{23}K_2K_3x^2$
2	${}^{11}\Psi_{24}(x) = 1 + (c_{124}K_1 + c_{234}K_3)c_{24}^{-1}x + c_{1234}c_{24}^{-1}K_1K_3x^2$	$c_{24}K_2K_4x^2$
2	${}^{11}\Psi_{34}(x) = 1 + (c_{134}K_1 + c_{234}K_2)c_{34}^{-1}x + c_{1234}c_{34}^{-1}K_1K_2x^2$	$c_{34}K_3K_4x^2$
3	${}^{000}\Psi_{123}(x) = 1 + K_4x$	1

Table 4.5. (*Cont.*)

Order	Contracted partition function	Weighting factor
3	$^{000}\Psi_{124}(x) = 1 + K_3 x$	1
3	$^{000}\Psi_{134}(x) = 1 + K_2 x$	1
3	$^{000}\Psi_{234}(x) = 1 + K_1 x$	1
3	$^{001}\Psi_{123}(x) = 1 + c_{34} K_4 x$	$K_3 x$
3	$^{001}\Psi_{124}(x) = 1 + c_{34} K_3 x$	$K_4 x$
3	$^{001}\Psi_{134}(x) = 1 + c_{24} K_2 x$	$K_4 x$
3	$^{001}\Psi_{234}(x) = 1 + c_{14} K_1 x$	$K_4 x$
3	$^{010}\Psi_{123}(x) = 1 + c_{24} K_4 x$	$K_2 x$
3	$^{010}\Psi_{124}(x) = 1 + c_{23} K_3 x$	$K_2 x$
3	$^{010}\Psi_{134}(x) = 1 + c_{23} K_2 x$	$K_3 x$
3	$^{010}\Psi_{234}(x) = 1 + c_{13} K_1 x$	$K_3 x$
3	$^{100}\Psi_{123}(x) = 1 + c_{14} K_4 x$	$K_1 x$
3	$^{100}\Psi_{124}(x) = 1 + c_{13} K_3 x$	$K_1 x$
3	$^{100}\Psi_{134}(x) = 1 + c_{12} K_2 x$	$K_1 x$
3	$^{100}\Psi_{234}(x) = 1 + c_{12} K_1 x$	$K_2 x$
3	$^{011}\Psi_{123}(x) = 1 + c_{234} c_{23}^{-1} K_4 x$	$c_{23} K_2 K_3 x^2$
3	$^{011}\Psi_{124}(x) = 1 + c_{234} c_{24}^{-1} K_3 x$	$c_{24} K_2 K_4 x^2$
3	$^{011}\Psi_{134}(x) = 1 + c_{234} c_{34}^{-1} K_2 x$	$c_{34} K_3 K_4 x^2$
3	$^{011}\Psi_{234}(x) = 1 + c_{134} c_{34}^{-1} K_1 x$	$c_{34} K_3 K_4 x^2$
3	$^{101}\Psi_{123}(x) = 1 + c_{134} c_{13}^{-1} K_4 x$	$c_{13} K_1 K_3 x^2$
3	$^{101}\Psi_{124}(x) = 1 + c_{134} c_{14}^{-1} K_3 x$	$c_{14} K_1 K_4 x^2$
3	$^{101}\Psi_{134}(x) = 1 + c_{124} c_{14}^{-1} K_2 x$	$c_{14} K_1 K_4 x^2$
3	$^{101}\Psi_{234}(x) = 1 + c_{124} c_{24}^{-1} K_1 x$	$c_{24} K_2 K_4 x^2$
3	$^{110}\Psi_{123}(x) = 1 + c_{124} c_{12}^{-1} K_4 x$	$c_{12} K_1 K_2 x^2$
3	$^{110}\Psi_{124}(x) = 1 + c_{123} c_{12}^{-1} K_3 x$	$c_{12} K_1 K_2 x^2$
3	$^{110}\Psi_{134}(x) = 1 + c_{123} c_{13}^{-1} K_2 x$	$c_{13} K_1 K_3 x^2$
3	$^{110}\Psi_{234}(x) = 1 + c_{123} c_{23}^{-1} K_1 x$	$c_{23} K_2 K_3 x^2$
3	$^{111}\Psi_{123}(x) = 1 + c_{1234} c_{123}^{-1} K_4 x$	$c_{123} K_1 K_2 K_3 x^3$
3	$^{111}\Psi_{124}(x) = 1 + c_{1234} c_{124}^{-1} K_3 x$	$c_{124} K_1 K_2 K_4 x^3$
3	$^{111}\Psi_{134}(x) = 1 + c_{1234} c_{134}^{-1} K_2 x$	$c_{134} K_1 K_3 K_4 x^3$
3	$^{111}\Psi_{234}(x) = 1 + c_{1234} c_{234}^{-1} K_1 x$	$c_{234} K_2 K_3 K_4 x^3$

Resolution of all independent site-specific parameters requires two additional constraints in the form of contracted partition functions. This follows directly from the properties of the site-specific affinity functions

$$^{x}K_1 = K_1 \lambda_1 / \mu_1 \tag{4.94a}$$

where $\lambda_1 = 1 + (c_{12} K_2 + c_{13} K_3 + c_{14} K_4)x + (c_{123} K_2 K_3 + c_{124} K_2 K_4 + c_{134} K_3 K_4)x^2 + c_{1234} K_2 K_3 K_4 x^3$ and $\mu_1 = 1 + (K_2 + K_3 + K_4)x + (c_{23} K_2 K_3 + c_{24} K_2 K_4 + c_{34} K_3 K_4)x^2 + c_{234} K_2 K_3 K_4 x^3$.

$$^xK_2 = K_2\lambda_2/\mu_2 \tag{4.94b}$$

where $\lambda_2 = 1 + (c_{12}K_1 + c_{23}K_3 + c_{24}K_4)x + (c_{123}K_1K_3 + c_{124}K_1K_4 + c_{234}K_3K_4)x^2 + c_{1234}K_1K_3K_4x^3$ and $\mu_2 = 1 + (K_1 + K_3 + K_4)x + (c_{13}K_1K_3 + c_{14}K_1K_4 + c_{34}K_3K_4)x^2 + c_{134}K_1K_3K_4x^3$.

$$^xK_3 = K_3\lambda_3/\mu_3 \tag{4.94c}$$

where $\lambda_3 = 1 + (c_{13}K_1 + c_{23}K_2 + c_{34}K_4)x + (c_{123}K_1K_2 + c_{134}K_1K_4 + c_{234}K_2K_4)x^2 + c_{1234}K_1K_2K_4x^3$ and $\mu_3 = 1 + (K_1 + K_2 + K_4)x + (c_{12}K_1K_2 + c_{14}K_1K_4 + c_{24}K_2K_4)x^2 + c_{124}K_1K_2K_4x^3$.

$$^xK_4 = K_4\lambda_4/\mu_4 \tag{4.94d}$$

where $\lambda_4 = 1 + (c_{14}K_1 + c_{24}K_2 + c_{34}K_3)x + (c_{124}K_1K_2 + c_{134}K_1K_3 + c_{234}K_2K_3)x^2 + c_{1234}K_1K_2K_3x^3$ and $\mu_4 = 1 + (K_1 + K_2 + K_3)x + (c_{12}K_1K_2 + c_{13}K_1K_3 + c_{23}K_2K_3)x^2 + c_{123}K_1K_2K_3x^3$.

Knowledge of X_1, X_2, X_3 and X_4 yields the values of K_1, K_2, K_3 and K_4, determined from the site-specific affinity functions in the limit $x \to 0$. This information is sufficient to resolve all site-specific binding constants and the coupling constant c_{1234}, since the value of $A_4 = c_{1234}K_1K_2K_3K_4$ is derived from the mean ligand activity of the overall binding curve $X = X_1 + X_2 + X_3 + X_4$. The ratios c_{1234}/c_{234}, c_{1234}/c_{134}, c_{1234}/c_{124} and c_{1234}/c_{123} are determined from the site-specific affinity functions in the limit $x \to \infty$ and yield the third-order coupling constants. However, the four site-specific binding curves do not provide enough constraints to resolve the six independent second-order coupling constants. Resolution of these parameters can only be obtained from contracted forms of the macromolecule where at least two sites are frozen in a particular ligation state. Two second-order contracted partition functions chosen properly, along with the four site-specific binding curves, are needed to solve the site-specific problem for $N = 4$. For example, $^{00}\Psi_{12}$ yields c_{34} and $^{00}\Psi_{34}$ yields c_{12}, thereby allowing resolution of all cs.

4.4 Site-specific properties of hemoglobin

The limitation arising in connection with the case of $N = 4$ applies to all other cases for larger N. Structural symmetry does not eliminate this problem and hemoglobin provides a convincing and pertinent example. Human hemoglobin is composed of two identical dimers of oxygen-binding subunits, the α and β chains, interacting across an extended interdimeric interface (Perutz, 1970, 1989; Baldwin and Chothia, 1979; Lesk, Janin, Wodak and Chothia, 1985). Each dimer is tightly held together by contacts at the $\alpha_1\beta_1$ or $\alpha_2\beta_2$ intradimeric interface. The partition function for hemoglobin in the site-specific description is (Di Cera, 1990, 1994a)

$$\Psi(x) = 1 + 2(K_\alpha + K_\beta)x + [c_{\alpha\alpha}K_\alpha^2 + c_{\beta\beta}K_\beta^2 + 2(c_{\alpha\beta} + c'_{\alpha\beta})K_\alpha K_\beta]x^2$$
$$+ 2(c_{\alpha\alpha\beta}K_\alpha + c_{\alpha\beta\beta}K_\beta)K_\alpha K_\beta x^3 + c_{\alpha\alpha\beta\beta}K_\alpha^2 K_\beta^2 x^4 \quad (4.95)$$

Here K_α and K_β denote the binding affinities of the two chains when all other chains are unligated, while the cs are the various coupling constants. A distinction between $c_{\alpha\beta}$ and $c'_{\alpha\beta}$ should be made since the interaction of the two chains across the interface between the dimers, $\alpha\beta$, is to be distinguished from that within each dimer, $\alpha\beta'$. All models proposed for hemoglobin cooperativity (Pauling, 1935; Wyman, 1948; Monod *et al.*, 1965; Koshland *et al.*, 1966; Szabo and Karplus, 1972; Herzfeld and Stanley, 1974; Johnson, Turner and Ackers, 1984; Di Cera, Robert and Gill, 1987b; Ackers *et al.*, 1992) can be derived from the model-independent partition function (4.95) as special cases.[†] The connection with the overall equilibrium constants is given by

$$A_1 = 2(K_\alpha + K_\beta) \quad (4.96a)$$
$$A_2 = c_{\alpha\alpha}K_\alpha^2 + c_{\beta\beta}K_\beta^2 + 2(c_{\alpha\beta} + c'_{\alpha\beta})K_\alpha K_\beta \quad (4.96b)$$
$$A_3 = 2(c_{\alpha\alpha\beta}K_\alpha + c_{\alpha\beta\beta}K_\beta)K_\alpha K_\beta \quad (4.96c)$$
$$A_4 = c_{\alpha\alpha\beta\beta}K_\alpha^2 K_\beta^2 \quad (4.96d)$$

The presence of two identical pairs of chains reduces the number of distinct configurations in the local description from sixteen to ten. Each configuration can be associated with a vector whose elements are in lexicographic order and refer to the chains in the order α_1, α_2, β_1, β_2, i.e.,

$$[0000]$$
$$[1000] \quad [0010]$$
$$[1100] \quad [1010] \quad [1001] \quad [0011] \quad\quad (4.97)$$
$$[1110] \quad [1011]$$
$$[1111]$$

[†] A few years after he had left Germany, Einstein was informed that Nazi scientists had put together 100 papers to prove his theories wrong. Asked whether he had any concerns, Einstein replied: 'These 100 papers must be all wrong. If right, one would have been sufficient.' One can hardly think of a better comment when it comes to the endless list of models for hemoglobin cooperativity. The exact thermodynamic treatment embodied by eq (4.95) is one and sufficient.

Symmetry of the hemoglobin tetramer imposes the following conditions

$$[1000] = [0100] \tag{4.98a}$$

$$[0010] = [0001] \tag{4.98b}$$

$$[1010] = [0101] \tag{4.98c}$$

$$[1001] = [0110] \tag{4.98d}$$

$$[1110] = [1101] \tag{4.98e}$$

$$[1011] = [0111] \tag{4.98f}$$

The unligated form of the macromolecule, M, is the same as the configuration M_{0000} labeled in a site-specific fashion and is used as reference in the definition of the partition function (4.95). The singly-ligated form of the macromolecule in the global description, MX, is the sum of two distinct site-specific configurations, M_{1000} and M_{0010}, corresponding to the intermediates with the α chain or the β chain ligated. In addition, there are four doubly-ligated configurations (M_{1100}, M_{1010}, M_{1001}, M_{0011}), two triply-ligated configurations (M_{1110}, M_{1011}) and the fully-ligated form M_{1111}. Since there are ten configurations in eq (4.95), one of which is used as reference, then there must be nine independent reactions among the possible intermediates and an equivalent number of independent equilibrium constants. The site-specific equilibrium constants K_α and K_β refer to the reactions

$$M_{0000} + X \Leftrightarrow M_{1000} \tag{4.99a}$$

$$M_{0000} + X \Leftrightarrow M_{0010} \tag{4.99b}$$

The coupling constants $c_{\alpha\alpha}$, $c_{\beta\beta}$, $c_{\alpha\beta}$, $c'_{\alpha\beta}$, $c_{\alpha\alpha\beta}$, $c_{\alpha\beta\beta}$ and $c_{\alpha\alpha\beta\beta}$ are associated with the dismutations

$$M_{1000} + M_{0100} \Leftrightarrow M_{0000} + M_{1100} \tag{4.100a}$$

$$M_{0010} + M_{0001} \Leftrightarrow M_{0000} + M_{0011} \tag{4.100b}$$

$$M_{1000} + M_{0001} \Leftrightarrow M_{0000} + M_{1001} \tag{4.100c}$$

$$M_{1000} + M_{0010} \Leftrightarrow M_{0000} + M_{1010} \tag{4.100d}$$

$$M_{1000} + M_{0100} + M_{0010} \Leftrightarrow 2M_{0000} + M_{1110} \tag{4.100e}$$

$$M_{1000} + M_{0010} + M_{0001} \Leftrightarrow 2M_{0000} + M_{1011} \tag{4.100f}$$

$$M_{1000} + M_{0100} +. M_{0010} + M_{0001} \Leftrightarrow 3M_{0000} + M_{1111} \tag{4.100g}$$

The site-specific binding curves of the two chains are

$$X_\alpha = K_\alpha x$$

$$\times \frac{1 + [c_{\alpha\alpha}K_\alpha + (c_{\alpha\beta} + c'_{\alpha\beta})K_\beta]x + (2c_{\alpha\alpha\beta}K_\alpha + c_{\alpha\beta\beta}K_\beta)K_\beta x^2 + c_{\alpha\alpha\beta\beta}K_\alpha K_\beta^2 x^3}{\Psi(x)}$$

$$\tag{4.101a}$$

$$X_\beta = K_\beta x$$

$$\times \frac{1 + [c_{\beta\beta}K_\beta + (c_{\alpha\beta} + c'_{\alpha\beta})K_\alpha]x + (c_{\alpha\alpha\beta}K_\alpha + 2c_{\alpha\beta\beta}K_\beta)K_\alpha x^2 + c_{\alpha\alpha\beta\beta}K_\alpha^2 K_\beta x^3}{\Psi(x)}$$

$$(4.101b)$$

with

$$X = 2(X_\alpha + X_\beta) \qquad (4.102)$$

as expected.

Complete resolution of the nine site-specific parameters in the partition function (4.95) demands measurements of at least three site-specific properties of the system in addition to the global properties. This is because A_2 contains four independent terms, thereby demanding three additional constraints. Although the intrinsic symmetry of the system has reduced the number of total configurations and independent parameters, the amount of information to be gathered from experimental measurements exceeds the limits accessible from measurements of X_α and X_β alone. These binding curves provide altogether one zero-order and two first-order contracted partition functions. An additional second-order contracted partition function is needed. The possible contracted partition functions, up to third order, and their proper weighting terms are listed in Table 4.6. As seen in the general case, the partition function Ψ can be written in a series of alternative forms using first-order contracted partition functions as follows

$$\Psi(x) = {}^0\Psi_\alpha(x) + {}^1\Psi_\alpha(x)K_\alpha x = {}^0\Psi_\beta(x) + {}^1\Psi_\beta(x)K_\beta x \quad (4.103)$$

Hence,

$$X_\alpha = K_\alpha x \frac{{}^1\Psi_\alpha(x)}{\Psi(x)} = 1 - \frac{{}^0\Psi_\alpha(x)}{\Psi(x)} \qquad (4.104a)$$

$$X_\beta = K_\beta x \frac{{}^1\Psi_\beta(x)}{\Psi(x)} = 1 - \frac{{}^0\Psi_\beta(x)}{\Psi(x)} \qquad (4.104b)$$

The properties of hemoglobin as a whole can be derived from those of the α or β chain if either system is known along with its interaction with the other chains. Each first-order contracted partition function can be written in terms of second-order contracted partition functions. If only the configurations pertaining to the α chain are considered we have

$${}^0\Psi_\alpha(x) = {}^{00}\Psi_{\alpha\alpha}(x) + {}^{01}\Psi_{\alpha\alpha}(x)K_\alpha x \qquad (4.105a)$$

$${}^1\Psi_\alpha(x) = {}^{10}\Psi_{\alpha\alpha}(x) + {}^{11}\Psi_{\alpha\alpha}(x)c_{\alpha\alpha}K_\alpha x \qquad (4.105b)$$

$$\Psi(x) = {}^{00}\Psi_{\alpha\alpha}(x) + 2{}^{10}\Psi_{\alpha\alpha}(x)K_\alpha x + {}^{11}\Psi_{\alpha\alpha}(x)c_{\alpha\alpha}K_\alpha^2 x^2 \quad (4.105c)$$

Therefore, the properties of hemoglobin as a whole can also be derived

Table 4.6. *Contracted partition functions and weighting factors for human hemoglobin.*

Order	Contracted partition function	Weighting factor
0	$\Psi(x) = 1 + 2(K_\alpha + K_\beta)x + (c_{\alpha\alpha}K_\alpha^2 + c_{\beta\beta}K_\beta^2 + 2c_{\alpha\beta}K_\alpha K_\beta$ $+ 2c'_{\alpha\beta}K_\alpha K_\beta)x^2 + 2(c_{\alpha\alpha\beta}K_\alpha + c_{\alpha\beta\beta}K_\beta)K_\alpha K_\beta x^3$ $+ c_{\alpha\alpha\beta\beta}K_\alpha^2 K_\beta^2 x^4$	1
1	${}^0\Psi_\alpha(x) = 1 + (K_\alpha + 2K_\beta)x + (c_{\beta\beta}K_\beta^2 + c_{\alpha\beta}K_\alpha K_\beta$ $+ c'_{\alpha\beta}K_\alpha K_\beta)x^2 + c_{\alpha\beta\beta}K_\alpha K_\beta^2 x^3$	1
1	${}^0\Psi_\beta(x) = 1 + (2K_\alpha + K_\beta)x + (c_{\alpha\alpha}K_\alpha^2 + c_{\alpha\beta}K_\alpha K_\beta$ $+ c'_{\alpha\beta}K_\alpha K_\beta)x^2 + c_{\alpha\alpha\beta}K_\alpha^2 K_\beta x^3$	1
1	${}^1\Psi_\alpha(x) = 1 + (c_{\alpha\alpha}K_\alpha + c_{\alpha\beta}K_\beta + c'_{\alpha\beta}K_\beta)x + (2c_{\alpha\alpha\beta}K_\alpha$ $+ c_{\alpha\beta\beta}K_\beta)K_\beta x^2 + c_{\alpha\alpha\beta\beta}K_\alpha K_\beta^2 x^3$	$K_\alpha x$
1	${}^1\Psi_\beta(x) = 1 + (c_{\beta\beta}K_\beta + c_{\alpha\beta}K_\alpha + c'_{\alpha\beta}K_\alpha)x + (c_{\alpha\alpha\beta}K_\alpha$ $+ 2c_{\alpha\beta\beta}K_\beta)K_\alpha x^2 + c_{\alpha\alpha\beta\beta}K_\alpha^2 K_\beta x^3$	$K_\beta x$
2	${}^{00}\Psi_{\alpha\alpha}(x) = 1 + 2K_\beta x + c_{\beta\beta}K_\beta^2 x^2$	1
2	${}^{00}\Psi_{\alpha\beta}(x) = 1 + (K_\alpha + K_\beta)x + c_{\alpha\beta}K_\alpha K_\beta x^2$	1
2	${}^{00}\Psi_{\alpha\beta'}(x) = 1 + (K_\alpha + K_\beta)x + c'_{\alpha\beta}K_\alpha K_\beta x^2$	1
2	${}^{00}\Psi_{\beta\beta}(x) = 1 + 2K_\alpha x + c_{\alpha\alpha}K_\alpha^2 x^2$	1
2	${}^{01}\Psi_{\alpha\alpha}(x) = 1 + (c_{\alpha\beta} + c'_{\alpha\beta})K_\beta x + c_{\alpha\beta\beta}K_\beta^2 x^2$	$K_\alpha x$
2	${}^{01}\Psi_{\alpha\beta}(x) = 1 + (c'_{\alpha\beta}K_\alpha + c_{\beta\beta}K_\beta)x + c_{\alpha\beta\beta}K_\alpha K_\beta x^2$	$K_\beta x$
2	${}^{01}\Psi_{\alpha\beta'}(x) = 1 + (c_{\alpha\beta}K_\alpha + c_{\beta\beta}K_\beta)x + c_{\alpha\beta\beta}K_\alpha K_\beta x^2$	$K_\beta x$
2	${}^{01}\Psi_{\beta\beta}(x) = 1 + (c_{\alpha\beta} + c'_{\alpha\beta})K_\alpha x + c_{\alpha\alpha\beta}K_\alpha^2 x^2$	$K_\beta x$
2	${}^{10}\Psi_{\alpha\beta}(x) = 1 + (c'_{\alpha\beta}K_\beta + c_{\alpha\alpha}K_\alpha)x + c_{\alpha\alpha\beta}K_\alpha K_\beta x^2$	$K_\alpha x$
2	${}^{10}\Psi_{\alpha\beta'}(x) = 1 + (c_{\alpha\beta}K_\beta + c_{\alpha\alpha}K_\alpha)x + c_{\alpha\alpha\beta}K_\alpha K_\beta x^2$	$K_\alpha x$
2	${}^{11}\Psi_{\alpha\alpha}(x) = 1 + 2c_{\alpha\alpha\beta}c_{\alpha\alpha}^{-1}K_\beta x + c_{\alpha\alpha\beta\beta}c_{\alpha\alpha}^{-1}K_\beta^2 x^2$	$c_{\alpha\alpha}K_\alpha^2 x^2$
2	${}^{11}\Psi_{\alpha\beta}(x) = 1 + (c_{\alpha\alpha\beta}K_\alpha + c_{\alpha\beta\beta}K_\beta)c_{\alpha\beta}^{-1}x + c_{\alpha\alpha\beta\beta}c_{\alpha\beta}^{-1}K_\alpha K_\beta x^2$	$c_{\alpha\beta}K_\alpha K_\beta x^2$
2	${}^{11}\Psi_{\alpha\beta'}(x) = 1 + (c_{\alpha\alpha\beta}K_\alpha + c_{\alpha\beta\beta}K_\beta)c_{\alpha\beta}'^{-1}x + c_{\alpha\alpha\beta\beta}c_{\alpha\beta}'^{-1}K_\alpha K_\beta x^2$	$c'_{\alpha\beta}K_\alpha K_\beta x^2$
2	${}^{11}\Psi_{\beta\beta}(x) = 1 + 2c_{\alpha\beta\beta}c_{\beta\beta}^{-1}K_\alpha x + c_{\alpha\alpha\beta\beta}c_{\beta\beta}^{-1}K_\alpha^2 x^2$	$c_{\beta\beta}K_\beta^2 x^2$
3	${}^{000}\Psi_{\alpha\alpha\beta}(x) = 1 + K_\beta x$	1
3	${}^{000}\Psi_{\alpha\beta\beta}(x) = 1 + K_\alpha x$	1
3	${}^{100}\Psi_{\alpha\alpha\beta}(x) = 1 + c_{\alpha\beta}K_\beta x$	$K_\alpha x$
3	${}^{010}\Psi_{\alpha\alpha\beta}(x) = 1 + c'_{\alpha\beta}K_\beta x$	$K_\alpha x$
3	${}^{001}\Psi_{\alpha\alpha\beta}(x) = 1 + c_{\beta\beta}K_\beta x$	$K_\beta x$
3	${}^{100}\Psi_{\alpha\beta\beta}(x) = 1 + c_{\alpha\alpha}K_\alpha x$	$K_\alpha x$
3	${}^{010}\Psi_{\alpha\beta\beta}(x) = 1 + c_{\alpha\beta}K_\alpha x$	$K_\beta x$
3	${}^{001}\Psi_{\alpha\beta\beta}(x) = 1 + c'_{\alpha\beta}K_\alpha x$	$K_\beta x$
3	${}^{110}\Psi_{\alpha\alpha\beta}(x) = 1 + c_{\alpha\alpha\beta}c_{\alpha\alpha}^{-1}K_\beta x$	$c_{\alpha\alpha}K_\alpha^2 x^2$
3	${}^{101}\Psi_{\alpha\alpha\beta}(x) = 1 + c_{\alpha\beta\beta}c_{\alpha\beta}'^{-1}K_\beta x$	$c'_{\alpha\beta}K_\alpha K_\beta x^2$
3	${}^{011}\Psi_{\alpha\alpha\beta}(x) = 1 + c_{\alpha\beta\beta}c_{\alpha\beta}^{-1}K_\beta x$	$c_{\alpha\beta}K_\alpha K_\beta x^2$
3	${}^{110}\Psi_{\alpha\beta\beta}(x) = 1 + c_{\alpha\alpha\beta}c_{\alpha\beta}'^{-1}K_\alpha x$	$c'_{\alpha\beta}K_\alpha K_\beta x^2$
3	${}^{101}\Psi_{\alpha\beta\beta}(x) = 1 + c_{\alpha\alpha\beta}c_{\alpha\beta}^{-1}K_\alpha x$	$c_{\alpha\beta}K_\alpha K_\beta x^2$
3	${}^{011}\Psi_{\alpha\beta\beta}(x) = 1 + c_{\alpha\beta\beta}c_{\beta\beta}^{-1}K_\alpha x$	$c_{\beta\beta}K_\beta^2 x^2$
3	${}^{111}\Psi_{\alpha\alpha\beta}(x) = 1 + c_{\alpha\alpha\beta\beta}c_{\alpha\alpha\beta}^{-1}K_\beta x$	$c_{\alpha\alpha\beta}K_\alpha^2 K_\beta x^3$
3	${}^{111}\Psi_{\alpha\beta\beta}(x) = 1 + c_{\alpha\alpha\beta\beta}c_{\alpha\beta\beta}^{-1}K_\alpha x$	$c_{\alpha\beta\beta}K_\alpha K_\beta^2 x^3$

from those of the two α chains, if their properties are known along with their interactions with the β chains. The structural symmetry of the hemoglobin molecule poses some precise limitations as to what contracted partition functions should be accessed experimentally to resolve the nine independent site-specific constants. Assume for instance that all contracted partition functions with the α chains unligated can be accessed, in addition to Ψ. This gives a total of three partition functions, Ψ, ${}^0\Psi_\alpha$ and ${}^{00}\Psi_{\alpha\alpha}$. We see from Table 4.6 that these partition functions are sufficient to resolve K_α, K_β, $c_{\alpha\alpha\beta\beta}$, $c_{\alpha\alpha\beta}$ and $c_{\alpha\beta\beta}$. Of the second-order interaction constants, $c_{\alpha\alpha}$ and $c_{\beta\beta}$ can be resolved separately, but only the sum of $c_{\alpha\beta} + c'_{\alpha\beta}$ can be obtained. The same result applies in the case where analogous partition functions are derived experimentally for the β chains. Information on the separate coupling constants $c_{\alpha\beta}$ and $c'_{\alpha\beta}$ must be obtained from contracted partition functions where one α chain and one β chain are frozen in particular ligation states.

The standard techniques developed for measuring oxygen binding equilibria of hemoglobin (Roughton, Oris and Lyster, 1955; Imai, 1974, 1982; Dolman and Gill, 1978; Gill *et al.*, 1987) yield global binding isotherms and binding capacity measurements. These techniques are not capable of resolving the separate contributions of the two chains, or the site-specific parameters involved in the partition function (4.95). Some qualitative information on the behavior of the two chains has been obtained by NMR measurements (Viggiano and Ho, 1979; Bondon, Petrinko, Sodano and Simonneaux, 1986; Simonneaux, Bondon, Brunel and Sodano, 1988; Bondon and Simonneaux, 1990) and by binding equilibria on hemoglobin hybrids where some of the chains carry a substitution of the FeII heme atom with Ni, Mn or Co (Ikeda-Saito and Yonetani, 1980; Inubushi, D'Ambrosio, Ikeda-Saito and Yonetani, 1986; Shibayama, Inubushi, Morimoto and Yonetani, 1987; Ackers and Smith, 1987; Spiros, LiCata, Yonetani and Ackers, 1991). The cryogenic quenching technique pioneered by Perrella (Perrella and Rossi-Bernardi, 1981; Perrella *et al.*, 1983, 1986) has provided a real breakthrough in the experimental approach to cooperative binding by hemoglobin. Using this technique, Perrella has been able to trap some of the key intermediates of ligation with CO (Perrella *et al.*, 1990, 1992, 1993) and to dissect the salient site-specific components of the linkage between heme ligation and proton release in hemoglobin (Perrella, Benazzi, Ripamonti and Rossi-Bernardi, 1994), also known as the Bohr effect. More recently, a detailed resolution of the CO-bound intermediates in scheme (4.97) has been obtained (Perrella, 1995). These measurements represent the first, most accurate

Table 4.7. *Distribution of ligated intermediates for CO binding to hemoglobin at 48% saturation (Perrella, 1995).*

Species	Fraction (%)	Associated site-specific term
[0000]	41.8 ± 0.2	$1/\Psi(x)$
[1000]	2.4 ± 0.1	$2K_\alpha x/\Psi(x)$
[0010]	4.9 ± 0.3	$2K_\beta x/\Psi(x)$
[1100]	1.0 ± 0.1	$c_{\alpha\alpha}K_\alpha^2 x^2/\Psi(x)$
[1010] + [1001]	2.9 ± 0.2	$2(c'_{\alpha\beta} + c_{\alpha\beta})K_\alpha K_\beta x^2/\Psi(x)$
[0011]	1.4 ± 0.2	$c_{\beta\beta}K_\beta^2 x^2/\Psi(x)$
[1110]	3.8 ± 0.5	$2c_{\alpha\alpha\beta}K_\alpha^2 K_\beta x^3/\Psi(x)$
[1011]	4.7 ± 0.7	$2c_{\alpha\beta\beta}K_\alpha K_\beta^2 x^3/\Psi(x)$
[1111]	37 ± 2	$c_{\alpha\alpha\beta\beta}K_\alpha^2 K_\beta^2 x^4/\Psi(x)$

Table 4.8. *Site-specific parameters for CO binding to hemoglobin.*

$K_\alpha = 1.5 \pm 0.1 \text{ Torr}^{-1}$	$c_{\beta\beta} = 10 \pm 2$
$K_\beta = 3.1 \pm 0.3 \text{ Torr}^{-1}$	$c_{\alpha\alpha\beta} = 942 \pm 186$
$c_{\alpha\alpha} = 29 \pm 4$	$c_{\alpha\beta\beta} = 570 \pm 108$
$c_{\alpha\beta} + c'_{\alpha\beta} = 21 \pm 4$	$c_{\alpha\alpha\beta\beta} = 3.1 \times 10^5 \pm 0.7 \times 10^5$

and relevant experimental determinations of the site-specific properties of the native hemoglobin molecule in its reaction with a gaseous ligand. The distribution of ligated intermediates in scheme (4.97) determined at 48% saturation with CO is given in Table 4.7. The parameters listed in Table 4.8, derived from the distribution shown in Table 4.7, make it possible to reconstruct the site-specific binding curves of the two chains. The fact that only the sum $c_{\alpha\beta} + c'_{\alpha\beta}$ could be obtained experimentally is inconsequential. It follows from eqs (4.101a) and (4.101b) and Table 4.6 that X_α and X_β depend on contracted partition functions that contain only the sum $c_{\alpha\beta} + c'_{\alpha\beta}$. The important response functions for the α and β chains in their reaction with CO in the hemoglobin tetramer are shown in Figure 4.10. Notwithstanding a significant difference in binding affinity, most evident in the affinity plot (see Figure 4.10(*c*)), the two chains are remarkably similar in their cooperative behaviour. The site-specific Hill coefficient (see Figure 4.10(*d*)) is practically identical for the two chains and indicates that up to 3.4 molecules of CO are bound when either chain is ligated. The communication among the chains is characterized by a significant degree of asymmetry. The second-order coupling constant between the α chains is three times stronger than that between the β chains. Likewise, the third-order coupling constant is more pronounced when the two α chains are ligated. The stronger coupling between the α chains compensates for their reduced affinity and results in a site-specific binding curve similar to that of the β chain (see Figures 4.10(a), (b)). It should be noted that coupling in the hemoglobin tetramer is always positive and therefore binding to any chain always promotes binding to unligated chains in a positively cooperative fashion.

The global description of CO binding to hemoglobin tells us nothing about the asymmetric interactions in the hemoglobin tetramer, or the chain heterogeneity, thereby missing a great deal of important information. To illustrate this point we compute the global binding parameters from the site-specific ones listed in Table 4.8, using eqs (4.71a)–(4.71d) and (4.80a)–(4.80d). The result is $A_1 = 9.2 \text{ Torr}^{-1}$, $A_2 = 357 \text{ Torr}^{-2}$, $A_3 = 3.0 \times 10^4 \text{ Torr}^{-3}$ and $A_4 = 6.7 \times 10^6 \text{ Torr}^{-4}$ for the overall constants and $k_1 = 2.3 \text{ Torr}^{-1}$, $k_2 = 26 \text{ Torr}^{-1}$, $k_3 = 126 \text{ Torr}^{-1}$ and $k_4 = 893 \text{ Torr}^{-1}$ for the stepwise constants. Binding of CO to hemoglobin in the global description is characterized by uniform positive cooperativity ($k_1 < k_2 < k_3 < k_4$) and symmetry ($k_1 k_4 \approx k_2 k_3$). These two features are easily accounted for by practically all models for hemoglobin cooperativity proposed to date, from the simplest ones (Pauling, 1935; Wyman, 1948) to the celebrated two-state MWC (Monod–Wyman–Changeux) model (Monod *et al.*, 1965). However, these models do not account for the distribution of intermediates as found by Perrella (1995) and listed in Table 4.7. As in the case discussed in Section 3.1, the important details of the local picture are swamped into the average global behavior of the macromolecule. The inadequacy of these models, including the MWC model that has been widely used in the analysis of hemoglobin cooperativity, is further demonstrated by consideration of the 'pairwise coupling' between sites within the general framework of site-specific thermodynamics. A detailed treatment of pairwise coupling is given in Section 4.6.

The coupling of a given chain to the rest of the macromolecule can be analyzed in terms of the contracted partition functions listed in Table 4.6. Likewise, for any given pair of chains, $\alpha_1 \alpha_2$, $\alpha_1 \beta_1$, $\alpha_1 \beta_2$ and $\beta_1 \beta_2$, the interaction with the complementary pair can be quantified in a direct way. Here we exploit the nature of the interaction constants as related to the free energy of coupling in a given thermodynamic cycle. Consider the thermodynamic cycle in Figure 4.11. Ligation of the $\alpha_1 \alpha_2$ pair is taken into consideration while the $\beta_1 \beta_2$ pair is kept in the unligated configuration.

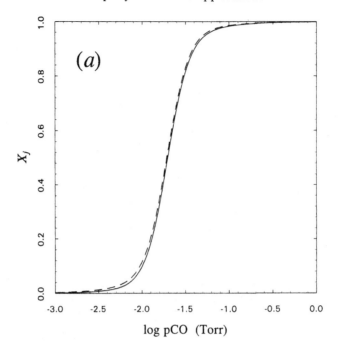

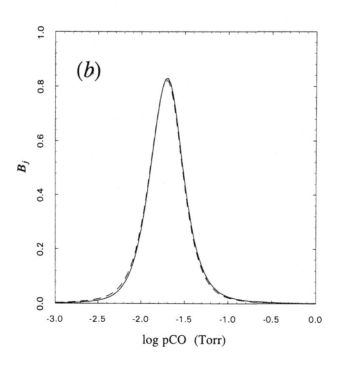

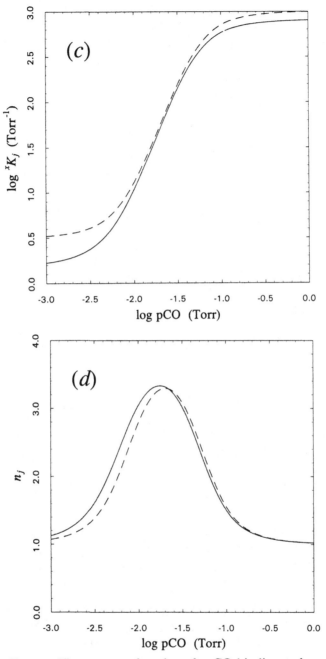

Figure 4.10   Site-specific response functions for CO binding to human hemoglobin derived from the distribution of intermediates given in Table 4.7. Continuous lines refer to the α chain and dashed lines refer to the β chain. Experimental conditions are (Perrella, 1995): 0.1 M Hepes, 0.1 M KCl, pH 7.0 at 20 °C: (*a*) site-specific binding curves, (*b*) binding capacities, (*c*) affinity functions and (*d*) Hill coefficients.

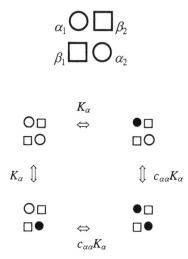

Figure 4.11   Schematic representation of the hemoglobin tetramer (top) and the thermodynamic cycle illustrating the pairwise coupling between the α chains when the β chains are kept unligated (bottom). Similar cycles can be constructed for all possible ligated configurations of the complementary pair and for all possible pairs of chains in the hemoglobin tetramer, as shown in Figure 4.12. The coupling free energy in the cycle is a model-independent measure of the communication between the α chains, given the particular configuration of the β chains.

The free energy of coupling (in kcal/mol) in the cycle is evidently

$$\Delta G_{\mathrm{c}} = -RT \ln c_{\alpha\alpha} \qquad (4.106)$$

This quantity reflects the linkage between the α chains when the β chains are unligated. As there are four possible configurations for the $\beta_1\beta_2$ pair, so are there four terms as eq (4.106). Furthermore, since there are four distinct pairs of chains to be considered in the hemoglobin tetramer, so are there a total of sixteen free energies of pairwise coupling as eq (4.106) (Di Cera, 1990). These terms are summarized in Table 4.9. The relevant thermodynamic cycles are sketched in Figure 4.12. For each pair, the four possible configurations of the complementary pair are listed in the usual vector form in lexicographic order. The relationships listed in Table 4.9 make it possible to quantify the linkage between any two chains in the hemoglobin tetramer as a function of the ligation state of the other two subunits. The coupling terms are independent of the binding affinities of the two chains, K_α and K_β, and depend solely on the interaction constants cs. They encapsulate the code for cooperative coupling among the subunits in a site-specific fashion.

The importance of defining pairwise coupling free energies stems from

Table 4.9. *Free energies of pairwise coupling in the hemoglobin tetramer.*

Pair	00	10	01	11
$\alpha_1\alpha_2$	$-RT\ln c_{\alpha\alpha}$	$-RT\ln(c_{\alpha\alpha\beta}/c_{\alpha\beta}c'_{\alpha\beta})$	$-RT\ln(c_{\alpha\alpha\beta}/c_{\alpha\beta}c'_{\alpha\beta})$	$-RT\ln(c_{\beta\beta}c_{\alpha\alpha\beta\beta}/c^2_{\alpha\beta\beta})$
$\alpha_1\beta_1$	$-RT\ln c'_{\alpha\beta}$	$-RT\ln(c_{\alpha\alpha\beta}/c_{\alpha\alpha}c_{\alpha\beta})$	$-RT\ln(c_{\alpha\beta\beta}/c_{\beta\beta}c_{\alpha\beta})$	$-RT\ln(c'_{\alpha\beta}c_{\alpha\alpha\beta\beta}/c_{\alpha\alpha\beta}c_{\alpha\beta\beta})$
$\alpha_1\beta_2$	$-RT\ln c_{\alpha\beta}$	$-RT\ln(c_{\alpha\alpha\beta}/c_{\alpha\alpha}c'_{\alpha\beta})$	$-RT\ln(c_{\alpha\beta\beta}/c_{\beta\beta}c'_{\alpha\beta})$	$-RT\ln(c_{\alpha\beta}c_{\alpha\alpha\beta\beta}/c_{\alpha\alpha\beta}c_{\alpha\beta\beta})$
$\beta_1\beta_2$	$-RT\ln c_{\beta\beta}$	$-RT\ln(c_{\alpha\beta\beta}/c_{\alpha\beta}c'_{\alpha\beta})$	$-RT\ln(c_{\alpha\beta\beta}/c_{\alpha\beta}c'_{\alpha\beta})$	$-RT\ln(c_{\alpha\alpha}c_{\alpha\alpha\beta\beta}/c^2_{\alpha\alpha\beta})$

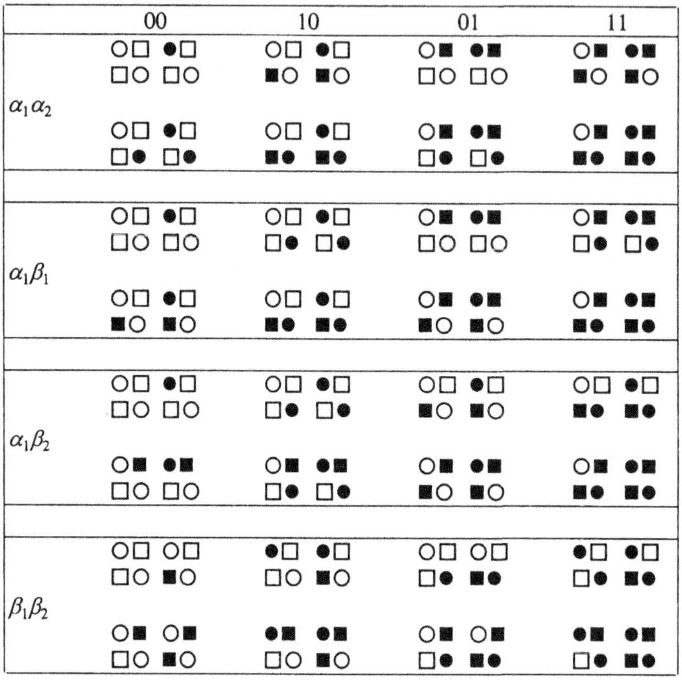

	00	10	01	11
$\alpha_1\alpha_2$	O□ ●□ □O □O	O□ ●□ ■O ■O	O■ ●■ □O □O	O■ ●■ ■O ■O
	O□ ●□ □● □●	O□ ●□ ■● ■●	O■ ●■ □● □●	O■ ●■ ■● ■●
$\alpha_1\beta_1$	O□ ●□ □O □O	O□ ●□ □● □●	O■ ●■ □O □O	O■ ●■ □● □●
	O□ ●□ ■O ■O	O□ ●□ ■● ■●	O■ ●■ ■O ■O	O■ ●■ ■● ■●
$\alpha_1\beta_2$	O□ ●□ □O □O	O□ ●□ □● □●	O□ ●□ ■O ■O	O□ ●□ ■● ■●
	O■ ●■ □O □O	O■ ●■ □● □●	O■ ●■ ■O ■O	O■ ●■ ■● ■●
$\beta_1\beta_2$	O□ O□ □O ■O	●□ ●□ □O ■O	O□ O□ □● ■●	●□ ●□ □● ■●
	O■ O■ □O ■O	●■ ●■ □O ■O	O■ O■ □● ■●	●■ ●■ □● ■●

Figure 4.12 Thermodynamic cycles for pairwise coupling in the hemoglobin tetramer. The configurations of the complementary pair are listed in lexicographic order. Symbols depict the various chains as in Figure 4.11 (top).

the possibility of deciphering the code for cooperativity in a model-independent fashion. Experimentally derived values for pairwise coupling can be compared directly with the predictions drawn from specific models. The KNF (Koshland–Nemethy–Filmer) model (Koshland *et al.*, 1966) based on nearest-neighbor interactions predicts the free energies of pairwise coupling to be independent of the ligation of the complementary pair (Di Cera, 1990). The MWC model predicts the free energies of pairwise coupling to be always negative and to change with ligation of the complementary pair (Di Cera, 1990). A detailed analysis of these interaction patterns is given in Section 4.6. The partition function for the two-state MWC model is[†] (Monod *et al.*, 1965)

$$\Psi(x) = \frac{L(1 + K_\mathrm{T}x)^N + (1 + K_\mathrm{R}x)^N}{L + 1} \qquad (4.107)$$

[†] It should be pointed out that the partition function of the MWC model was first derived by Terrell Hill in an analogous context in 1960 (Hill, 1960; see eq (7–68) at p. 143), i.e., five years before publication of the MWC paper (Monod *et al.*, 1965). Hill's derivation was ignored in the MWC paper and in practically all subsequent treatments of allosteric models.

The various terms in eq (4.107) represent the binding affinities of the T and R states, K_T and K_R, and the allosteric constant, L, that gives the ratio between the concentration of unligated molecules in the T state relative to the R state. Cooperativity arises from the fact that the macromolecule is predominantly in the low affinity T state at low saturation and progressively switches to the high affinity R state as ligation proceeds. Both states are assumed to be reference systems. The T $\rightarrow$ R transition involves all sites in a concerted fashion and brings about cooperative behavior, as in the example dealt with in Section 2.6 for the linkage between two ligands (see eq (2.162)). The free energies of pairwise coupling for the MWC model are derived from (4.107) as (Di Cera, 1990)

$$\Delta G_{00} = -RT \ln \frac{(Lc^2 + 1)(L + 1)}{(Lc + 1)^2} \qquad (4.108a)$$

$$\Delta G_{10} = \Delta G_{01} = -RT \ln \frac{(Lc^3 + 1)(Lc + 1)}{(Lc^2 + 1)^2} \qquad (4.108b)$$

$$\Delta G_{11} = -RT \ln \frac{(Lc^4 + 1)(Lc^2 + 1)}{(Lc^3 + 1)^2} \qquad (4.108c)$$

where $c = K_T/K_R < 1$ by definition. The various free energy terms in eqs (4.108a)–(4.108c) give the pairwise coupling between any two chains when the other two chains are unligated, singly-ligated or doubly-ligated. The free energy terms are always negative because

$$(Lc^j + 1)(Lc^{j-2} + 1) > (Lc^{j-1} + 1)^2 \qquad (4.109)$$

for any value of j, as long as $c > 0$. Furthermore, pairwise coupling is large at the switchover point, where $Lc^s \approx 1$, and tends to zero for values of $j \neq s$.

The global CO binding curve constructed from Perrella's data (Perrella, 1995) can be represented very well in terms of the MWC model with $c \approx 0.0026$ and $L \approx 9.5 \times 10^4$. This gives a switchover point of $s \approx 2$, i.e., the T $\rightarrow$ R transition takes place after two CO molecules are bound to the T state, as expected for a symmetric binding curve in the case $N = 4$. The free energies of pairwise coupling derived from these parameter values are: $\Delta G_{00} = -0.5$ kcal/mol, $\Delta G_{10} = \Delta G_{01} = -2.6$ kcal/mol, $\Delta G_{11} = -0.3$ kcal/mol. As expected, the coupling between any two chains is maximum when the T $\rightarrow$ R transition is bridged in going from the singly-ligated to the triply-ligated species. The values for the pairwise coupling derived from Perrella's data are listed in Table 4.10, assuming that $c_{\alpha\beta} = c'_{\alpha\beta} = 10.5$, or $c_{\alpha\beta} = 1$ and $c'_{\alpha\beta} = 20$ as extreme cases. The free

Table 4.10. *Free energies of pairwise coupling (in kcal/mol) for CO binding to hemoglobin.*

Pair	00	10	01	11
$c_{\alpha\beta} = c'_{\alpha\beta} = 10.5$				
$\alpha_1\alpha_2$	-2.0 ± 0.2	-1.2 ± 0.2	-1.2 ± 0.2	-1.3 ± 0.2
$\alpha_1\beta_1$	-1.4 ± 0.2	-0.7 ± 0.2	-1.0 ± 0.2	-1.0 ± 0.2
$\alpha_1\beta_2$	-1.4 ± 0.2	-0.7 ± 0.2	-1.0 ± 0.2	-1.0 ± 0.2
$\beta_1\beta_2$	-1.3 ± 0.2	-1.0 ± 0.2	-1.0 ± 0.2	-1.3 ± 0.2
$c_{\alpha\beta} = 1, c'_{\alpha\beta} = 20$				
$\alpha_1\alpha_2$	-2.0 ± 0.2	-2.2 ± 0.2	-2.2 ± 0.2	-1.3 ± 0.2
$\alpha_1\beta_1$	-1.7 ± 0.2	-2.0 ± 0.2	-2.3 ± 0.2	-1.4 ± 0.2
$\alpha_1\beta_2$	0.0 ± 0.2	-0.3 ± 0.2	-0.6 ± 0.2	0.3 ± 0.2
$\beta_1\beta_2$	-1.3 ± 0.2	-1.9 ± 0.2	-1.9 ± 0.2	-1.3 ± 0.2

energies of pairwise coupling are practically all negative, indicating positive coupling between pairs at all stages of ligation. However, in no case is the pattern predicted by the MWC model observed. In the case $c_{\alpha\beta} = c'_{\alpha\beta} = 10.5$, the pattern is almost independent of the ligation of the complementary pair and suggests the presence of direct interactions of the nearest-neighbor type. In the case $c_{\alpha\beta} = 1$ and $c'_{\alpha\beta} = 20$, coupling tends to increase slightly at intermediate saturation and remains strong at low and high saturation for all pairs but $\alpha_1\beta_2$, contrary to the predictions of the MWC model. Although a conclusive answer as to the precise code for cooperative CO binding to hemoglobin must await the resolution of $c_{\alpha\beta}$ and $c'_{\alpha\beta}$ separately, it is clear from the data in Table 4.10 that the MWC model cannot account, even qualitatively, for the precise mechanism of hemoglobin cooperativity. There is no evidence from Perrella's data of the existence of a concerted transition between T and R. The transition probably takes place gradually with saturation, with the chains communicating directly through nearest-neighbor interactions, as predicted by the KNF model.

The conclusion drawn for CO binding provides important insights into the mechanism of hemoglobin cooperativity and is anticipated by the results obtained by Ackers on the distribution of ligation intermediates using model systems (Ackers *et al.*, 1992). Construction of these systems has been made possible by the fundamental discovery due to Bucci and Fronticelli (1965) on the separation of the hemoglobin chains. In these systems, the ligated or unligated states of the heme FeII are mimicked by *ad hoc* substitutions. A remarkable finding on these systems is that the intermediates of ligation distribute themselves according to a 'symmetry rule' for cooperative switching. The rule is cast within an MWC-like framework and states that any intermediate can exist in either one of two quaternary states, T (low affinity) or R (high affinity). Switching from T to R is combinatorial and intermediates with ligated chains across the inter-dimeric $\alpha\beta$ interface are in the R state, while those with ligated chains on the same side of the interdimeric $\alpha\beta$ interface are in the T state (Daugherty *et al.*, 1991; Ackers *et al.*, 1992; Doyle and Ackers, 1992; Ackers and Hazzard, 1993; LiCata, Dalessio and Ackers, 1993). The elegance of this rule stems from its simplicity. Since ligation must eventually involve chains across the interdimeric $\alpha\beta$ interface, the T → R transition becomes an 'inevitable' consequence of the rule and so is cooperativity. The ensemble of intermediates partitioned according to the symmetry rule is shown in Figure 4.13. Resolution of the free energy levels relative to all intermediates allows for the calculation of all the free energies of pairwise coupling. The code for cooperativity is then 'broken' by inspection of the pairwise coupling pattern, independent of any mechanism. The results are given in Table 4.11 for three different model systems.

In the FeII/FeIII–CN system, the bound state of the heme FeII is mimicked by a cyanide bound to the oxidized FeIII (Smith and Ackers, 1985). In the FeII/MnIII system, MnIII mimics the ligated heme FeII (Smith, Gingrich, Hoffman and Ackers, 1987). In the system CoII/FeII–CO, the unligated FeII is mimicked by CoII (Spiros *et al.*, 1991). Despite

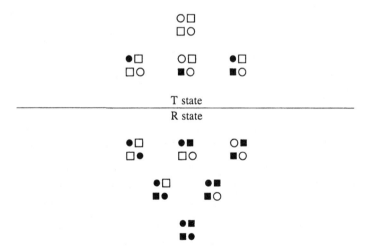

Figure 4.13 Partitioning of the ten ligated intermediates of hemoglobin according to the symmetry rule (Ackers *et al.*, 1992). Ligation within the $\alpha_1\beta_1$ or $\alpha_2\beta_2$ dimers does not affect the quaternary state. On the other hand, ligation across the $\alpha\beta$ interdimeric interface induces the T → R transition, regardless of the ligation state. Symbols depict the various chains as in Figure 4.11 (top).

Table 4.11. *Free energies of pairwise coupling (in kcal/mol) for hemoglobin model systems.*

Pair	00	10	01	11
FeII/FeIII–CN (Huang and Ackers, 1995)				
$\alpha_1\alpha_2$	-0.3 ± 0.2	0.0 ± 0.2	0.0 ± 0.2	-0.2 ± 0.2
$\alpha_1\beta_1$	-3.2 ± 0.2	-2.9 ± 0.2	-3.0 ± 0.2	-3.2 ± 0.2
$\alpha_1\beta_2$	-0.1 ± 0.2	0.2 ± 0.2	0.1 ± 0.2	-0.1 ± 0.2
$\beta_1\beta_2$	-0.2 ± 0.2	0.0 ± 0.2	0.0 ± 0.2	-0.3 ± 0.2
FeII/MnIII (Ackers, 1990)				
$\alpha_1\alpha_2$	1.0 ± 0.5	0.2 ± 0.6	0.2 ± 0.6	0.1 ± 0.6
$\alpha_1\beta_1$	-3.2 ± 0.5	-4.0 ± 0.6	-2.6 ± 0.6	-2.7 ± 0.6
$\alpha_1\beta_2$	0.0 ± 0.5	-0.8 ± 0.6	0.6 ± 0.6	0.5 ± 0.6
$\beta_1\beta_2$	-1.2 ± 0.5	-0.6 ± 0.6	-0.6 ± 0.6	0.7 ± 0.6
CoII/FeII–CO (Spiros *et al.*, 1991)				
$\alpha_1\alpha_2$	0.1 ± 0.4	0.0 ± 0.6	0.0 ± 0.6	-0.2 ± 0.4
$\alpha_1\beta_1$	-1.4 ± 0.5	-1.5 ± 0.5	-1.2 ± 0.6	-1.4 ± 0.4
$\alpha_1\beta_2$	-0.5 ± 0.6	-0.6 ± 0.4	-0.3 ± 0.5	-0.5 ± 0.5
$\beta_1\beta_2$	-0.8 ± 0.6	-0.6 ± 0.5	-0.6 ± 0.5	-0.5 ± 0.4

the substantial differences in chemical perturbation of the heme sites and the ligands mimicking the bound state of FeII, the free energies of pairwise coupling show, with few exceptions, some definite and general trends. Unlike the case of CO binding, the $\alpha_1\alpha_2$ pair does not show appreciable coupling and therefore the α chains behave as independent in these model systems. The $\beta_1\beta_2$ pair shows small but significant cooperativity and the same is seen for the $\alpha_1\beta_2$ pair. The $\alpha_1\beta_1$ pair is the most cooperative one in all systems. A striking observation is that in both the FeII/FeIII–CN and CoII/FeII–CO systems the free energies of pairwise coupling seem to be independent of the ligation state of the complementary pair. In the FeII/MnIII system some differences are observed, most notably in the $\alpha_1\alpha_2$ pair that shows negative coupling at low saturation. In all cases, the pattern of pairwise coupling rules out the simplistic assumptions of the MWC model and echoes the results obtained in the case of CO binding. Direct nearest-neighbor interactions of the KNF-type dominate the scenario of site–site communication in these model systems also.

Daugherty, Shea and Ackers (1994) have resolved all free energy levels for the FeII/FeIII–CN systems as a function of pH. The free energies of pairwise coupling derived from their results are listed in Table 4.12 and sketched in Figure 4.14. Pairwise coupling is independent of ligation over the entire pH range 7.0–8.0. At pH > 8.0 pairwise coupling is strong at low saturation and tends to disappear at intermediate and high saturation.

Table 4.12. *Free energies of pairwise coupling (in kcal/mol) for the FeII/ FeIII–CN hemoglobin system as a function of pH (Daugherty et al., 1994).*

Pair	00	10	01	11	
$\alpha_1\alpha_2$	-0.3 ± 0.2	0.1 ± 0.3	0.1 ± 0.3	-0.5 ± 0.2	pH 7.0
$\alpha_1\beta_1$	-4.0 ± 0.3	-3.6 ± 0.3	-3.6 ± 0.3	-4.2 ± 0.2	
$\alpha_1\beta_2$	-0.2 ± 0.3	0.2 ± 0.2	0.2 ± 0.3	-0.4 ± 0.3	
$\beta_1\beta_2$	-0.2 ± 0.3	0.2 ± 0.3	0.2 ± 0.3	-0.4 ± 0.2	
$\alpha_1\alpha_2$	-0.3 ± 0.2	0.0 ± 0.2	0.0 ± 0.2	-0.2 ± 0.2	pH 7.4
$\alpha_1\beta_1$	-3.2 ± 0.2	-2.9 ± 0.2	-3.0 ± 0.2	-3.2 ± 0.2	
$\alpha_1\beta_2$	-0.1 ± 0.2	0.2 ± 0.2	0.1 ± 0.2	-0.1 ± 0.2	
$\beta_1\beta_2$	-0.2 ± 0.2	0.0 ± 0.2	0.0 ± 0.2	-0.3 ± 0.2	
$\alpha_1\alpha_2$	-0.9 ± 0.4	-0.4 ± 0.4	-0.4 ± 0.4	-0.2 ± 0.3	pH 8.0
$\alpha_1\beta_1$	-2.7 ± 0.3	-2.2 ± 0.4	-2.3 ± 0.3	-2.1 ± 0.4	
$\alpha_1\beta_2$	-0.7 ± 0.3	-0.2 ± 0.4	-0.3 ± 0.3	-0.1 ± 0.4	
$\beta_1\beta_2$	-0.7 ± 0.2	-0.3 ± 0.4	-0.3 ± 0.4	-0.2 ± 0.4	
$\alpha_1\alpha_2$	-1.1 ± 0.3	-1.6 ± 0.4	-0.6 ± 0.4	-0.1 ± 0.3	pH 8.5
$\alpha_1\beta_1$	-2.1 ± 0.4	-1.6 ± 0.4	-1.6 ± 0.4	-1.1 ± 0.4	
$\alpha_1\beta_2$	-1.0 ± 0.4	-0.5 ± 0.4	-0.5 ± 0.4	0.0 ± 0.4	
$\beta_1\beta_2$	-1.0 ± 0.3	-0.5 ± 0.4	-0.5 ± 0.4	0.0 ± 0.3	
$\alpha_1\alpha_2$	-1.8 ± 0.4	-0.5 ± 0.5	-0.5 ± 0.5	-0.3 ± 0.3	pH 8.8
$\alpha_1\beta_1$	-2.4 ± 0.4	-1.1 ± 0.5	-1.1 ± 0.4	-0.9 ± 0.4	
$\alpha_1\beta_2$	-1.6 ± 0.4	-0.3 ± 0.5	-0.3 ± 0.4	-0.1 ± 0.4	
$\beta_1\beta_2$	-1.6 ± 0.3	-0.3 ± 0.5	-0.3 ± 0.5	-0.1 ± 0.4	
$\alpha_1\alpha_2$	-2.1 ± 0.4	-0.4 ± 0.5	-0.4 ± 0.5	-0.1 ± 0.3	pH 9.5
$\alpha_1\beta_1$	-2.5 ± 0.5	-0.8 ± 0.5	-1.1 ± 0.4	-0.8 ± 0.5	
$\alpha_1\beta_2$	-1.8 ± 0.4	-0.1 ± 0.5	-0.4 ± 0.4	-0.1 ± 0.4	
$\beta_1\beta_2$	-1.6 ± 0.3	-0.2 ± 0.5	-0.2 ± 0.5	-0.2 ± 0.4	

This would suggest that a concerted transition takes place after the first ligation step, thereby supporting an MWC-like mechanism of cooperativity at high pH. However, the behavior of the $\alpha_1\beta_1$ pair provides a notable exception. Coupling within this pair remains strong and positive even at high pH. In the complex hierarchy of site–site communications of the hemoglobin tetramer in this model system, the $\alpha_1\beta_1$ and $\alpha_2\beta_2$ dimers clearly stand out for the strength of cooperative coupling. The other pairs show a small coupling in the pH range 7.0–8.0, which, however, tends to increase significantly with pH. Also, the linkage between pairwise coupling and proton binding is eminently different for the $\alpha_1\beta_1$ and $\alpha_2\beta_2$ dimers. The slope in the plots in Figure 4.14, divided by $2.303RT$, gives the net number of protons exchanged in the thermodynamic cycle relative to the pairwise coupling free energy. Protons are taken up as a result of

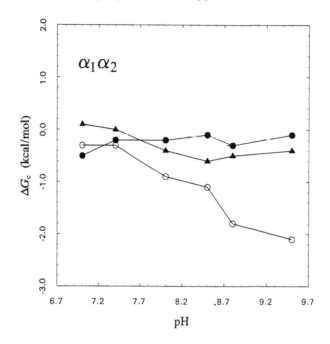

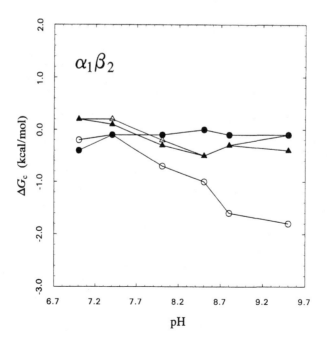

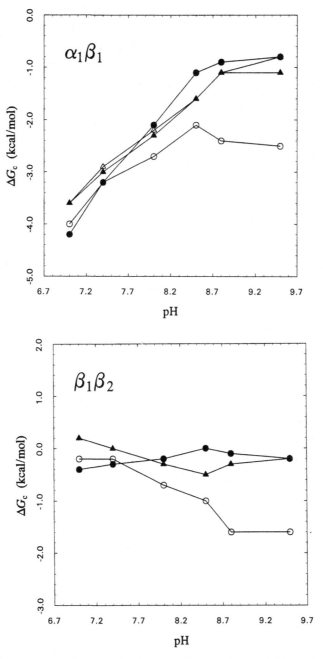

Figure 4.14   Free energies of pairwise coupling for hemoglobin pairs of chains, as indicated, as a function of pH. The values are listed in Table 4.12 and are computed from the free energy levels for the FeII/FeIII–CN model system (Daugherty *et al.*, 1994). Symbols refer to different ligated configurations of the complementary pair as follows (see also Table 4.12): (open circles) 00; (open triangles) 10; (full triangles) 01; (full circles) 11.

the interaction between α_1 and β_1, but are released in the interaction between α_1 and α_2, α_1 and β_2, or β_1 and β_2. This result also deserves much consideration in view of the fact that the overall linkage with protons reconstructed from the data on the intermediates for the FeII/FeIII–CN system (Daugherty *et al.*, 1994) is practically identical to the Bohr effect of the native hemoglobin (Chu *et al.*, 1982). The Bohr effect is the linkage between O_2 binding and proton release that allows hemoglobin to accomplish its important role in the blood (Bunn and Forget, 1986). The structural origin of the Bohr effect was intensely debated until the elegant work of Ho (1992) proved that the effect arises from the contribution of many ionizable groups. The findings by Daugherty *et al.* (1994) suggest the existence of two classes of Bohr groups: one linked to the coupling *within* dimers and the other linked to the coupling *between* dimers. The foregoing analysis based on site-specific thermodynamics suggests that the FeII/FeIII–CN system and its intermediates should be used in NMR studies to identify the Bohr groups directly.

The importance of a site-specific analysis of hemoglobin stems also from the possibility of defining a mechanism for cooperativity from model-independent considerations. Here we show how this can be done in practice. From the definitions given in Table 4.7 and the experimental results given in Tables 4.10–4.12, we assume that the free energies of coupling are independent of the ligation state of the complementary pair. This implies that the third-order and fourth-order interaction constants can be expressed in terms of the second-order interaction constants as follows

$$c_{\alpha\alpha\beta} = c_{\alpha\alpha}c_{\alpha\beta}c'_{\alpha\beta} \tag{4.110a}$$

$$c_{\alpha\beta\beta} = c_{\beta\beta}c_{\alpha\beta}c'_{\alpha\beta} \tag{4.110b}$$

$$c_{\alpha\alpha\beta\beta} = c_{\alpha\alpha}c_{\beta\beta}c^2_{\alpha\beta}c'^2_{\alpha\beta} \tag{4.110c}$$

The partition function for our model takes the form

$$\Psi(x) = 1 + 2(K_\alpha + K_\beta)x + [c_{\alpha\alpha}K^2_\alpha + c_{\beta\beta}K^2_\beta + 2(c_{\alpha\beta} + c'_{\alpha\beta})K_\alpha K_\beta]x^2$$
$$+ 2c_{\alpha\beta}c'_{\alpha\beta}(c_{\alpha\alpha}K_\alpha + c_{\beta\beta}K_\beta)K_\alpha K_\beta x^3 + c_{\alpha\alpha}c_{\beta\beta}c^2_{\alpha\beta}c'^2_{\alpha\beta}K^2_\alpha K^2_\beta x^4 \tag{4.111}$$

The overall equilibrium constants written in terms of the model parameters are

$$A_1 = 2(K_\alpha + K_\beta) \tag{4.112a}$$

$$A_2 = c_{\alpha\alpha}K^2_\alpha + c_{\beta\beta}K^2_\beta + 2(c_{\alpha\beta} + c'_{\alpha\beta})K_\alpha K_\beta \tag{4.112b}$$

$$A_3 = 2c_{\alpha\beta}c'_{\alpha\beta}(c_{\alpha\alpha}K_\alpha + c_{\beta\beta}K_\beta)K_\alpha K_\beta \tag{4.112c}$$

$$A_4 = c_{\alpha\alpha}c_{\beta\beta}c^2_{\alpha\beta}c'^2_{\alpha\beta}K^2_\alpha K^2_\beta \tag{4.112d}$$

The site-specific binding curves of the two chains are

$$X_\alpha = K_\alpha x \lambda_\alpha / \Psi(x) \tag{4.113a}$$

where $\lambda_\alpha = 1 + [c_{\alpha\alpha}K_\alpha + (c_{\alpha\beta} + c'_{\alpha\beta})K_\beta]x + c_{\alpha\beta}c'_{\alpha\beta}(2c_{\alpha\alpha}K_\alpha + c_{\beta\beta}K_\beta)K_\beta x^2 + c_{\alpha\alpha}c_{\beta\beta}c^2_{\alpha\beta}c'^2_{\alpha\beta}K_\alpha K^2_\beta x^3$.

$$X_\beta = K_\beta x \lambda_\beta / \Psi(x) \tag{4.113b}$$

where $\lambda_v = 1 + [c_{\beta\beta}K_\beta + (c_{\alpha\beta} + c'_{\alpha\beta})K_\alpha]x + c_{\alpha\beta}c'_{\alpha\beta}(c_{\alpha\alpha}K_\alpha + 2c_{\beta\beta}K_\beta)K_\alpha x^2 + c_{\alpha\alpha}c_{\beta\beta}c^2_{\alpha\beta}c'^2_{\alpha\beta}K^2_\alpha K_\beta x^3$.

This model easily accounts for the results obtained in the FeII/FeIII–CN system in the pH range 7.0–8.0. Also, it accounts for the high cooperativity in the global CO binding curve reconstructed from Perrella's results shown in Figure 4.10(*a*) and almost completely explains the distribution of intermediates for CO binding. Extension of the model can readily account for the results obtained with model systems at pH > 8.0. Pairwise coupling patterns can be used to formulate a revised two-state model where both the T and R states contain second-order interactions between chain pairs. Specifically,

$$\begin{aligned}
\Psi_T(x) = {}& 1 + 2(K_{\alpha T} + K_{\beta T})x \\
& + [c_{\alpha\alpha T}K^2_{\alpha T} + c_{\beta\beta T}K^2_{\beta T} + 2(c_{\alpha\beta T} + c'_{\alpha\beta T})K_{\alpha T}K_{\beta T}]x^2 \\
& + 2c_{\alpha\beta T}c'_{\alpha\beta T}(c_{\alpha\alpha T}K_{\alpha T} + c_{\beta\beta T}K_{\beta T})K_{\alpha T}K_{\beta T}x^3 \\
& + c_{\alpha\alpha T}c_{\beta\beta T}c^2_{\alpha\beta T}c'^2_{\alpha\beta T}K^2_{\alpha T}K^2_{\beta T}x^4
\end{aligned} \tag{4.114a}$$

$$\begin{aligned}
\Psi_R(x) = {}& 1 + 2(K_{\alpha R} + K_{\beta R})x \\
& + [c_{\alpha\alpha R}K^2_{\alpha R} + c_{\beta\beta R}K^2_{\beta R} + 2(c_{\alpha\beta R} + c'_{\alpha\beta R})K_{\alpha R}K_{\beta R}]x^2 \\
& + 2c_{\alpha\beta R}c'_{\alpha\beta R}(c_{\alpha\alpha R}K_{\alpha R} + c_{\beta\beta R}K_{\beta R})K_{\alpha R}K_{\beta R}x^3 \\
& + c_{\alpha\alpha R}c_{\beta\beta R}c^2_{\alpha\beta R}c'^2_{\alpha\beta R}K^2_{\alpha R}K^2_{\beta R}x^4
\end{aligned} \tag{4.114b}$$

$$\Psi(x) = \frac{L\Psi_T(x) + \Psi_R(x)}{L + 1} \tag{4.114c}$$

and all parameters are self-explanatory. The partition function for the symmetry rule (Ackers *et al.*, 1992) is embodied by eqs (4.114a)–(4.114c) as a special case, and so are several other models of hemoglobin cooperativity. Specific assumptions on the strength of pairwise interactions within either quaternary state can be made to simplify eq (4.114c) for practical applications.

Models of cooperativity can be developed only after detailed information is obtained on the properties of individual sites and intermediates of ligation. The information contained in the global picture, say the O_2 or CO binding curve of hemoglobin, is minimal and many different interpretations yield equally acceptable results. In the local description, on the other hand, the ambiguity is resolved and only a few models survive the test of experimental data. It is noteworthy that the simple model based solely on pairwise coupling constants, given in eq (4.111), explains most of the results on the intermediates of ligation in model systems at pH <8.0 and the CO intermediates. In these systems, cooperativity is not dominated by a concerted transition between two states, as implied by the central idea of the MWC model (Monod *et al.*, 1965). Rather, the heme sites communicate directly in a pairwise fashion. Ligation at one site affects all other sites in contact with it through direct interactions. These ideas were first put forward by Pauling in 1935 (Pauling, 1935) and redefined in a more systematic way by Koshland *et al.* (1966). The existence of two distinct structures for the T and R states of hemoglobin (Perutz, 1970; Baldwin and Chothia, 1979; Shaanan, 1983; Fermi, Perutz and Shaanan, 1984) has widely been accepted, almost dogmatically, as the most direct demonstration of the validity of the MWC model for hemoglobin cooperativity. Detailed stereochemical and statistical thermodynamic mechanisms have been derived from analysis of these limiting intermediates of ligation (Perutz, 1970; Szabo and Karplus, 1972). More recently, new structures of fully-ligated hemoglobin have been resolved under low salt conditions (Smith, Lattman and Carter, 1991; Silva, Roger and Arnone, 1992; Smith and Simmons, 1994). The R2 structure that emerged from these studies is quite different from the R structure and is not intermediate on the pathway of the $T \rightarrow R$ transition (Janin and Wodak, 1993). Srinivasan and Rose (1994) have carried out a very thorough analysis of the T, R and R2 structures and demonstrated unequivocally that R is an intermediate between T and R2. The R structure is nothing but a 'handicapped' R2 structure, an artifact of crystallization conditions (high salt) that freeze the molecule in an intermediate state that does not exist under physiological salt conditions and prevents hemoglobin assuming its 'natural' R2 form. The interesting analysis by Srinivasan and Rose indicates that the detailed mechanisms of hemoglobin cooperativity based on the T and R structures may be flawed. Furthermore, if T and R2 are the two limiting states, the transition between them can take place through several intermediates that can be populated by direct site–site interactions, as predicted by the KNF model.

The lesson to be learned from hemoglobin is far more general. Mechanisms of protein function inferred from crystal structures alone may be very misleading. It is not possible to understand mechanisms of cooperative transitions without gathering information on ligation intermediates under conditions that are relevant to physiological function (i.e., low salt). The landmark work of Perrella (Perrella and Rossi-Bernardi, 1981; Perrella *et al.*, 1983) and Ackers (Smith and Ackers, 1985) should be properly credited in this regard. Like the beautiful structural work of Perutz (1970), these functional studies have contributed enormously to our understanding of cooperative interactions in biology and will serve as a paradigm for any future studies on allosteric systems.

We conclude our discussion of hemoglobin with the problem of the symmetry of the binding curve. It has been known for a long time that the O_2 binding curve of human hemoglobin under physiological conditions is strongly asymmetric, since it becomes steeper at high saturation (Roughton *et al.*, 1955). Various attempts to relate this feature to a detailed mechanism of interaction among the chains have appeared in the literature (Peller, 1982; Weber, 1982). These approaches based on specific mechanisms are of limited validity. Site-specific thermodynamics has the advantage of being model-independent and the conclusions derived from this approach are valid in general. As shown in Section 2.3, symmetry of the binding curve X for hemoglobin demands $A_4 = (A_3/A_1)^2$. This condition is independent of A_2 and therefore symmetry, or lack of it, cannot depend on the interaction constants $c_{\alpha\alpha}$, $c_{\beta\beta}$, $c_{\alpha\beta}$ or $c'_{\alpha\beta}$. Symmetry requires (Di Cera *et al.*, 1992)

$$c_{\alpha\alpha\beta\beta}(K_\alpha + K_\beta)^2 = (c_{\alpha\alpha\beta}K_\alpha + c_{\alpha\beta\beta}K_\beta)^2 \qquad (4.115)$$

and depends on the association constants of the two chains and the third- and fourth-order coupling constants. If $K_\alpha = K_\beta$, asymmetry is observed whenever $4c_{\alpha\alpha\beta\beta} \neq (c_{\alpha\alpha\beta} + c_{\alpha\beta\beta})^2$, which can be satisfied even if $c_{\alpha\alpha\beta} = c_{\alpha\beta\beta}$, provided $c_{\alpha\alpha\beta\beta} \neq c^2_{\alpha\alpha\beta}$. Therefore, asymmetry of the O_2 binding curve under physiological conditions demands neither subunit heterogeneity nor asymmetric interactions, contrary to the conclusions derived by Weber (1982) and Peller (1982). It is interesting to note the effect of symmetry on the site-specific binding curves X_α and X_β of the two chains. If these curves are symmetric, then one necessarily finds from eq (3.125) that the two centers of symmetry are

$$x_\alpha = \frac{1}{K_\alpha}\sqrt{\frac{c_{\alpha\beta\beta}}{c_{\alpha\alpha\beta\beta}}} \qquad (4.116a)$$

$$x_\beta = \frac{1}{K_\beta}\sqrt{\frac{c_{\alpha\alpha\beta}}{c_{\alpha\alpha\beta\beta}}} \qquad (4.116b)$$

These centers are identical to the mean ligand activities of the chains and therefore

$$\sqrt{x_\alpha x_\beta} = \frac{1}{\sqrt[4]{c_{\alpha\alpha\beta\beta} K_\alpha^2 K_\beta^2}} = x_m \qquad (4.117)$$

where x_m is the mean ligand activity of the system as a whole. Hence, it follows from eqs (4.116a), (4.116b) and (4.117) that $c_{\alpha\alpha\beta\beta} = c_{\alpha\alpha}c_{\beta\beta}$. Substitution in eq (4.115) shows that asymmetry necessarily demands $c_{\alpha\alpha\beta} \neq c_{\alpha\beta\beta}$, independent of the value of K_α and K_β. In the case of interest for hemoglobin, the left-hand side of eq (4.115) exceeds the right-hand side, which implies either $(K_\alpha/K_\beta)^2 > c_{\alpha\beta\beta}/c_{\alpha\alpha\beta} > 1$, or $(K_\alpha/K_\beta)^2 < c_{\alpha\beta\beta}/c_{\alpha\alpha\beta} < 1$. Since $K_\alpha/K_\beta \approx 4$ under physiological conditions (Viggiano and Ho, 1979), then $c_{\alpha\beta\beta}/c_{\alpha\alpha\beta} < 16$ and also $x_\beta > x_\alpha$. The binding curve X_β is shifted to the right with respect to X_α. If X_α and X_β are not symmetric, there is no need to invoke asymmetric interactions to obtain a resulting global binding curve that is itself asymmetric. Knowledge of the shape of chain-specific oxygen binding curves under physiological conditions is therefore necessary to establish unequivocally whether the asymmetric oxygen binding curve of human hemoglobin is due to asymmetric interactions, or to other factors not accounted for by current mechanistic models.

4.5 Site-specific binding curves from global properties of structurally perturbed systems

If only global properties of a macromolecular system can be accessed experimentally, is it possible to derive relevant information on site-specific properties? This is a question of considerable importance in practical applications. In this section we show a possible approach to the problem, which combines the powerful tools of molecular biology with the principles of site-specific thermodynamics illustrated so far. This approach has the potential to broaden our understanding of cooperative phenomena in biology considerably by revealing the site-specific components of global effects from analysis of *ad hoc* perturbed systems. The practical advantage of this approach is obvious, since global properties of macromolecular systems are far easier to obtain experimentally than their site-specific components.

We start with the basic relationship (Di Cera, 1990, 1994b)

$$X_j = K_j x \frac{^1\Psi_j(x)}{\Psi(x)} = 1 - \frac{^0\Psi_j(x)}{\Psi(x)} \qquad (4.118)$$

The physical significance of eq (4.118) has been discussed at length. Here we are concerned with the operational significance of this equation. Note that the site-specific binding curve, X_j, can be constructed from two global quantities: the partition function of the system as a whole, Ψ, and its contracted analogues $^0\Psi_j$, obtained when site j is kept unligated, or $^1\Psi_j$, obtained when site j is kept ligated. This holds true in general, regardless of the number of sites, or the cooperativity pattern. Hence, if a chemical perturbation of site j is applied in such a way as essentially to abolish binding to that site or to mimick its bound state, then information on the binding properties of site j can be derived. The perturbation acts as the source of information.

Critical to the practical applicability of this approach is the assumption that the perturbation itself does not alter the network of interaction among the sites, nor does it change the binding affinities of the remaining sites. The assumption is tenable for simple systems. Indeed this strategy, cast in a different conceptual framework, was first introduced in 1895 by the Austrian chemist Wegscheider in the study of ionization reactions of polybasic acids (Wegscheider, 1895). Wegscheider's idea was to replace the ionizable carboxylate with a nonionizable group, an ethyl ester, that could mimick the protonated form of the 'wild-type' COO^- group. By blocking carboxyl groups, ionization reactions at remaining groups could be studied in 'reduced' systems. Such 'reduced' systems are merely contracted forms of the system as a whole and are therefore amenable to description in terms of contracted partition functions. Neuberger (1936), among many others (Edsall and Blanchard, 1933), exploited Wegscheider's idea to resolve completely the three ionization reactions of glutamic acid into their site-specific components. The partition function for such a system is

$$\Psi(x) = 1 + (K_1 + K_2 + K_3)x + (c_{12}K_1K_2 + c_{13}K_1K_3 + c_{23}K_2K_3)x^2$$
$$+ c_{123}K_1K_2K_3x^3 \tag{4.119}$$

where the suffixes refer to the α-COO^- (1), NH_2 (2) and γ-COO^- (3) groups, as indicated in Figure 4.15. The variable x denotes the proton activity. Neuberger synthesized esters of either the α- or γ-carboxyl

| Site 1 | Site 2 | | Site 3 |

$$COO^- - CHNH_2 - CH_2 - CH_2 - COO^-$$

Figure 4.15 Schematic representation of glutamic acid in its fully deprotonated form. The three ionizable groups, or proton binding sites, are indicated.

Table 4.13. pK_as *of glutamic acid and its ethyl esters (Neuberger, 1936).*

	^1pK_a	^2pK_a	^3pK_a
Glutamic acid	2.155	4.324	9.960
α-Ethyl glutamate	3.846	7.838	
γ-Ethyl glutamate	2.148	9.19	
Ethyl glutamate	7.035		

groups and titrated the other groups. The results of his experiments are summarized in Table 4.13. The assumption embodied by the Wegscheider principle is that the ethyl ester effectively mimics the protonated state of the carboxyl group. Hence, the data in Table 4.13 are *de facto* measurements of the equilibrium constants for the system as a whole and its contracted forms. The partition function for α-ethyl glutamate is derived from eq (4.119) and Section 4.2 as follows

$$^1\Psi_1(x) = 1 + (c_{12}K_2 + c_{13}K_3)x + c_{123}K_2K_3x^2 \qquad (4.120)$$

In this derivative, proton binding to the amino and γ-carboxyl groups is measured while the α-carboxyl group is protonated. Likewise, the partition function for γ-ethyl glutamate is

$$^1\Psi_3(x) = 1 + (c_{13}K_1 + c_{23}K_2)x + c_{123}K_1K_2x^2 \qquad (4.121)$$

Equations (4.119)–(4.121) are sufficient to resolve the site-specific parameters for glutamic acid. In the case of the 'wild-type' molecule we have (see Table 4.13)

$$10^{9.960} = K_1 + K_2 + K_3 \qquad (4.122a)$$

$$10^{4.324+9.960} = c_{12}K_1K_2 + c_{13}K_1K_3 + c_{23}K_2K_3 \qquad (4.122b)$$

$$10^{2.155+4.324+9.960} = c_{123}K_1K_2K_3 \qquad (4.122c)$$

In the case of α-ethyl glutamate we have

$$10^{7.838} = c_{12}K_2 + c_{13}K_3 \qquad (4.123a)$$

$$10^{3.846+7.838} = c_{123}K_2K_3 \qquad (4.123b)$$

and for γ-ethyl glutamate we have

$$10^{9.19} = c_{13}K_1 + c_{23}K_2 \qquad (4.124a)$$

$$10^{2.148+9.19} = c_{123}K_1K_2 \qquad (4.124b)$$

The resulting values of the site-specific parameters for the ionization reactions of glutamic acid are given in Table 4.14, along with their pK estimates. The results are identical to those obtained by Hill using a

Table 4.14. *Site-specific parameters for proton binding to glutamic acid.*

	K	pK
Site-specific binding constants (M^{-1})		
K_1	5.69×10^4	4.755
K_2	9.12×10^9	9.960
K_3	1.26×10^5	5.101
Interaction constants		
c_{12}	7.55×10^{-3}	-2.122
c_{13}	3.53×10^{-1}	-0.452
c_{23}	1.70×10^{-1}	-0.770
c_{123}	4.20×10^{-4}	-3.377

different approach (Hill, 1944). Note that the pK_a of the amino group in ethyl glutamate can be predicted from the data in Table 4.14 to be 7.035, as found experimentally. In fact, titration of the amino group is expected to obey the second-order contracted partition function

$$^{11}\Psi_{13}(x) = 1 + \frac{c_{123}}{c_{13}} K_2 x \qquad (4.125)$$

with a pK_a given by $pK_a = \log_{10}(c_{123}c_{13}^{-1}K_2)$.

The site-specific analysis of glutamic acid yields useful information on the energetics of protonation from which structural information can be inferred. This point illustrates the importance of resolving site-specific energetics. When all groups are deprotonated, binding of the proton occurs with high affinity to the amino group and with low affinity to the carboxyl groups. However, the α-carboxyl group has an affinity lower than the γ-carboxyl group, due to the proximity of the amino group. The interaction constants are all less than one, indicating the presence of negative cooperativity in the protonation reactions. This effect acts in concert with the extreme heterogeneity of the sites to produce a strongly negatively cooperative proton binding curve for glutamic acid, as shown in Figure 4.16. Negative coupling among the sites is expected from electro-static considerations. Again, the role of the amino group in differentially affecting the protonation properties of the two carboxyl groups should be emphasized. The two groups are located far enough away that $c_{13} \approx 1$ when the amino group is deprotonated. Protonation of this group has an effect almost three orders of magnitude larger on the α- than the γ-car-boxyl group. Interestingly, the third-order interaction constant is practic-ally the same as the product of all three second-order constants. This

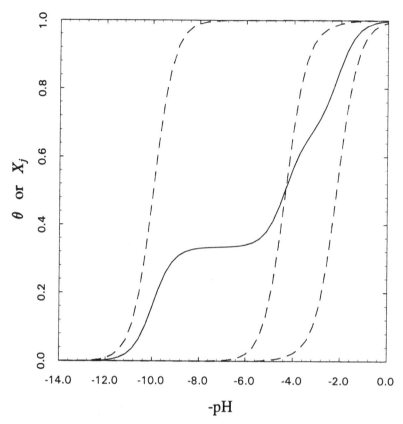

Figure 4.16 Proton binding curve of glutamic acid (continuous line) shown as fractional saturation, along with the site-specific proton binding curves (dashed lines) of the α-carboxyl (right), α-NH$_2$ (left) and γ-carboxyl (middle) groups. Curves were drawn using the parameter values listed in Table 4.14.

indicates that pairwise coupling in glutamic acid is direct, a result that echoes the hemoglobin scenario dealt with in the previous section.

Can the Wegscheider principle be used in the case of complex macromolecular systems? Consider for instance the long standing problem of calculating the pK_a of ionizable groups in proteins (Linderstrøm-Lang, 1924; Tanford and Kirkwood, 1957), where the site-specific proton binding curve of each ionizable group must be determined experimentally. For a protein containing twenty such groups, there are a total of $2^{20} \approx 10^6$ configurations to be taken into account in the partition function Ψ and no experimental technique can provide such information. However, many residues like Asp, Glu, Arg, Lys and Tyr do not ionize in the pH range of physiological interest and can be treated as fully protonated or deprotonated. The remaining groups again form a 'reduced' system in the Wegscheider sense and the contracted partition function associated with them contains a drastically reduced number of configurations. This idea has been implemented in the analysis of site-specific titration curves and is known as the 'reduced-site model' (Bashford and Karplus, 1991). Given these assumptions, we can determine the site-specific titration curve of a His residue in a protein in the pH range of physiological interest as follows. First, the titration curve of the protein, X, is determined in the pH range of interest to obtain Ψ. Here Ψ denotes the partition function of all residues that ionize in the pH range under study. Then, a site-directed mutant of the His residue is made and the titration curve of the mutant is measured over the same pH range. The substitutions His $\rightarrow$ Lys and His $\rightarrow$ Val may be used to mimic the protonated and unprotonated states of the His residue. In the former case the titration curve of the mutant yields ^1X_j (j labels the His residue), while in the latter it yields ^0X_j. These global binding curves yield the partition functions

$$\ln {}^1\Psi_j(x) = \int_{-\infty}^{\ln x} {}^1X_j\, d\ln x' \tag{4.126a}$$

$$\ln {}^0\Psi_j(x) = \int_{-\infty}^{\ln x} {}^0X_j\, d\ln x' \tag{4.126b}$$

The partition function of the wild-type is derived from X as follows

$$\ln \Psi(x) = \int_{-\infty}^{\ln x} X\, d\ln x' \tag{4.127}$$

and application of eq (4.118) yields the site-specific proton binding curve for the His as follows

$$X_j = K_j x \exp\left[\int_{-\infty}^{\ln x}({}^1X_j - X)\,d\ln x'\right] = 1 - \exp\left[\int_{-\infty}^{\ln x}({}^0X_j - X)\,d\ln x'\right] \tag{4.128}$$

Integration of the difference titration curve gives a direct measure of X_j. Analysis of X_j reveals important properties of the system, e.g., whether the His is coupled to other residues and the nature of the coupling, positive or negative. The apparent pK_a can be resolved from the mean ligand activity of the binding curve as $2.303 pK_a = -\ln x_{mj}$. However, it may be reasonably objected that the His $\rightarrow$ Lys or His $\rightarrow$ Val substitutions may affect other properties of the macromolecule, i.e., the ionization of other groups, or the coupling of the His with other residues. In general, the assumption that a given substitution may actually mimic a particular ligation state for the His may be questioned. More importantly, how does this strategy apply to ligands such as metal ions, organic molecules, peptides and nucleic acids, for which binding sites in proteins involve large structural domains? For example, in the specific case of Ca^{2+} binding to an EF-hand (Kretsinger, 1980), what kind of substitutions are to be made to mimick the Ca^{2+}-bound state of the site, or the Ca^{2+}-free form? When binding of a ligand involves several protein residues, it becomes difficult to perturb the site as is done for proton binding reactions. In this case, the perturbation is likely to be more extensive and therefore the assumption that it mimicks a specific ligation state becomes untenable.

We now show that site-specific binding curves can be obtained from eq (4.118) in an alternative way, using *ad hoc* perturbations of the binding affinity of the site of interest. This approach has a considerable practical advantage, since it is relatively easy to perturb binding sites structurally only slightly using recombinant DNA technology. The key advantage of our approach is that the requirement for the perturbation to mimick either the free or bound form of the site is no longer necessary. Also, it uses exclusively information derived from global properties of the system, thereby eliminating the need for directly monitoring site-specific effects at individual sites.

Consider the partition function of the system as a whole written in terms of the first-order contracted partition functions involving site j

$$\Psi(x) = {}^0\Psi_j(x) + {}^1\Psi_j(x)K_jx \qquad (4.129)$$

Consider then a perturbation of site j, say a chemical modification induced by a site-directed mutation, as illustrated schematically in Figure 4.17. The perturbation can either affect the properties of the site where it applies, or it can carry over to other sites. Assume that the perturbation remains localized at site j, and therefore it affects only K_j. We speak in this case of a 'first-order perturbation' that changes the value of K_j to K'_j, while leaving the value of all other parameters unchanged. Clearly, the smaller the perturbation, the more likely that it will cause a first-order effect. In general, a first-order perturbation encompasses all cases where the change has its primary effect at the site where it applies and only a negligible secondary effect on other sites. Then, for the perturbed system we have

$$\Psi(x)' = {}^0\Psi_j(x) + {}^1\Psi_j(x)K'_jx \qquad (4.130)$$

The contracted partition functions are the same as those for the wild-type, or unperturbed system, since they do not depend on K_j. Given that both Ψ and Ψ' are accessible experimentally as global properties of the system, the function

$$f_j = 1 - \frac{\Psi(x)'}{\Psi(x)} = \left(1 - \frac{K'_j}{K_j}\right)X_j = \eta_j X_j \qquad (4.131)$$

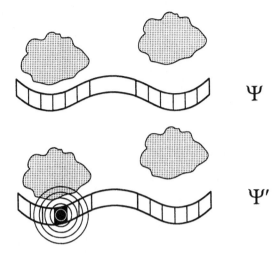

Figure 4.17 Schematic representation of a perturbation applied to a specific site (bottom) of a biological macromolecule. The unperturbed system (top) binds two molecules at sites 1 (left) and 2 (right). The mutation at site 1 can either remain localized at this site (first-order perturbation), or carry over to site 2 (second-order perturbation). If the perturbation is first-order, the site-specific properties of site 1 can be derived from knowledge of the partition functions Ψ and Ψ' according to eq (4.131).

can be constructed. Except for a constant factor η_j, this function is the same as the quantity of interest X_j in the unperturbed, wild-type system. The constant factor η_j is easily derived from f_j in the limit $x \to \infty$. In the limiting case where mutation abolishes binding to site j, eq (4.131) yields eq (4.118). In general, eq (4.131) only requires η_j to be finite. The approach based on eq (4.131) should be tested with a number of mutants at site j, in order to guarantee the uniqueness of the function X_j. Different mutants may yield different perturbations of K_j, but they should all give the same function X_j from eq (4.131). Only in this case are the perturbations at site j truly first-order.

A simple way to implement this strategy is to construct the 'double-ratio' plot in Figure 4.18. Once the partition function is resolved in the global description for the wild-type and a number of mutants, the overall equilibrium constants As can be derived. The first and last such constants are of particular importance. In fact,

$$A_1 = K_1 + K_2 + \ldots K_j \ldots + K_N \tag{4.132a}$$

$$A_1' = K_1 + K_2 + \ldots K_j' \ldots + K_N \tag{4.132b}$$

$$A_N = c_{12\ldots j\ldots N} K_1 K_2 \ldots K_j \ldots K_N \tag{4.132c}$$

$$A_N' = c_{12\ldots j\ldots N} K_1 K_2 \ldots K_j' \ldots K_N \tag{4.132d}$$

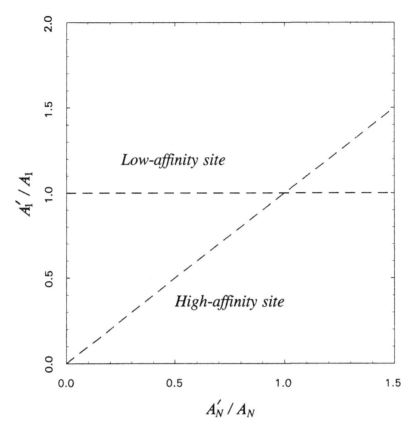

Figure 4.18 Double-ratio plot for the analysis of first-order perturbations. The values of A'_N for the mutants, relative to A_N for the wild-type, are plotted versus the ratio A'_1/A_1. Two limiting cases in this plot denote perturbation of a high-affinity (slope = 1), or low-affinity (slope = 0) site. Points outside these boundaries denote violations of the first-order perturbation hypothesis. Points following a straight line inside the boundaries support the validity of the first-order perturbation hypothesis.

where the prime refers to the mutant of site j. Hence,

$$\frac{A'_1}{A_1} = \frac{K_1 + K_2 + \ldots + K_N}{K_1 + K_2 + \ldots K_j \ldots + K_N} + \frac{K_j}{K_1 + K_2 + \ldots K_j \ldots + K_N} \frac{A'_N}{A_N}$$

$$= (1 - \rho) + \rho \frac{A'_N}{A_N} \qquad (4.133)$$

If several mutants perturbing site j are analyzed and the results plotted, a straight line should be observed in the double-ratio plot if the first-order approximation is obeyed. The relationship (4.133) deserves consideration. The intercept and slope are correlated and sum to one. The slope, ρ, is bounded from zero to one and reflects the value of K_j in the unperturbed system relative to A_1. The intercept, $1 - \rho$, is likewise bounded from zero to one and reflects the value of the sum of all site-specific binding constants, but K_j, relative to A_1. The first-order approximation imposes a constraint on the double-ratio plot. Two boundaries define the limiting behavior of the system in response to the perturbation. The boundary with zero slope defines the case in which A_N is perturbed with practically no change in A_1. This case applies when K_j is negligibly small compared to all other Ks and the site being perturbed has low affinity. The boundary with unit slope defines the case in which A_1 changes linearly with A_N. This is the case in which K_j is large compared to all other Ks and the site being perturbed has high affinity. The two boundaries cross over at the point reflecting the wild-type behavior. The necessary condition for the first-order approximation to be valid is that the experimental points lie within the foregoing boundaries, as illustrated in Figure 4.18. The condition is also sufficient if data points are within the boundaries and obey eq (4.133). In this case, the value of K_j can be derived directly from the slope in the plot. If points lie outside the boundaries, then they violate the first-order approximation. This case provides a direct and simple demonstration that the mutation has carried over to other sites. The double-ratio plot is therefore extremely informative.

Assume, then, that the perturbation at site j carries over to other sites. We speak in this case of second-order, third-order or nth-order perturbation, depending on how many sites are involved. The approach based on eq (4.131) is easily generalized. For example, in the case of a second-order perturbation the partition function of the wild-type is written in terms of second-order contracted partition functions as follows

$$\Psi(x) = {}^{00}\Psi_{ij}(x) + ({}^{10}\Psi_{ij}K_i + {}^{01}\Psi_{ij}K_j)x + {}^{11}\Psi_{ij}c_{ij}K_iK_jx^2 \quad (4.134)$$

If the mutation affects K_i and K_j, but leaves all other site-specific terms unaltered, then

$$\Psi(x)' = {}^{00}\Psi_{ij}(x) + ({}^{10}\Psi_{ij}K_i' + {}^{01}\Psi_{ij}K_j')x + {}^{11}\Psi_{ij}c_{ij}K_i'K_j'x^2 \quad (4.135)$$

The function f_{ij}, analogous to f_j in the case of a first-order perturbation, in this case yields

$$f_{ij} = 1 - \frac{\Psi(x)'}{\Psi(x)} = \eta_iX_i + \eta_jX_j - \eta_i\eta_jc_{ij}K_iK_jx^2\frac{{}^{11}\Psi_{ij}(x)}{\Psi(x)}$$

$$= \eta_iX_i + \eta_jX_j - {}^xc_{ij}\eta_i\eta_jX_iX_j \quad (4.136)$$

where

$$^xc_{ij} = \frac{{}^{11}\Psi_{ij}(x)\Psi(x)}{{}^1\Psi_i(x){}^1\Psi_j(x)} \quad (4.137)$$

is the pairwise coupling function between sites i and j, and the ηs have the same form as in eq (4.131). As expected, the function f_{ij} contains terms relative to sites i and j, along with the coupling between them. The interaction constant c_{ij} can be derived in the limit $x \to \infty$ once η_i and η_j are known from first-order perturbation experiments, where K_i and K_j are perturbed separately. Likewise, measurements of Ψ and Ψ' can be used to access $^{11}\Psi_{ij}$ using eqs (4.136) and (4.137) and to resolve all the overall binding constants of a second-order contracted partition function of the wild-type system. Similar arguments apply to higher-order perturbations, or those cases where interaction constants can also be affected.

We now show an application of the foregoing approach to the analysis of cooperative Ca^{2+} binding to calbindin, one of the smallest members of the calmodulin superfamily (Skelton *et al.*, 1994). Calbindin is composed of two helix–loop–helix motifs responsible for Ca^{2+} binding, as shown in Figure 4.19. The C-terminal site (site 2) has the amino acid sequence and fold of an archetypal EF-hand, while the N-terminal site (site 1) differs from the usual EF-hand in that it contains two additional residues (Szebenyi and Moffat, 1986). The protein binds Ca^{2+} with positive

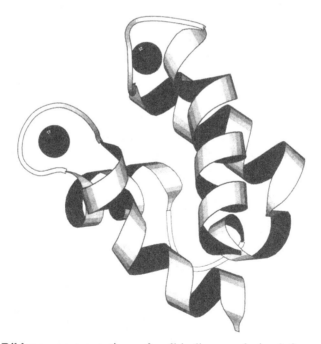

Figure 4.19 Ribbon representation of calbindin, as derived from the crystal structure (Szebenyi and Moffat, 1986). The bound Ca^{2+} ions are shown by circles. Site 1, shaped as a pseudo EF-hand, is at right. Site 2, shaped as an archetypal EF-hand, is at left.

cooperativity and a Hill coefficient of about 1.5 (Linse *et al.*, 1987, 1988, 1991). Due to the structural differences between the binding sites, it is expected that the true coupling between them is partially reduced by site heterogeneity. The arguments discussed in Section 3.1 become particularly relevant in this respect. In fact, calbindin provides a sort of 'hydrogen atom' for site-specific thermodynamics, being the simplest cooperative system with site heterogeneity.

Experimental evidence provided by NMR suggests that Cd^{2+} binds with higher affinity to site 2 (Akke, Forsén and Chazin, 1991) and that the binding pathways $[00] \rightarrow [10] \rightarrow [11]$ and $[00] \rightarrow [01] \rightarrow [11]$ induce distinct structural transitions in the molecule (Carlström and Chazin, 1993). Spectroscopic and kinetic studies on the global Ca^{2+} binding properties of calbindin suggest that site heterogeneity may be a factor of 4–6 (Linse *et al.*, 1987, 1991; Brodin *et al.*, 1990; Martin *et al.*, 1990). However, no direct evidence has been provided for Ca^{2+} binding with higher affinity to site 1 or site 2. Electrostatic calculations suggest that site 2 may have only a slightly higher affinity (Ahlström *et al.*, 1989). Valence maps indicate that site 1, rather than site 2, may bind Ca^{2+} with higher affinity (Nayal and Di Cera, 1994). Furthermore, these maps suggest that Cd^{2+} should not be used as a model for the energetics of Ca^{2+} binding, since the parameters relating bond strength to bond order are significantly different for these metal ions (Pauling, 1929, 1947; Brown and Wu, 1976). A number of mutants of residues in and around site 1, and partially site 2, have been made to assess the role of electrostatic contributions to the binding of Ca^{2+} and the global binding parameters have been resolved for all of them (Linse *et al.*, 1987, 1988, 1991; Martin *et al.*, 1990). Some of the residues mutated in the wild-type are shown in Figure 4.20. Particularly interesting is the observation that mutations around site 1 remain localized at this site and do not propagate to site 2 (Linse *et al.*, 1987; Martin *et al.*, 1990). The global binding parameters for the wild-type calbindin and a number of mutants are summarized in Table 4.15. Mutations significantly affect the binding affinity and cooperativity. We will use this set of mutants to resolve the site-specific parameters for calbindin.

We assume that the single mutants around site 1 provide a small perturbation of the affinity of site 1, while leaving the affinity of site 2 and the interaction between the sites unchanged. Our assumption of the first-order perturbation is supported by NMR data showing that the mutants E17Q, E26Q and E60Q have practically the same structure as the wild-type protein (Akke *et al.*, 1991; Carlström and Chazin, 1993). The mutant P20G and the deletion mutants P20G, ΔN21 and ΔP20 show

Table 4.15. *Global binding parameters for Ca²⁺ binding to calbindin and its mutants (Linse et al., 1987, 1988, 1991).*

	A_1 (M^{-1})	A_2 (M^{-2})	k_1 (M^{-1})	k_2 (M^{-1})	n_H
Wild-type	2.0×10^8	7.9×10^{16}	1.0×10^8	7.9×10^8	1.5
P20G	6.5×10^7	5.2×10^{14}	3.3×10^7	1.6×10^7	0.8
P20G, ΔN21	1.7×10^8	8.5×10^{13}	8.5×10^7	1.0×10^6	0.2
ΔP20	8.0×10^7	4.0×10^{13}	4.0×10^7	1.0×10^6	0.3
Y13F	3.7×10^8	9.3×10^{16}	1.9×10^8	4.9×10^8	1.2
E17Q	2.5×10^7	3.2×10^{15}	1.3×10^7	2.5×10^8	1.6
D19N	4.0×10^7	4.0×10^{15}	2.0×10^7	2.0×10^8	1.5
E26Q	6.3×10^7	1.6×10^{16}	3.2×10^7	5.0×10^8	1.6
E60Q	1.0×10^8	3.2×10^{16}	5.0×10^7	6.4×10^8	1.6
E17Q/D19N	2.5×10^7	3.2×10^{14}	1.3×10^7	2.5×10^7	1.2
E17Q/E26Q	6.3×10^6	4.0×10^{14}	3.2×10^6	1.3×10^8	1.7
D19N/E26Q	7.9×10^6	3.2×10^{14}	4.0×10^6	8.0×10^7	1.6
E17Q/D19N/E26Q	6.3×10^6	3.2×10^{13}	3.2×10^6	1.0×10^7	1.3

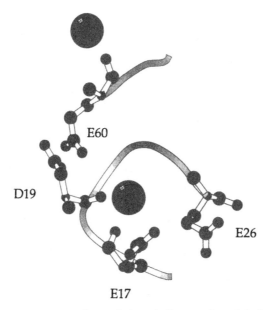

Figure 4.20   Ribbon representation of sites 1 (bottom) and 2 (top) of calbindin, indicating the side chains of residues whose mutations obey the first-order perturbation hypothesis. The bound Ca²⁺ ions are shown by circles.

drastic perturbation of the macroscopic cooperativity pattern. Nonetheless, there is evidence that the structural perturbation is confined to site 1 (Linse *et al.*, 1987). The perturbation of the global binding constants in these mutants may indeed reflect a perturbation of the site-specific binding constant K_1 only. Construction of the double-ratio plot in Figure 4.21 confirms this expectation. The mutant Y13F clearly lies outside the boundaries for first-order perturbation. Replacement of Tyr13 with a Phe induces changes that propagate to site 2 and may also involve the communication between the sites. Other mutations lie within the boundaries, but do not give rise to a linear dependence of the perturbed binding constants. On the other hand, there are five mutants whose perturbed constants clearly follow a straight line in the double-ratio plot. The values of K_1, K_2 and c_{12} are readily derived from this plot, since the ratios K_1/A_1

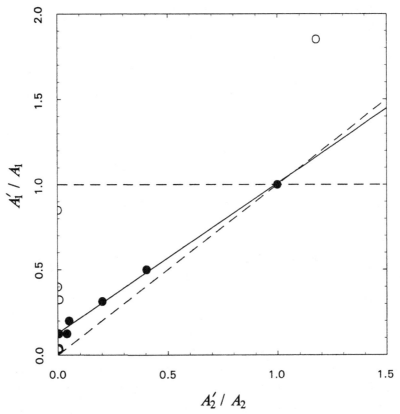

Figure 4.21 Double-ratio plot for calbindin. The point outside the boundaries refers to the mutant Y13F, which clearly violates the first-order perturbation hypothesis. The results pertaining to five mutants (filled circles) follow a straight line according to eq (4.133). These mutants are given in the key on Figure 4.22. The point of coordinates 1,1 refers to the wild-type. Since the straight line is closer to the high-affinity boundary (see also Figure 4.18), mutations affect predominantly the high-affinity site 1. The value of K_1 is derived from the value of A_1 and the slope in the plot and is 1.7×10^8 M^{-1}. The values of K_2 and c_{12} are 2.7×10^7 M^{-1} and 18.

and K_2/A_1 are given by the slope and intercept respectively, while A_1 and A_2 are known. From the results in Figure 4.21 it also follows that the site being perturbed, site 1, is the high-affinity one since the straight line is closer to the high-affinity boundary (see also Figure 4.18).

A striking observation is that the mutants behaving according to the first-order perturbation are all isosteric substitutions of negatively charged residues around site 1. The site-specific binding curves of the two sites can be reconstructed from application of eq (4.131). The partition function of unperturbed calbindin is

$$\Psi(x) = 1 + A_1 x + A_2 x^2 = 1 + (K_1 + K_2)x + c_{12}K_1 K_2 x^2 \quad (4.138)$$

For the mutant obeying first-order perturbation of site 1 we have

$$\Psi(x)' = 1 + A_1' x + A_2' x^2 = 1 + (K_1' + K_2)x + c_{12}K_1' K_2 x^2 \quad (4.139)$$

Application of eq (4.131) yields X_1, and X_2 is derived from the global binding curve of the wild-type as $X_2 = X - X_1$. The results are shown by symbols in Figure 4.22 for the five mutants on the straight line of Figure 4.21. The predicted site-specific binding curves are practically identical for all these mutants, although the perturbation of the overall binding constants is quite different in each case, as shown in Table 4.15. Site 1 binds Ca^{2+} with an affinity of about $1.7 \times 10^8 \, M^{-1}$, which is nearly six times the affinity of site 2. As a result of cooperative coupling between the sites, the affinity of either site increases by a factor of eighteen when the other site is ligated. This analysis provides a solution for the site-specific problem for calbindin and indicates that site 1 is probably the high-affinity binding site, contrary to the conclusion drawn from Cd^{2+} binding studies (Akke *et al.*, 1991). The results are also consistent with the structure of calbindin. The crystal structure of site 1 shown in Figure 4.20 suggests that it is unlikely that a mutation of Glu26 will be carried over to site 2, and the same applies to Glu17. These residues are about 10 Å closer to the Ca^{2+} in site 1 than that in site 2. Asp19 and Glu60 are close to both sites and, in principle, mutations of these residues should perturb sites 1 and 2. In practice, however, mutation of these residues yields effects similar to mutation of Glu26 and Glu17. A molecular dynamics simulation of calbindin shows that Glu60 may actually be part of the coordination sphere of Ca^{2+} in site 1 (Ahlström *et al.*, 1989). This residue, and possibly Asp19, may be closer to site 1 in solution, contrary to the conclusions drawn from the crystal structure (Szebenyi and Moffat, 1986). It may also be hypothesized that these residues may be more strongly coupled to site 1 than site 2 so that although perturbation of site 2 cannot be ruled out in principle, it is likely to be insignificant in practice.

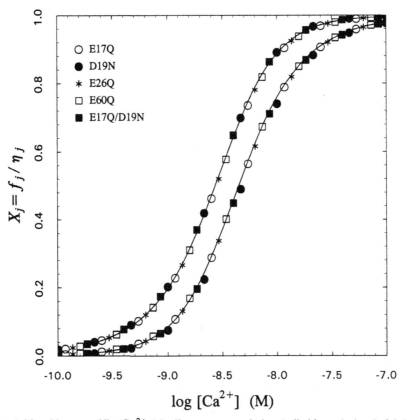

Figure 4.22 Site-specific Ca²⁺ binding curves of site 1 (left) and site 2 (right) of calbindin, as derived from the global binding parameters of the wild-type and the mutants as listed. Curves were drawn from eqs (4.27a) and (4.27b) using parameter values: $K_1 = 1.7 \times 10^8$ M⁻¹, $K_2 = 2.7 \times 10^7$ M⁻¹ and $c_{12} = 18$.

For a two-site macromolecule, knowledge of the site-specific binding curve of either site and the overall binding constants is sufficient to resolve all site-specific binding parameters. In the case of calbindin these parameters can be derived directly in the first-order perturbation using the following transformations of the overall binding constants

$$K_1 = \frac{A_2 A_1' - A_1 A_2}{A_2' - A_2} \tag{4.140a}$$

$$K_2 = \frac{A_2' A_1 - A_1' A_2}{A_2' - A_2} \tag{4.140b}$$

$$c_{12} = \frac{A_2}{K_1 K_2} \tag{4.140c}$$

The results are listed in Table 4.16 for all mutants. The mutants on the

Table 4.16. *Site-specific binding parameters for*
Ca²⁺ binding to calbindin.

Derived from	K_1 (M^{-1})	K_2 (M^{-1})	c_{12}
P20G	1.4×10^8	6.2×10^7	9
P20G, ΔN21	4.1×10^7	1.6×10^8	12
ΔP20	1.2×10^8	7.9×10^7	8
Y13F	7.7×10^8		
E17Q	1.8×10^8	1.8×10^7	24
D19N	1.7×10^8	3.1×10^7	15
E26Q	1.7×10^8	2.9×10^7	16
E60Q	1.7×10^8	3.4×10^7	14
E17Q/D19N	1.8×10^8	2.4×10^7	19
E17Q/E26Q	1.9×10^8	5.3×10^6	77
D19N/E26Q	1.9×10^8	7.2×10^6	58
E17Q/D19N/E26Q	1.9×10^8	6.2×10^6	66

straight line of Figure 4.21 yield very similar estimates for the site-specific
parameters, as expected. These estimates differ slightly for the mutants
P20G and ΔP20 that are also in close proximity to site 1. However, the
difference may not be statistically significant, since the error on the global
parameters reported for these mutants is large (Linse *et al.*, 1987). The
double and triple mutants, on the other hand, provide very different
estimates for the site-specific parameters, and so do the mutants P20G,
ΔN21 and Y13F. The double mutant E17Q, D19N provides a notable
exception. It may be concluded that neutralization of two or three surface
charges induces changes that propagate to site 2 and possibly affect the
interaction between the sites as well. Deletion of Gln21 may drastically
affect the overall conformation of calbindin and the substitution Y13F
clearly affects K_1 and K_2, as well as c_{12}, since A_1 increases while A_2 stays
constant (see also Figure 4.18). No physically meaningful values for K_2
and c_{12} can be derived from this mutant in the first-order perturbation
approach, since eq (4.140b) returns a negative value.

4.6 Pairwise coupling

We now focus on the concept of 'pairwise coupling' already introduced in
Section 4.4. This concept has an applicability that goes well beyond
site-specific binding cooperativity and extends to practically all cooperative processes that can be accessed experimentally in terms of site-specific
effects. The relevance of this concept stems from the possibility of

'breaking' the code for cooperativity in a model-independent fashion. This is an aspect of central importance to a variety of processes as diverse as ligand binding, protein folding and mutational perturbations. All these processes can be cast in terms of an underlying conceptual framework based on site-specific thermodynamics. The advantage of this approach is that the specific details of the system under investigation become of marginal interest, since the energetics are treated in their most general form.

Analysis of a cooperative process involving a macromolecular system entails the following three steps:

(1) Identification of structural domains important for energetic coupling.
(2) Assessment of the coupling among different domains.
(3) Elucidation of the precise mechanism of coupling.

These steps encapsulate the relevant aspects of the problem and its solution, regardless of the specific process under study, whether it involves ligand binding, protein folding or mutational perturbations. The first step determines the 'degrees of freedom' of the system, like the number of binding sites in the case of ligand binding. In the case of protein folding studies, binding sites are replaced by folding domains, like helices and sheets, that can be treated as elementary units. In mutational perturbation studies, the first step is equivalent to identifying those residues whose mutation significantly affects the energetics under investigation, e.g., substrate specificity or protein stability. The second step establishes a connection among sites, domains or residues from analysis of coupling effects. Examples are given by site–site communication leading to co-operative ligand binding, coupling of helices and sheets that determines the correct folding pattern, or coupling of individual residues as revealed by double mutations. The third step is the most important and challenging, since it provides a direct connection between structure and energetics. Once residues, sites or domains important for protein function are identified, along with interaction patterns, the key problem is to decipher the exact mechanism of communication among individual elements. Such mechanisms are often assumed *a posteriori* and modeled to fit the specific properties of the system. Site-specific thermodynamics provides a model-independent strategy to 'break' cooperativity codes that can be used in general, without formulation of *ad hoc* assumptions. This strategy is based on the properties of the coupling free energy in a thermodynamic cycle.

A cooperative system composed of N elements, like binding sites, protein domains or residues, can be associated with an N-dimensional vector of binary digits, 0 and 1. The entries 0 and 1 stand for free and bound in the case of binding sites, unfolded and folded in the case of protein domains, or wild-type and mutated in the case of mutational perturbations. There are 2^N possible configurations for the system, as already pointed out in the case of ligand binding. Pairwise coupling between any two elements, i and j, is defined as the coupling free energy in the thermodynamic cycle involving the $0 \to 1$ switching reactions of these elements, as illustrated in Figure 4.23. Keeping the remaining $N - 2$ elements in a given configuration, the cycle in Figure 4.23 has a coupling free energy

$$\Delta G_c = {}^1\Delta G_i - {}^0\Delta G_i = {}^1\Delta G_j - {}^0\Delta G_j \qquad (4.141)$$

The elements are positively linked if the coupling free energy is negative, and negatively linked otherwise. The case $\Delta G_c = 0$ indicates the absence of coupling. Since there are $N - 2$ other elements, besides i and j, then there must be a total of 2^{N-2} possible configurations for which the coupling between i and j can be computed. Hence, the total number of coupling free energies in the system is

$$v = \binom{N}{2} 2^{N-2} = N(N - 1)2^{N-3} \qquad (4.142)$$

The binomial term enumerates the possible ways of selecting any two elements among N, while the power law term reflects the number of thermodynamic cycles that can be constructed for any given pair of elements by exhausting the configurations of all other $N - 2$ elements. There is obviously no coupling free energy for $N = 1$ and only one value for $N = 2$, as expected. The value of v increases rapidly with N.

The approach based on the thermodynamic cycle in Figure 4.23 has been widely used in the analysis of mutational perturbations (Ackers and

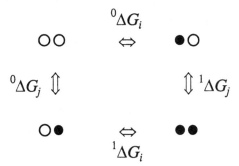

Figure 4.23 Thermodynamic cycle for the coupling between elements i and j. Symbols denote states 0 (open circles) and 1 (filled circles). The free energy of coupling between i and j is given by eq (4.141).

Smith, 1985; Horovitz and Fersht, 1990; Wells, 1990) and binding cooperativity (Jencks, 1981; Di Cera, 1990). Here we extend previous approaches by drawing attention to an important property of the cycle. Once coupling free energies are calculated for all possible configurations of the system, it is possible to decipher the code for cooperativity using the following properties of pairwise coupling:

Property 1 If the coupling interactions involving two elements are *direct*, the free energy of pairwise coupling is *independent* of the configuration of the remaining $N - 2$ elements.

Property 2 If the coupling interactions involving two elements are *indirect*, the free energy of pairwise coupling is *dependent* on the configuration of the remaining $N - 2$ elements.

We now prove the first property in a rigorous manner. Consider the elements i and j and assume that they are coupled directly to each other and to all other elements. This guarantees that the coupling between i and j is not mediated by indirect interactions. To compute the coupling free energy in the thermodynamic cycle in Figure 4.23 for an arbitrary configuration of the other $N - 2$ elements, we have to evaluate the contribution of all interactions to the free energy terms. When i and j are in state 0, the free energy associated with the system is

$$^{ij}G_{00} = G_{i,0} + G_{j,0} + \sum_{l \neq i,j}^{N} G_{l,\oplus} + {}^{ij}U_{00} + \sum_{l \neq i,j}^{L_i} {}^{il}U_{00} + \sum_{l \neq i,j}^{M_i} {}^{il}U_{01} + \sum_{l \neq i,j}^{L_j} {}^{jl}U_{00}$$

$$+ \sum_{l \neq i,j}^{M_j} {}^{jl}U_{01} + \Lambda \tag{4.143}$$

The terms in the foregoing expression refer to the free energy of elements i and j in state 0, $G_{i,0}$ and $G_{j,0}$, the free energy of element $l \neq i, j$ whether it is in state 0 or 1, $G_{l,\oplus}$, the interaction free energy between i and j when both are in state 0, $^{ij}U_{00}$, and the interaction free energies between i and any other element, and likewise for j and any other element. These interaction terms can be partitioned into two groups, one containing all elements other than i and j in state 0 and the other containing all elements other than i and j in state 1. There are L_i elements in state 0 and M_i elements in state 1 that interact with i. Likewise, there are L_j elements in state 0 and M_j elements in state 1 that interact with j. The term Λ contains all interaction terms among the $N - 2$ elements other than i and j. It is inconsequential to enumerate these terms, as well as to specify whether the interactions are direct or indirect, since the term Λ is by definition independent of the state of elements i and j and hence makes no contribution to the coupling free energy between these elements. When i switches to state 1, the free energy of the system is

$$
{}^{ij}G_{10} = G_{i,1} + G_{j,0} + \sum_{l \neq i,j}^{N} G_{l,\oplus} + {}^{ij}U_{10} + \sum_{l \neq i,j}^{P_i} {}^{il}U_{10} + \sum_{l \neq i,j}^{Q_i} {}^{il}U_{11}
$$

$$
+ \sum_{l \neq i,j}^{L_j} {}^{jl}U_{00} + \sum_{l \neq i,j}^{M_j} {}^{jl}U_{01} + \Lambda \tag{4.144}
$$

The free energy now contains new terms. Specifically, the free energy of element i in state 1, $G_{i,1}$, the interaction term between i and j with i in state 1 and j in state 0, ${}^{ij}U_{10}$, and the interaction terms between i and all other elements with i in state 1. Again, these terms are partitioned into two groups, P_i involving elements in state 0 and Q_i involving elements in state 1. Clearly,

$$
L_i = P_i \tag{4.145a}
$$

$$
M_i = Q_i \tag{4.145b}
$$

because the number of direct interactions involving i is fixed. The free energies pertaining to all elements other than i and j remain unchanged, and so are the interaction terms for element j. Likewise, when j switches to state 1, the free energy of the system is

$$
{}^{ij}G_{01} = G_{i,0} + G_{j,1} + \sum_{l \neq i,j}^{N} G_{l,\oplus} + {}^{ij}U_{01} + \sum_{l \neq i,j}^{L_i} {}^{il}U_{00} + \sum_{l \neq i,j}^{M_i} {}^{il}U_{01} + \sum_{l \neq i,j}^{P_j} {}^{jl}U_{10}
$$

$$
+ \sum_{l \neq i,j}^{Q_j} {}^{jl}U_{11} + \Lambda \tag{4.146}
$$

with

$$
L_j = P_j \tag{4.147a}
$$

$$
M_j = Q_j \tag{4.147b}
$$

Finally, when both i and j are in state 1, the free energy is

$$
{}^{ij}G_{11} = G_{i,1} + G_{j,1} + \sum_{l \neq i,j}^{N} G_{l,\oplus} + {}^{ij}U_{11} + \sum_{l \neq i,j}^{P_i} {}^{il}U_{10} + \sum_{l \neq i,j}^{Q_i} {}^{il}U_{11} + \sum_{l \neq i,j}^{P_j} {}^{jl}U_{10}
$$

$$
+ \sum_{l \neq i,j}^{Q_j} {}^{jl}U_{11} + \Lambda \tag{4.148}
$$

The coupling free energy in the thermodynamic cycle in Figure 4.23 is

$$
\Delta G_c = {}^1\Delta G_i - {}^0\Delta G_i = {}^1\Delta G_j - {}^0\Delta G_j = {}^{ij}G_{00} + {}^{ij}G_{11} - {}^{ij}G_{10} - {}^{ij}G_{01}
$$

$$
\tag{4.149}
$$

Substitution of eqs (4.143), (4.144), (4.146) and (4.148) into eq (4.149) yields

$$\Delta G_c = {}^{ij}U_{00} + {}^{ij}U_{11} - {}^{ij}U_{10} - {}^{ij}U_{01} \qquad (4.150)$$

The foregoing expression is independent of the particular state of elements other than i and j. Hence, if the coupling interactions involving i and j are direct, the coupling free energy is independent of the configuration of all other elements. This result is general and holds regardless of the specific nature of the direct interactions involving i and j. Also, it is independent of the nature of the interaction among all other elements, whether they are direct or indirect.

The second property is easier to prove. It is intuitively obvious that if the coupling between i and j is indirect, then there must be other elements responsible for the linkage and therefore the coupling free energy will necessarily depend on the state of these elements. The free energy of the system in the presence of indirect interactions, when i and j are in state 0, is given by

$$^{ij}G_{00} = G_{i,0} + G_{j,0} + U_{00}(\sigma) \qquad (4.151)$$

where $U_{00}(\sigma)$ is the interaction between i and j, given the particular configuration σ of the other $N - 2$ elements. The other free energy terms in the thermodynamic cycle in Figure 4.23 are given by

$$^{ij}G_{10} = G_{i,1} + G_{j,0} + U_{10}(\sigma) \qquad (4.152)$$

$$^{ij}G_{01} = G_{i,0} + G_{j,1} + U_{01}(\sigma) \qquad (4.153)$$

$$^{ij}G_{11} = G_{i,1} + G_{j,1} + U_{11}(\sigma) \qquad (4.154)$$

Hence,

$$\Delta G_c = U_{00}(\sigma) + U_{11}(\sigma) - U_{10}(\sigma) - U_{01}(\sigma) \qquad (4.155)$$

and the coupling free energy clearly depends on the configuration of the elements other than i and j.

A specific example is provided by a two-state model where each element can exist in two possible energetic states, T and R. It is assumed that the T$\rightarrow$R transition is positively linked to the $0 \rightarrow 1$ transition, and that the transition is global and affects all elements in a concerted fashion. This is essentially a generalization of the MWC model (Monod *et al.*, 1965) that allows for treatment of cooperative processes other than ligand binding, like protein folding or mutational perturbations. The various free energy terms involved in the thermodynamic cycle in Figure 4.23 for such a generalized model are

$$
^{ij}G_{00} = {}^{R}G_{i,0} + {}^{R}G_{j,0} + \sum_{l \neq i,j}^{N} {}^{R}G_{l,\oplus}
$$
$$
- RT \ln \frac{Lc_{i,0}c_{j,0} \exp\left[\sum_{l \neq i,j}^{N}\left(\dfrac{^{R}G_{l,\oplus} - {}^{T}G_{l,\oplus}}{RT}\right)\right] + 1}{L + 1} \tag{4.156a}
$$

$$
^{ij}G_{10} = {}^{R}G_{i,1} + {}^{R}G_{j,0} + \sum_{l \neq i,j}^{N} {}^{R}G_{l,\oplus}
$$
$$
- RT \ln \frac{Lc_{i,1}c_{j,0} \exp\left[\sum_{l \neq i,j}^{N}\left(\dfrac{^{R}G_{l,\oplus} - {}^{T}G_{l,\oplus}}{RT}\right)\right] + 1}{L + 1} \tag{4.156b}
$$

$$
^{ij}G_{01} = {}^{R}G_{i,0} + {}^{R}G_{j,1} + \sum_{l \neq i,j}^{N} {}^{R}G_{l,\oplus}
$$
$$
- RT \ln \frac{Lc_{i,0}c_{j,1} \exp\left[\sum_{l \neq i,j}^{N}\left(\dfrac{^{R}G_{l,\oplus} - {}^{T}G_{l,\oplus}}{RT}\right)\right] + 1}{L + 1} \tag{4.156c}
$$

$$
^{ij}G_{11} = {}^{R}G_{i,1} + {}^{R}G_{j,1} + \sum_{l \neq i,j}^{N} {}^{R}G_{l,\oplus}
$$
$$
- RT \ln \frac{Lc_{i,1}c_{j,1} \exp\left[\sum_{l \neq i,j}^{N}\left(\dfrac{^{R}G_{l,\oplus} - {}^{T}G_{l,\oplus}}{RT}\right)\right] + 1}{L + 1} \tag{4.156d}
$$

where the ^{R}Gs are the free energy levels for state 0 in the R state, $^{R}G_{l,\oplus}$ and $^{T}G_{l,\oplus}$ denote the free energy levels for states 0 or 1 in R and T for the elements other than i and j, L is the allosteric constant reflecting the relative concentrations of T and R when all elements are in state 0, and

$$
c_{i,0} = \exp\left(\frac{^{R}G_{i,0} - {}^{T}G_{i,0}}{RT}\right) \tag{4.157a}
$$

$$
c_{i,1} = \exp\left(\frac{^{R}G_{i,1} - {}^{T}G_{i,1}}{RT}\right) \tag{4.157b}
$$

$$
c_{j,0} = \exp\left(\frac{^{R}G_{j,0} - {}^{T}G_{j,0}}{RT}\right) \tag{4.157c}
$$

$$c_{j,1} = \exp\left(\frac{{}^{R}G_{j,1} - {}^{T}G_{j,1}}{RT}\right) \tag{4.157d}$$

The free energy of pairwise coupling for this model is

$\Delta G_c =$

$$-RT\ln\frac{\left\{Lc_{i,0}c_{j,0}\exp\left[\sum_{l\neq i,j}^{N}\left(\frac{{}^{R}G_{l,\oplus} - {}^{T}G_{l,\oplus}}{RT}\right)\right] + 1\right\}\left\{Lc_{i,1}c_{j,1}\exp\left[\sum_{l\neq i,j}^{N}\left(\frac{{}^{R}G_{l,\oplus} - {}^{T}G_{l,\oplus}}{RT}\right)\right] + 1\right\}}{\left\{Lc_{i,1}c_{j,0}\exp\left[\sum_{l\neq i,j}^{N}\left(\frac{{}^{R}G_{l,\oplus} - {}^{T}G_{l,\oplus}}{RT}\right)\right] + 1\right\}\left\{Lc_{i,0}c_{j,1}\exp\left[\sum_{l\neq i,j}^{N}\left(\frac{{}^{R}G_{l,\oplus} - {}^{T}G_{l,\oplus}}{RT}\right)\right] + 1\right\}}$$

$$\tag{4.158}$$

It is straightforward to prove that eq (4.158) is always negative, thereby implying that the coupling between any two elements in the system is always positive. In fact, eq (4.158) can be rewritten in the simpler form

$$\Delta G_c = -RT\ln\frac{1 + (c_{i,0}c_{j,0} + c_{i,1}c_{j,1})fL + c_{i,0}c_{j,0}c_{i,1}c_{j,1}f^2L^2}{1 + (c_{i,1}c_{j,0} + c_{i,0}c_{j,1})fL + c_{i,0}c_{j,0}c_{i,1}c_{j,1}f^2L^2} \tag{4.159}$$

where f replaces the exponential term. The numerator and denominator of eqs (4.158) and (4.159) are second-order polynomials in L that differ only in the coefficient of L. Simple transformations prove that

$$c_{i,0}c_{j,0} + c_{i,1}c_{j,1} > c_{i,1}c_{j,0} + c_{i,0}c_{j,1} \tag{4.160}$$

since

$$(c_{i,1} - c_{i,0})(c_{j,1} - c_{j,0}) > 0 \tag{4.161}$$

This is because, by hypothesis, the $T \to R$ transition is positively linked to the $0 \to 1$ transition. If the R state has a lower free energy than the T state, then state 1 must have a lower free energy than state 0. Hence, $c_{i,1}$ is necessarily less than one, but is also smaller than $c_{i,0}$, because it is more likely to find element i in the T state when it is in state 0. The same applies to element j, so that eq (4.161) is satisfied. Complex patterns of coupling free energies can be obtained in this model if each element shows a different linkage between the $T \to R$ and $0 \to 1$ transitions. In general, the free energy of coupling will depend on the state of the other elements, contrary to the case of direct interactions in eq (4.150).

Pairwise coupling profiles can be constructed for a variety of cooperative processes. An application to ligand binding is given in Section 4.4. The approach is general enough to be extended to other relevant processes, like protein folding and mutational perturbations. Mutational perturbations are discussed in some detail, due to the widespread use of site-directed mutagenesis studies in the analysis of structure–function relationships in macromolecules (Knowles, 1987). Mutational perturbations are currently used in a variety of systems to explore folding pathways (Horovitz and Fersht, 1992), protein stability (Lyu, Liff, Marky and Kallenbach, 1990; Padmanabhan *et al.*, 1990; Matthews, 1993), enzyme specificity (Carter and Wells, 1988; Scrutton, Berry and Perham, 1990; Wells, 1990; Cornish and Schultz, 1994) and molecular recognition (Ebright, 1986; Lesser, Kurpiewski and Jen-Jacobson, 1991; Draper, 1993). In most of these studies early events of protein folding or important aspects of protein stability have been dissected using single and double mutations. The approach based on pairwise coupling requires the construction of triple, quadruple and higher-order mutants. For the approach to be exploited fully, it is necessary to construct a complete set of mutations that includes at least three elements. This is because the number of thermodynamic cycles for pairwise coupling must be greater than one, which is seen if and only if $N \geqslant 3$ in eq (4.142). Hence, mechanisms of cooperativity cannot be dissected in a model-independent fashion using only single and double mutations.

Assume that the problem at hand is to determine the mechanism through which the four helices A, D, F and H of apomyoglobin communicate and give rise to protein stability. First, the free energy of unfolding is derived for the wild-type. This is the quantity to be used in the thermodynamic cycle in Figure 4.23. Second, single site-directed mutations are made in each of the helices, as shown in Figure 4.24, to identify residues that perturb the stability of apomyoglobin. Third, once the best candidate is identified for each helix from single mutations, the entire set of mutants up to the quadruple mutant is constructed. This set can be depicted in vectorial form as

$$
\begin{array}{ccccccc}
 & & [0000] & & & & \\
 & [1000] & [0100] & [0010] & [0001] & & \\
[1100] & [1010] & [1001] & [0110] & [0101] & [0011] & \quad (4.162) \\
 & [1110] & [1101] & [1011] & [0111] & & \\
 & & [1111] & & & &
\end{array}
$$

and is identical to the ligand binding case dealt with in the scheme (4.72). The vector [0000] denotes the wild-type, [1000], [0100], [0010] and [0001] pertain to single mutants on helices A, D, F and H, and so forth for the double, triple and quadruple mutants. The pairwise coupling pattern is constructed by measuring the free energy of unfolding for all mutants. This yields the stability of the four species in Figure 4.23 and hence the

Figure 4.24   Ribbon representation of apomyoglobin. Circles depict residues on the A, D, F and H helices that are assumed to be critical for protein stability.

coupling free energy from the general eq (4.149). There are a total of twenty-four such values and analysis of the pattern reveals whether the four helices are coupled directly or indirectly. The same approach can be used to assess the mechanism of coupling within each structural element. For example, the stability of helix A can be studied by multiple substitutions within the helix and subsequent analysis of the pairwise coupling pattern.

As an example of the approach based on pairwise coupling we analyze the mutations of the salt-bridged triad Asp8, Asp12 and Arg110 of barnase (Horovitz *et al.*, 1990, 1991; Horovitz and Fersht, 1992). All the

Table 4.17. *Free energies of pairwise coupling (in kcal/mol) for the stability of barnase (Horovitz and Fersht, 1992).*

Pair	0	1
Asp8–Asp12	-0.44 ± 0.06	0.33 ± 0.06
Asp8–Arg110	-0.98 ± 0.06	-0.21 ± 0.06
Asp12–Arg110	-1.25 ± 0.06	-0.48 ± 0.06

D8D12R110
0.00 ± 0.03

ADR	**DAR**	**DDA**
0.99 ± 0.03	0.34 ± 0.03	0.46 ± 0.03

AAR	**ADA**	**DAA**
0.89 ± 0.03	0.47 ± 0.03	-0.45 ± 0.03

AAA
-0.11 ± 0.03

Figure 4.25   Schematic representation of the complete set of mutants of the salt-bridged triad Asp8, Asp12 and Arg110 of barnase (Horovitz and Fersht, 1992). The free energy of unfolding (in kcal/mol) relative to the wild-type, as derived from eq (4.163), is shown for each mutant. The pairwise coupling patterns derived from the set of mutations are given in Table 4.17.

residues of this triad have been mutated to Ala in a complete set and the free energy of unfolding has been measured for all members of the set. The results are given in Figure 4.25, as calculated relative to the wild-type using the expression

$$\Delta\Delta G_u = {}^{wt}\Delta G_u - {}^{mut}\Delta G_u \qquad (4.163)$$

where $\Delta\Delta G_u$ is the difference in the unfolding free energy between the wild-type and the mutant. A positive value of this parameter indicates that the mutant is less stable than the wild-type and *vice versa* for negative values. The pairwise coupling patterns for each of the three pairs are listed in Table 4.17. It is evident from the table that the three residues are indirectly linked in the wild-type and contribute to the stability of the protein as a 'unit'. Interestingly, mutations tend to interact with positive cooperativity in the stabilization of the protein. The close proximity of the residues explains their behavior as a 'concerted switch', so that the state of any residue is 'felt' energetically by the other two upon mutation.

Table 4.18. *Free energies of pairwise coupling (in kcal/mol) for the stability of the transition state of subtilisin (Carter and Wells, 1988).*

Pair	0	1
Asp32–His64	-7.88 ± 0.06	-0.11 ± 0.06
Asp32–Ser221	-6.59 ± 0.06	1.18 ± 0.06
His64–Ser221	-8.94 ± 0.06	-1.17 ± 0.06

D32H64S221
0.00 ± 0.03

AHS	**DAS**	**DHA**
6.52 ± 0.03	8.84 ± 0.03	8.93 ± 0.03

AAS	**AHA**	**DAA**
7.48 ± 0.03	8.86 ± 0.03	8.83 ± 0.03

AAA
8.65 ± 0.03

Figure 4.26 Schematic representation of the complete set of mutants of the catalytic triad Asp32, His64 and Ser221 of subtilisin (Carter and Wells, 1988). The free energy in the transition state (in kcal/mol) relative to the wild-type, as derived from eq (4.164), is shown for each mutant. The pairwise coupling patterns derived from the set of mutations are given in Table 4.18.

A similar effect is observed in mutational studies of subtilisin, where the spatially adjacent residues of the catalytic triad, Asp32, His64 and Ser221, have been mutated to Ala in a complete set (Carter and Wells, 1988). In this case, the free energy measured in the thermodynamic cycle in Figure 4.23 is the difference in the transition state free energy between the mutant and the wild-type, i.e.,

$$\Delta \Delta G_T^{\ddagger} = {}^{mut}\Delta G_T^{\ddagger} - {}^{wt}\Delta G_T^{\ddagger} \tag{4.164}$$

and the results are given in Figure 4.26. A positive value of this parameter indicates that the mutation has reduced the specificity of the enzyme and *vice versa* for negative values. The pairwise coupling pattern for each of the three possible pairs is given in Table 4.18. A drastic difference is observed upon mutation of the third residue, thereby indicating that the members of the catalytic triad act as a 'concerted switch' where each residue is influenced energetically by the others.

The criterion of spatial proximity linked to indirect coupling of the concerted type, as envisioned by the picture dealt with in eqs (4.156)–(4.161), is, however, not valid in general. We have shown in Section 4.5 that the protonation of the spatially close ionizable groups of glutamic acid can be explained fully in terms of direct coupling among the groups. A similar coupling mechanism is also seen for the four hemes of hemoglobin that are quite far apart from each other (see Section 4.4). Mutational studies on glutathione reductase have been carried out to perturb Ala179, Arg198 and Arg204, three residues not in close contact (Scrutton *et al.*, 1990). The set of mutations is given in Figure 4.27, along with the relevant free energy values for the perturbation of the transition state in the oxidation of glutathione by NADH and NADPH. The pairwise coupling pattern given in Table 4.19 shows that the three residues are linked indirectly, although they are not in spatial proximity. The rules for communication in a biological macromolecule may be quite subtle. More information on higher-order perturbations is to be gathered and analyzed in a model-independent way using pairwise coupling patterns. The combination of massive mutational analysis of proteins and the principles of site-specific thermodynamics will eventually unravel general rules for gaining a predictive understanding of protein function and cooperative transitions from structure.

Another application of the approach based on pairwise coupling can be developed in the case of studies on molecular recognition. In this case, the goal is to identify regions that are important for specificity and the

A179R198R204
0.00,0.00

GRR	**AMR**	**ARL**
−1.10,0.08	−0.62,2.68	0.41,2.42

GMR	**GRL**	**AML**
−1.32,2.11	−1.54,0.87	−0.51,3.70

GML
−1.72,2.22

Figure 4.27   Schematic representation of the complete set of mutants of residues Ala179, Arg198 and Arg204 of glutathione reductase (Scrutton *et al.*, 1990). The free energy in the transition state (in kcal/mol) relative to the wild-type, as derived from eq (4.164), is shown for each mutant. The two values refer to oxidation in the presence of NADH (left) or NADPH (right). The pairwise coupling patterns derived from the set of mutations are given in Table 4.19.

Table 4.19. *Free energies of pairwise coupling (in kcal/mol) for the stability of the transition state of glutathione reductase (Scrutton* et al., *1990).*

Pair	0	1
NADH		
Ala179–Arg198	0.40 ± 0.06	0.74 ± 0.06
Ala179–Arg204	-0.85 ± 0.06	-0.51 ± 0.06
Arg198–Arg204	-0.30 ± 0.06	0.04 ± 0.06
NADPH		
Ala179–Arg198	-0.65 ± 0.06	0.07 ± 0.06
Ala179–Arg204	-1.63 ± 0.06	-0.91 ± 0.06
Arg198–Arg204	-1.40 ± 0.06	-0.68 ± 0.06

coupling among the various structural elements. The analysis can be carried out on either the 'macromolecule' or the 'ligand'. First, the subsites involved in the recognition mechanism need to be identified by single site-directed mutations. The quantity of interest can be the binding affinity of the ligand or the k_{cat}/K_m ratio for the substrate. The perturbation of these signals due to mutation is used as a measure of the appropriate ΔG in the thermodynamic cycle of Figure 4.23. Second, a set of mutants is constructed to map the entire specificity domain and to derive the pairwise coupling pattern for the system. Assume that the system of interest is the specificity pocket of thrombin, a serine protease involved in blood coagulation. This pocket entails three subsites, as shown in Figure 4.28, and its structural organization sets the preference for substrates with Arg at P1, Pro at P2 and Phe at P3 (Bode, Turk and Kanshikov, 1992). The coupling mechanism among these subsites can be explored with mutants of thrombin, or using synthetic substrates that carry mutations at P1, P2 and P3. The latter alternative is clearly more convenient in practice. The set of synthetic substrates can be cast in vectorial form

$$-S3-S2-S1-S1'-S2'-S3'-$$

$$-X-X-X-P3-P2-P1-P1'-P2'-P3'-X-X-X-$$

Figure 4.28 Active-site mapping of a serine protease, using the strategy based on pairwise coupling. The specificity subsites of the enzyme S1, S2 and S3 (Schechter and Berger, 1967), interact with the residues P1, P2 and P3 of the substrate. Similar interactions involve the primed subsites. A complete set of mutants of either the S or P subsites yields the precise mechanism of coupling in the transition state.

as follows

$$[000]$$

$$[100] \quad [010] \quad [001]$$

$$[110] \quad [101] \quad [011] \tag{4.165}$$

$$[111]$$

and is analogous to the ligand binding case dealt with in eq (4.39), and the examples given in Figures 4.25–4.27. The wild-type sequence FPR is perturbed stepwise to generate a triple mutant VAK, where Phe at P3 is replaced by Val, Pro at P2 is replaced by Ala, and Arg at P1 is replaced by Lys. Hence,

$$XXFPRXX$$

$$XXVPRXX \quad XXFARXX \quad XXFPKXX$$

$$XXVARXX \quad XXVPKXX \quad XXFAKXX \tag{4.166}$$

$$XXVAKXX$$

is the set of substrates that 'chemically' maps the manifold in eq (4.165). X refers to an arbitrary residue. Once the ratio k_{cat}/K_m is determined for each substrate, the pairwise coupling free energy between any two subsites is derived from the transition state free energies of the four species in the thermodynamic cycle in Figure 4.23. For example, the quantity

$$\Delta G_c = -RT \ln \frac{(k_{cat}/K_m)_{FPR}(k_{cat}/K_m)_{VPK}}{(k_{cat}/K_m)_{VPR}(k_{cat}/K_m)_{FPK}} \tag{4.167}$$

gives the coupling free energy in the transition state between P1 and P3, when there is no mutation at P2. The pairwise coupling pattern reveals how the three residues of the substrate are coupled in the transition state. This provides direct information on how thrombin subsites recognize these residues and are themselves coupled. A similar strategy can be used with protein–protein, protein–DNA or protein–RNA interactions. In the case of enzymatic cleavage, the quantity of interest is the transition state free energy. In the case of equilibrium binding, the signal is provided by the binding free energy. A combined approach with mutations involving the macromolecule and ligands probing different domains can yield all the information necessary to break the code for the communication among structural components. The higher the dimensionality of the functional space generated by the mutations with triple, quadruple and higher-order substitutions, the more complete the picture we obtain of macromolecular interactions using the approach based on pairwise coupling.

5

Site-specific effects in Ising networks

In this chapter we deal with the thermodynamics of site-specific effects in Ising networks (Newell and Montroll, 1953; Hill, 1960; Stanley, 1971; Domb and Green, 1972; Itzykson and Drouffe 1989) of particular biological relevance. The basic results of the phenomenological theory described in the foregoing chapters can be merged with those of the statistical thermodynamics of model systems that have set conceptual landmarks in our understanding of cooperative phenomena. A number of important properties of biological macromolecules have been approached in terms of lattice statistics, e.g., helix-coil transitions (Zimm and Bragg, 1959; Glauber, 1963; Cohen and Penrose, 1970; Poland and Scheraga 1970), protein folding (Dill *et al.*, 1995) and ligand-induced conformational transitions (Herzfeld and Stanley, 1974; Hermans and Premilat, 1975). These processes entail cooperative transitions involving a number of subsystems. Molecular recognition, or the process that allows proteins to interact specifically with small ligands, other proteins, RNA and DNA molecules, can be modeled in terms of the contribution of several individual recognition subsites that give rise to the macroscopic energetics measured experimentally. The properties of subsites can be modeled using Ising networks (Keating and Di Cera, 1993). A mechanistic description of cooperative processes should capture essential features which can be used eventually to develop more elaborate and accurate descriptions. An exact treatment of cooperative transitions in systems composed of a large number of interacting components remains, at present, computationally unfeasible. Interest in the analysis of site-specific effects in Ising networks also arises from the fact that the site-specific description can shed new light on outstanding problems related to properties of these networks in two, three and higher dimensions. Furthermore, these networks can be used to model site-specific effects under nonequilibrium conditions, such as those pertaining to the hydrogen exchange properties of peptides and proteins (Englander and Kallenbach, 1984; Bai, Milne, Mayne and Englander, 1994; Zhou *et al.*, 1994). It seems, therefore, quite appropriate to conclude the treatment of site-specific effects with a mechanistic description that complements and enriches the phenomenological analysis dealt with in previous chapters. Much of our understanding of site-specific thermodynamics of extended systems, along with their nonequilibrium properties, will depend on our ability to exploit the results of the phenomenological theory in the analysis of the Ising problem.

5.1 Transition modes

An Ising network consists of a set of N units connected through a given geometry in arbitrary dimensions (Newell and Montroll, 1953; Domb and Green, 1972; Itzykson and Drouffe, 1989; Keating and Di Cera, 1993). Each unit exists in two distinct states, which can be denoted by 0 and 1 without loss of generality. The network is called *homogeneous* if the units are equivalent and *heterogeneous* for all other cases encompassing differences among the units or asymmetric interaction geometries. The ring shown in Figure 5.1 is an example of a homogeneous network where all units are identical and equivalent since they contact two other units. The linear chain shown in Figure 5.1 is an example of a heterogeneous network. Although its units are identical, the end units differ insofar as they contact only one other unit. Heterogeneous networks in two or three dimensions are generated by randomly connecting an arbitrary set of identical or different units. If two units contact each other in the network,

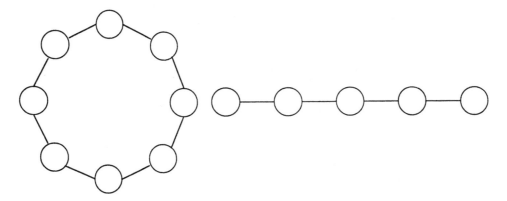

Figure 5.1 Examples of Ising networks. The ring (left) is a homogeneous network since all units are identical and interact equally. The linear chain (right) is heterogeneous. Although its units are identical, they do not interact equally.

there is a nearest-neighbor interaction which depends only on the states of the units. Two ways of defining the rules for interaction have historically received much attention. The spin-lattice type of interaction assigns an interaction energy to two neighbor units in unlike states (Onsager, 1944; Lee and Yang, 1952; Newell and Montroll, 1953). The lattice-gas type of interaction assigns an interaction energy to two neighbor units in state 1 (Mayer and Mayer, 1940; Yang and Lee, 1952). Each treatment has advantages and fundamental properties. The spin-lattice rule encapsulates some general symmetry properties, while the lattice-gas rule simplifies considerably the mathematics when it comes to dealing with site-specific effects.

A rather basic quantity in an Ising network is the energy spent per unit in the $0 \to 1$ transition. Intuitively, if a given unit is decoupled from the rest of the network, then the energy spent in the $0 \to 1$ transition for this unit is equal to the intrinsic energy for switching from state 0 to state 1. However, in the more interesting and realistic case where the unit is coupled to the rest of the network, the energy spent in the $0 \to 1$ transition will also depend on the state of all other units. This conclusion is reached independently of the nature of the two states, 0 and 1, and the driving force responsible for the $0 \to 1$ transition. If the two states represent two different conformational states of individual domains of a macromolecular system at constant P, and the temperature of the heat bath provides the driving force for the $0 \to 1$ transition, then the network can be treated as a mechano-thermal ensemble. Alternatively, the $0 \to 1$ transition can model the binding of a ligand to a specific site of the macromolecule, at constant P and T, with states 0 and 1 denoting the free and bound forms. In this case the driving force for the transition is the chemical potential of the ligand and the network can be treated as a generalized ensemble. In the following, we will refer to states 0 and 1 as the free and bound forms of a site, modeled as a unit in the network. We assume T and P constant and use the generalized ensemble. We know from site-specific thermodynamics that for a system of N sites existing in two states there are 2^N possible configurations, each of which is associated with a N-dimensional vector

$$[\sigma] = [\theta_1 \theta_2 \ldots \theta_j \ldots \theta_N] \tag{5.1}$$

where $\theta_j = 0$, 1 labels the state of the jth site in the network. Each configuration is characterized by a Gibbs free energy specified by the states 0 and 1, and the interaction network among the sites.

If the configuration with all sites in state 0 is taken as reference, then a total of $2^N - 1$ independent free energy terms characterize the network, each corresponding to the work spent in the transition from the reference configuration to any one of the remaining configurations. These terms completely define the partition function of the system in the site-specific scenario. The energetics of the network can also be cast in terms of the free energy spent per site in the $0 \to 1$ (free $\to$ bound) transition. It follows from the phenomenological theory of global effects dealt with in Section 2.3 that the mean ligand activity is related to the average work spent in ligating one site of the macromolecule. This quantity refers to the work spent in the transition from the fully-unligated to the fully-ligated form, divided by the number of sites. However, the average work spent in ligating one site, or in the $0 \to 1$ transition, can be defined from many other transitions. This point can be illustrated with the Ising networks depicted in Figure 5.2 in the form of graphs (Gould, 1988). The possible configurations are denoted by vertices and labeled accordingly. For $N = 1$, the free energy spent per site in the $0 \to 1$ transition coincides with the mean free energy of ligation. For $N = 2$, on the other hand, the free

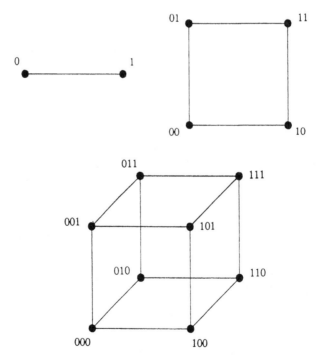

Figure 5.2 Graphs representing the transitions of Ising networks of 1, 2 or 3 sites. Each possible configuration is labeled by a binary code and is depicted as a vertex in the graph. The connection between two vertices indicates the existence of a one-step transition that generates one configuration from the other by switching the state of a single site.

energy of transition can be defined in a number of ways. The transitions $00 \rightarrow 01$ and $00 \rightarrow 10$ provide two possible values. The transitions $10 \rightarrow 11$ and $01 \rightarrow 11$ provide two other values. The transition $00 \rightarrow 11$ defines a free energy change for switching two sites to state 1 and when this value is divided by two, it gives yet another measure of the mean free energy change spent per site in the $0 \rightarrow 1$ transition. This particular value is the mean free energy of ligation defined in the global description. The five possible values listed above need not be the same, unless the sites are identical and independent. In the case of $N = 3$, there are transitions like $000 \rightarrow 001$, $001 \rightarrow 011$ and $011 \rightarrow 111$ that involve one site, those like $000 \rightarrow 011$ and $001 \rightarrow 111$ that involve two sites and the transition $000 \rightarrow 111$ which involves all sites. For arbitrary N, there are transitions involving $j = 1, 2, \ldots N$ sites, each providing in principle a different value for the mean free energy change spent per site in the $0 \rightarrow 1$ transition. Given any two configurations in the network, $[\sigma]$ and $[\sigma']$, the number of digits that one needs to change in $[\sigma']$ to obtain $[\sigma]$ is the 'Hamming distance', $H_{\sigma\sigma'}$, between $[\sigma]$ and $[\sigma']$ (Hamming, 1986). Hence, the mean free energy spent per site in the $0 \rightarrow 1$ transition, $\Delta G_{\mathrm{m}}(\sigma, \sigma')$, can be written as

$$\Delta G_{\mathrm{m}}(\sigma, \sigma') = \frac{\Delta G_{\sigma\sigma'}}{H_{\sigma\sigma'}} \tag{5.2}$$

Given any two configurations in the network, $[\sigma]$ and $[\sigma']$, such that $[\sigma']$ can be obtained from $[\sigma]$ by switching n sites from 0 to 1, the value of $\Delta G_{\mathrm{m}}(\sigma, \sigma')$ is the free energy change for the $[\sigma] \rightarrow [\sigma']$ transition divided by n, i.e., the Hamming distance $H_{\sigma\sigma'}$ between $[\sigma]$ and $[\sigma']$.

The transition free energy mode, or simply the *transition mode* (Keating and Di Cera, 1993) as defined in eq (5.2), is a sort of local 'intensive' property of the system which labels the energetics in a site-specific manner. A pictorial representation of transition modes is given in Figure 5.3 for the case $N = 3$. The configurations $[\sigma]$ and $[\sigma']$ are depicted by filled circles and the Hamming distance is the number of edges connecting them. Each edge is a valid transition step provided it increases the number of sites in state 1 by one. The transition mode is the total free energy spent in the trajectory, divided by the number of steps. When all possible transitions between allowable pairs of $[\sigma]$ and $[\sigma']$ are considered, a distribution of transition mode values is obtained. This distribution reflects the energetic landscape of the network, a sort of spectral decomposition of the free energy values for the $0 \rightarrow 1$ transition. Essentially, a transition mode is the mean free energy of $0 \rightarrow 1$ switching when the state of up to

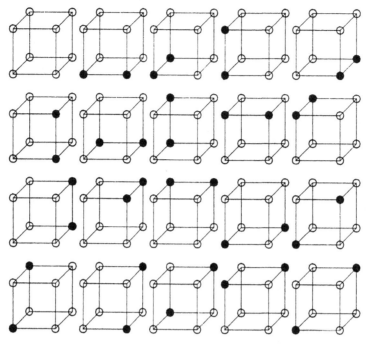

Figure 5.3　Transition diagrams for $N = 3$. The vertices of the cubes correspond to the same configurations shown in Figure 5.2 and the cube at the top left is the same as the one in Figure 5.2. Filled circles in the other cubes label the configurations $[\sigma]$ and $[\sigma']$. The Hamming distance between two configurations depicted as closed circles is the minimum number of edges joining them. The total number of transition modes is derived from eq (5.3) and corresponds to the total number of contracted partition functions for a macromolecule with $N = 3$.

$N - 1$ sites is held fixed. The operation of fixing M, $0 \leqslant M \leqslant N - 1$, sites in a given state has already been dealt with in the phenomenological theory in Section 3.2 and is called a *contraction*. The set of sites whose state is fixed is the *contraction domain* and the set of unconstrained sites is the *free domain*. Among the M sites in the contraction domain, a number j, $0 \leqslant j \leqslant M$, may be fixed in state 1. The total number of distinct contractions, and hence transition modes, that can be generated in this way is

$$v_N = 3^N - 2^N \tag{5.3}$$

This is because the number of ways of selecting M contracted sites is $C_{M,N} = N!/(N - M)!M!$, and likewise the number of ways of selecting j sites in state 1 is $C_{j,M} = M!/(M - j)!j!$. Hence the value of v_N is given by the double sum over M and j

$$v_N = \sum_{M=0}^{N-1} \sum_{j=0}^{M} C_{M,N} C_{j,M} = 3^N - 2^N \tag{5.4}$$

Alternatively one can observe that each of the N sites has three possible states: contracted in state 0, contracted in state 1 and free, giving 3^N possibilities in all. Of these, 2^N where all N units are contracted in either state 0 or 1 must be eliminated. Hence, the transition modes are merely the mean free energies of ligation derived from the ensemble of contracted partition functions of the system. This is easily understood from Figure 5.2. In the case $N = 1$ there is only one contracted partition function of order zero and hence only one possible mean free energy of ligation. In the case $N = 2$ there is one zero-order contracted partition function and four first-order contracted partition functions, giving a total of five possible values of the mean free energy of ligation per site. In the case $N = 3$ there are one zero-order, six first-order and twelve second-order contracted partition functions, giving a total of nineteen possible values of the mean free energy of ligation per site. Contracted partition functions of the same order as the number of sites are not included in eq (5.4), since they are all equal to one and do not depend on the ligand activity (see Section 3.2).

As to the free domain, we note that a transition mode can obviously be defined for any subsystem where the number of free sites is $N - M$. Consider an arbitrary configuration with M sites in the contraction domain, j of which are in state 1, while the remaining $N - M$ sites in the free domain are all in state 0. Let $[\tau]$ be the label for the vector associated with the configuration of the contracted sites. Then, $^{[\tau]}\Delta G_{\mathrm{m}}(N, M, j)$ is the transition mode associated with $[\tau]$ and represents the free energy change for switching all sites in the free domain from state 0 to 1, divided by $N - M$. For a fixed value of M and j, there are as many as $C_{M,N} C_{j,M}$ possible values of $^{[\tau]}\Delta G_{\mathrm{m}}(N, M, j)$. We are interested in the distribution of $\Delta G_{\mathrm{m}}(N, M, j)$ values in the space of the v_N possible contractions on the network and in how many of the contractions contribute to each transition mode. In principle, for a system where all sites are different and interact differently, such as the one dealt with in the phenomenological description, as many as v_N different values can be expected. The transition modes for CO binding to human hemoglobin are shown in Figure 5.4. The number of modes is very close to the actual number of contracted partition functions listed in Table 4.6. In the case of Ising networks, a high degeneracy of the transition modes is to be expected due to symmetry.

In a homogeneous network where all sites are equivalent, each

ΔG_{m} (kcal/mol)

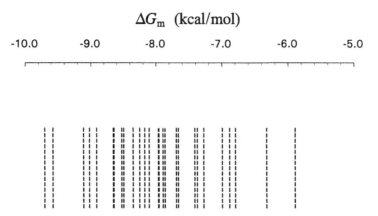

Figure 5.4 Transition modes for CO binding to human hemoglobin, obtained from the mean ligand activities associated with the contracted partition functions listed in Table 4.6. The values for the site-specific binding constants are listed in Table 4.8 and were derived from analysis of CO intermediates (Perrella, 1995).

transition mode has a degeneracy of $w(N, M, j) = C_{M,N} C_{j,M}$. The unligated state 0 of a site is used as reference in the free energy scale, and ΔG_{s} is the free energy of $0 \to 1$ switching of an isolated site. Two sites contact one another when in the graph associated with the network they are connected through a single edge. Specifically, any given site in the ring shown in Figure 5.1 contacts two sites, while the end sites in the linear chain contact only one site. If a site contacts r others in the network, its connectivity is r. In a homogeneous Ising network, r is uniform everywhere and is a measure of the connectivity of the network as a whole. In a heterogeneous network the connectivity can only be defined as an average, since different sites may contact a different number of other sites.

We now prove a theorem on the distribution of transition modes that holds for any Ising network. We choose the spin-lattice interaction rule, so that two sites in unlike states contribute an additional interaction free energy U to the total. The sign of U defines the nature of cooperativity, positive ($U > 0$) or negative ($U < 0$). Consider an arbitrary configuration of the sites in the contraction domain, with all sites in the free domain in state 0. The free energy associated with this configuration of N sites, relative to the reference with all sites unligated, is defined by the number of sites in state 1 and the number of 01 contact pairs in the network. Since each site in state 1 contributes ΔG_{s} to the free energy, and each 01 pair contributes U, a given configuration with M units in the contraction domain and $N - M$ units in state 0 in the free domain is characterized by a free energy

$$\Delta G_0(N, M, j) = j\Delta G_s + [\chi_0(N,M, j) + \psi(N, M, j)]U \quad (5.5)$$

where $\chi_0(N, M, j)$ is the number of 01 pairs across the boundary between the free and contraction domains, and the suffix zero labels the state of all sites in the free domain. The coefficient $\psi(N, M, j)$ gives the number of 01 pairs inside the contraction domain. Upon switching all sites in the free domain to state 1 one has

$$\Delta G_1(N, M, j) = (N - M + j)\Delta G_s + [\chi_1(N, M, j) + \psi(N, M, j)]U$$

$$(5.6)$$

where $\chi_1(N, M, j)$ denotes the number of 01 pairs across the boundary between the free and contraction domains when all sites in the free domain are in state 1. The transition mode associated with the foregoing transformation is

$$\Delta G_m(N, M, j) = \Delta G_s + \frac{\chi_1(N, M, j) - \chi_0(N, M, j)}{N - M}U \quad (5.7)$$

and is independent of $\psi(N, M, j)$. In the special case where $M = 0$, there is no contraction domain and eq (5.7) gives the mean free energy change per unit in switching from the reference configuration to that where all sites are in state 1. We know from the phenomenological theory that this transition mode is related to the mean ligand activity, x_m, and we will refer to it henceforth as the *standard transition mode* $\Delta G_m^\circ = \Delta G_s$. The sum of the possible 01 pairs $\chi_1(N, M, j) + \chi_0(N, M, j)$ is the total number, $\chi_{tot}(N, M)$, of physical contacts between the domains. Hence (Keating and Di Cera, 1993),

Theorem: The distribution of transition modes for an Ising network is the sum of subdistributions symmetric about the standard transition mode ΔG_m° and itself possesses this symmetry.

To prove this theorem let us introduce the dual domain, defined as that obtained from the contraction domain by switching the state of all the contracted sites. All those in state 1 switch to state 0 and *vice versa*. This leaves M unchanged and results in the replacement $j \to M - j$. The sum of the possible 01 pairs χ_0 and χ_1 for a given contraction domain and its dual is the total number of interdomain contacts, i.e.,

$$\chi_0(N, M, j) + \chi_0(N, M, M - j) = \chi_{tot}(N, M) \quad (5.8a)$$

$$\chi_1(N, M, j) + \chi_1(N, M, M - j) = \chi_{tot}(N, M) \quad (5.8b)$$

The sum of the transition modes for any contraction domain and its dual is

$$\Delta G_\mathrm{m}(N, M, j) + \Delta G_\mathrm{m}(N, M, M - j) = 2\Delta G_\mathrm{m}^\circ +$$

$$\frac{\chi_1(N, M, j) + \chi_1(N, M, M - j) - \chi_0(N, M, j) - \chi_0(N, M, M - j)}{N - M} U$$

$$(5.9)$$

and from eqs (5.8a) and (5.8b) it follows that

$$\Delta G_\mathrm{m}(N, M, j) + \Delta G_\mathrm{m}(N, M, M - j) = 2\Delta G_\mathrm{m}^\circ \qquad (5.10)$$

Hence, the transition modes of a given contraction domain and its dual are equidistant from the standard transition mode. Furthermore, since $w(N, M, j) = C_{M,N} C_{j,M} = C_{M,N} C_{M-j,M} = w(N, M, M - j)$, the intrinsic weighting of any contraction, and the degeneracy of the transition mode associated to it, is equal to that of its dual. For M fixed, the components in the distribution of transition modes generated by changing j occur in pairs symmetrically disposed about the standard transition mode $\Delta G_\mathrm{m}^\circ$. The self-dual contraction, $M = 0$, also obeys this rule as a special case. Thus, the subdistribution for any given M is symmetric about $\Delta G_\mathrm{m}^\circ$. The overall distribution of transition modes, which is the sum of these subdistributions, must also possess this symmetry. Hence the theorem.

It should be mentioned that if the lattice-gas rule of interaction is employed, then a weaker form of the foregoing theorem holds, requiring that the network be homogeneous. The generality of the theorem in the case of spin-lattice interaction rule is quite interesting and bears on the results embodied by the Lee–Yang theorem of statistical mechanics (Lee and Yang, 1952). This theorem asserts that the partition function of an Ising network under the spin-lattice interaction rule is always symmetric, with the roots distributed on the unit circle in complex plane. The theorem on the transition mode distribution provides further information on the properties of the partition function by stating that all of its contracted forms are also symmetric. This corollary to the theorem follows from the fact that if a given configuration enters the definition of a contracted partition function, so does its dual with identical degeneracy. As for the Lee–Yang theorem, our results are independent of the dimensional embedding of the network, since the analysis is based entirely on combinatorial arguments.

The connectivity of the network, r, and hence its dimensionality, comes into play when the range of each subdistribution for fixed M is to be determined. This quantity gives a measure of how the transition modes spread over the ΔG axis for a given value of M. Any given transition mode can be written as

$$\Delta G_m(N, M, j) = 2\Delta G_m^0 - \Delta G_m(N, M, M - j)$$

$$= \Delta G_m^0 - \frac{\chi_{tot}(N, M) - 2\chi_1(N, M, j)}{N - M} U \qquad (5.11)$$

The number of 01 pairs, $\chi_1(N, M, j)$ varies from 0 for $j = M$, to $\chi_{tot}(N, M)$ for $j = 0$. Hence, the limits for the transition modes for a given value of M are

$$\pm\Delta G_m(N, M, j) = \Delta G_m^0 \pm \frac{\chi_{tot}(N, M)}{N - M} U \qquad (5.12)$$

and we seek the maximum value of $\chi_{tot}(N, M)$ for a given M. Suppose that $M \leqslant N/r$. Then there is a disposition of the contracted sites such that there is no contact within the contraction domain, in which case $\chi_{tot}(N, M) = rM$ and $\pm\Delta G_m(N, M, j) = \Delta G_m^\circ \pm M(N - M)^{-1}rU$. Likewise, for $M \geqslant N/r$ there is a disposition of the contracted sites such that there is no contact within the free domain, so that $\chi_{tot}(N, M) = r(N - M)$ and $\pm\Delta G_m(N, M, j) = \Delta G_m^\circ \pm rU$. The transition modes of the overall distribution satisfy the condition

$$\Delta G_m^\circ - rU \leqslant \Delta G_m(N, M, j) \leqslant \Delta G_m^\circ + rU \qquad (5.13)$$

Since the term $\chi_{tot}(N, M) - 2\chi_1(N, M, j)$ in eq (5.11) must be an integer, each subdistribution for M fixed has a characteristic spacing of $U/(N - M)$. This is the minimum possible distance between two transition modes within the subdistribution. The set of values that $\chi_{tot}(N, M) - 2\chi_1(N, M, j)$ can take on depends on the details of the network, such as the value of r and the dimensional embedding.

To illustrate the properties of the transition mode distribution we consider the special case of an Ising network where each site contacts every other site. In this case the connectivity is $r = N - 1$ and both $\chi_{tot}(N, M)$ and $\chi_1(N, M, j)$ are independent of the disposition of the contracted sites and are uniquely determined once N, M and j are specified. For any given contraction one has $\chi_{tot}(N, M) = M(N - M)$ and $\chi_1(N, M, j) = (M - j)(N - M)$, so that

$$\Delta G_m(N, M, j) = \Delta G_m^\circ + (M - 2j)U = \Delta G_m^\circ + nU \qquad (5.14)$$

where $n = M - 2j$ goes from $-(N - 1)$ to $(N - 1)$, since $M = 0, 1, \ldots N - 1$ and $j = 0, 1, \ldots M$. The transition modes for this network are uniformly distributed in the range $\pm(N - 1)U$, with a spacing of U. For fixed M, the distribution is symmetric around $j = M/2$, as implied by the theorem, has a spacing of $2U$ and follows the binomial $C_{j,M}$. For N large, M is also large and the binomial distribution $C_{j,M}$ goes into a Gaussian with mean $M/2$ and variance $M/4$, as implied by the De Moivre–Laplace theorem (Wilson, 1911; Feller, 1950). The Gaussian approximation is remarkably good even for small values of N. Hence, each subdistribution for fixed M can be approximated by a Gaussian distribution with mean and variance given by $\mu(N, M) = \Delta G_{\mathrm{m}}^{\circ}$ and $\sigma^2(N, M) = MU^2$. The mean is independent of M, while the variance grows linearly with the size of the contraction domain, independent of N. This is equivalent to a Brownian motion in one dimension around the origin $x = \Delta G_{\mathrm{m}}^{\circ}$, where M plays the role of time and the diffusion coefficient D is proportional to U^2 (Chandrasekhar, 1943). In the absence of interactions, as expected for a reference system, $U = 0$ and no diffusion of modes is observed around the standard transition mode. The overall distribution is a δ-function peaked at the origin. For finite U, and regardless of whether interactions stabilize $(U < 0)$ or destabilize $(U > 0)$ 01 pairs in contact, the mode distribution spreads out with increasing M under the effect of a 'drift' provided by the interaction energy U. The further away two configurations $[\sigma]$ and $[\sigma']$ in the ensemble are, the more likely that the transition mode would be close to the standard transition mode $\Delta G_{\mathrm{m}}^{\circ}$. The greater the Hamming distance between two configurations, the smaller is the variance of the distribution of transition modes associated with it. Each subdistribution has a variance of MU^2 and an intrinsic weight of 2^M. The overall distribution of transition modes is the sum of all subdistributions obtained for fixed M, times the number of ways of fixing M units among N, $C_{M,N}$.

Some overlap exists among the subdistributions since all values of M and j such that $M - 2j = n = $ constant contribute to the same mode. It is quite instructive to derive analytically the degeneracy, $P(n)$, of a given mode as follows. By definition

$$P(n) = \sum_{M=0}^{N-1} \sum_{j=0}^{M} C_{M,N} C_{j,M} \delta_{mn} \tag{5.15}$$

where the Kronecker delta, δ_{mn}, filters out all values of $m = M - 2j$ not equal to the prescribed value of n. The form of $P(n)$ can be found by introducing the integral representation of δ_{mn} (Pearson, 1990)

$$\delta_{mn} = \frac{1}{2\pi} \int_0^{2\pi} \exp\left[i(n - m)\varphi\right] d\varphi \tag{5.16}$$

Interchanging eq (5.16) with the double sum in eq (5.15) yields

$$P(n) = \frac{1}{2\pi} \int_0^{2\pi} \exp\left(in\varphi\right) \sum_{M=0}^{N-1} \sum_{j=0}^{M} C_{N,M} C_{M,j} \exp\left[i(2j - M)\varphi\right] d\varphi \tag{5.17}$$

Elementary transformations lead to

$$P(n) = \frac{1}{2\pi} \int_0^{2\pi} \exp(in\varphi)[(1 + 2\cos\varphi)^N - 2^N \cos^N \varphi]\,d\varphi \quad (5.18)$$

and symmetry of the integrand gives the equivalent form

$$P(n) = \frac{1}{\pi} \int_0^{\pi} \exp(in\varphi)[(1 + 2\cos\varphi)^N - 2^N \cos^N \varphi]\,d\varphi \quad (5.19)$$

In the limit of large N, the expression in square brackets has a dominant peak with maximum 3^N at $\varphi = 0$. The value of the maximum is the expected value of v_N in the same limit. The Taylor expansion around this point yields

$$(1 + 2\cos\varphi)^N - 2^N \cos^N \varphi \approx 3^N \left(1 - \frac{\varphi^2}{3}\right)^N \approx 3^N \exp\left(-\frac{N\varphi^2}{3}\right) \quad (5.20)$$

In this limit the upper bound of the integral can be extended to ∞ and substitution of eq (5.20) into eq (5.19) yields

$$P(n) = \frac{3^N}{\pi} \int_0^{\infty} \exp(in\varphi) \exp\left(-\frac{N\varphi^2}{3}\right)\,d\varphi = \frac{3^N}{2\sqrt{\pi N/3}} \exp\left(-\frac{3n^2}{4N}\right)$$
$$(5.21)$$

The distribution $P(n)$ of n values is the Fourier transform of a Gaussian distribution and is therefore itself a Gaussian (Pearson, 1990), with a mean of zero and a variance equal to $2N/3$. The overall distribution of transition modes for this Ising network is a Gaussian with mean and variance equal to $\mu(N) = \Delta G_{\mathrm{m}}^{\circ}$ and $\sigma^2(N) = 2NU^2/3$. This distribution is shown in Figure 5.5. The distribution spans the expected range $\Delta G_{\mathrm{m}}^{\circ} \pm (N - 1)U$ with the $2N - 1$ distinct modes being uniformly distributed with a spacing of U.

It is of interest to compare the uniform distribution of modes in the range $\Delta G_{\mathrm{m}}^{\circ} \pm (N - 1)U$ for the foregoing Ising network with that actually obtained from the intermediates of the FeII/FeIII–CN system of hemoglobin as a function of pH (Daugherty *et al.*, 1994). This system has been dealt with in detail in Section 4.4. The spectra of transition modes are shown in Figure 5.6 as a function of pH. The modes are expressed relative to the standard value $\Delta G_{\mathrm{m}}^{\circ}$ for the sake of comparison. There are a number of striking features revealed by these spectra. First, the range spanned decreases with increasing pH, thereby indicating that cooperativity decreases with pH. Second, the pattern observed at pH < 8.0 is distinctly more 'discrete' than that observed at pH > 8.0. Three main clusters of modes are clearly distinguishable at pH < 8.0 and tend to merge at higher pH. Third, in no case is the pattern consistent with that of

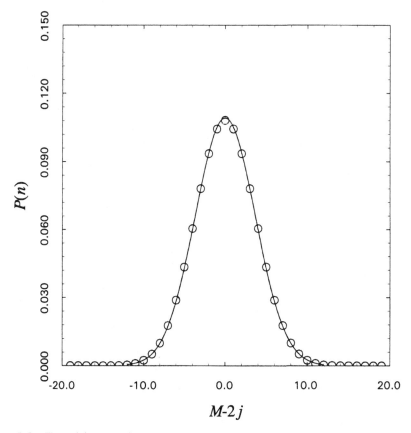

Figure 5.5   Transition mode distribution for a homogeneous Ising network with $N = 20$ and $r = N - 1$. The continuous curve was drawn according to eq (5.21), while points were calculated by computer analysis of the network exhausting all possible configurations (Keating and Di Cera, 1993).

an Ising network with all sites coupled to the same extent, thereby suggesting that coupling in the hemoglobin molecule involves interactions of different strength that change drastically at pH > 8.0. These conclusions concur with the results of our analysis in Section 4.4.

5.2 A combinatorial model for molecular recognition

The concept of transition mode can be exploited in the analysis of molecular recognition, a basic problem in physical and computational biochemistry that is tantamount to predicting free energies of interaction from structural components. Here we discuss a simple model that reduces molecular recognition to a combinatorial problem (Keating and Di Cera,

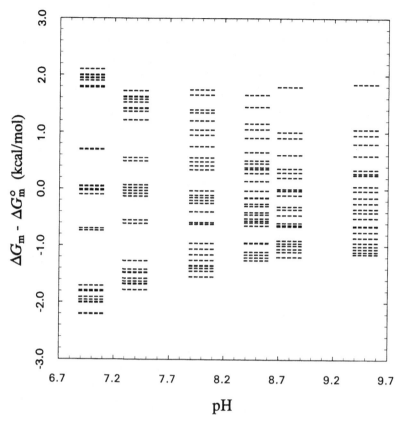

Figure 5.6 Spectral representation of the transition modes for the FeII/FeIII–CN hemoglobin system as a function of pH. The modes were obtained from the mean ligand activities associated with the contracted partition functions listed in Table 4.6, using values for the site-specific binding constants derived by Daugherty *et al.* (1994).

1993). Let us consider two surfaces, A and B, with A containing N acceptor subsites for N donor subsites of B as shown in Figure 5.7. For the sake of simplicity surface A is assumed to be inert, i.e., its energetic state is the same when free or bound to B. The donors of B can exist in two energetically equivalent states, 0 and 1. When surface B is free, all donors are in state 0. When surface B is bound to surface A all donors switch to state 1 in a concerted fashion. Each pair 'donor in state 1 : acceptor' provides a free energy of stabilization, $\Delta G°$, across the A–B interface of the bound complex. State 0 can be interpreted as a neutral state with the pair 'donor in state 0 : acceptor' providing neutral stabilization to the bound complex. The free energy of binding of surface B to A is simply $\Delta G_0 = N\Delta G°$. This is a very simple model for the interaction of

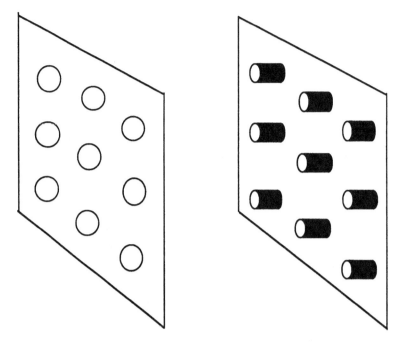

Figure 5.7 Schematic representation of molecular recognition as a combinatorial problem. Two surfaces, A (left) and B (right), form a complex involving the interaction of a number of complementary subsites. Each ensemble of subsites can be modeled as a two-dimensional Ising network and the free energy distribution for the binding reaction can be derived from the properties of the transition modes of the networks.

two macromolecules and can also be used in the analysis of the protein folding problem by allowing the two surfaces A and B to be distinct domains of the same macromolecule. Protein folding is, in fact, a problem of molecular self-recognition, as illustrated by the elegant work of Rose and his colleagues (Presta and Rose, 1988; Harper and Rose, 1993; Lattman and Rose, 1993; Aurora, Srinivasan and Rose, 1994).

Assume now that a number of site-specific 'mutations' are made on surface B. We assume that any mutation may have the effect of 'freezing' one or more donors in either one of the allowable energy states, 0 or 1. In principle, there is no way one can predict the effect of the mutation on the free energy value for the formation of the AB complex, since the mutation can act locally at the subsite where it applies and/or propagate to distant sites. However, regardless of the extent and precise nature of the perturbation, we know that surface B free can assume only one of the 2^N possible configurations allowed for the manifold of donors. Assume that the mutation freezes M donors, j of which are in state 1, and that the probabilities of freezing a donor in state 0 or state 1 are the same. In mathematical terms, the effect of the mutation is equivalent to generating a contraction domain in the system. The free energy of binding of this mutant to surface A can be computed as follows. A total of $N - M$ donors are free to switch to state 1 when bound to surface A. These donors provide a free energy change $(N - M)\Delta G°$. Among the frozen donors, $M - j$ remain in state 0 even when bound to surface A. These donors provide no contribution to the binding free energy. On the other hand, the j donors frozen in state 1 provide an additional term $j\Delta G°$ to the total free energy of binding, which is then

$$\Delta G(N, M, j) = (N - M + j)\Delta G° \tag{5.22}$$

The binding free energy difference between the mutant and the wild-type surface B is evidently

$$\Delta\Delta G(N, M, j) = -(M - j)\Delta G° \tag{5.23}$$

In this simple model any mutation is expected to reduce the binding affinity, unless the number of donors frozen in state 1 coincides with the total number of frozen donors M. The entire spectrum of $\Delta\Delta G$ values for the system is generated by changing M and j. For a fixed value of M, the values of $\Delta G(N, M, j)$ are distributed as a Gaussian with mean $\mu(N, M) = (N - M/2)\Delta G°$ and variance $\sigma^2(N, M) = (M/4)\Delta G°^2$. The probability that the binding free energy for any mutant is x is given by the expression

$$P(x) = \sum_{M=0}^{N-1} w(N, M) \exp\left\{-\frac{1}{2}\left[\frac{x - \mu(N, M)}{\sigma(N, M)}\right]^2\right\} \tag{5.24}$$

where $w(N, M)$ is an *ad hoc* weighting factor for the subdistribution of free energy values obtained for a given M. The basic assumption of the foregoing model is that the system (surface B) explores a unique 'free energy space' when structural perturbations are considered. Single- or multiple-site mutations can be analyzed using the same framework. Essentially, the number of allowable free energy values for binding is finite and is determined by the combinatorics of the network of donors.

The predictions of this model can be tested by analyzing free energy profiles obtained for macromolecular systems and their mutant derivatives. The relevant parameters of the model can be obtained by analyzing the distribution of $\Delta G(N, M, j)$ values obtained experimentally in terms of a combination of Gaussian terms, whose mean and variance are to be considered as independent parameters. Each Gaussian term should be weighted properly using another set of independent parameters for the filters $w(N, M)$. Once the best-fit values of μ and σ^2 are determined for all Gaussian terms, then a plot of σ^2 versus μ should be constructed. The model predicts that such a plot would be a straight line and specifically

$$\sigma^2 = \Delta G^\circ \frac{N \Delta G^\circ - \mu}{2} \tag{5.25}$$

This plot yields the number of subsites N and the free energy of binding provided by a single 'donor in state 1 : acceptor' pair across the A–B interface. Both parameters are of considerable interest in relating perturbed energetics to structural features.

In the case where interactions exist among the donor subsites of surface B, the treatment is more elaborate. We make use of the results derived in Section 5.1 and assume that each donor is in contact with $r = N - 1$ other donors. Two donors in contact interact according to an energy U only if they are in different states. The interaction network of this fully connected Ising network resembles a 'molecular field' (Hill, 1960; Itzykson and Drouffe, 1989), since the state of any given donor in the surface will be felt by all other donors in unlike states. The effect of a mutation on surface B is more complex than that discussed in the case of independent donors. When M units are frozen, j of which are in state 1, the energetic state of the surface can be predicted from the number of donors in states 0 and 1 and the interactions among them as $G_B(\text{free}) = j(N - j)U$. When the surface is bound, all donors in the free domain switch to state 1, while the donors in the contracted domain remain frozen. The new energetic state of the surface is $G_B(\text{bound}) = (N - M + j)(M - j)U$. The energetic state of the interface between A and B is computed by counting the number of 'donor in state 1 : acceptor' pairs. The free energy of binding of the two surfaces is evidently

$$\Delta G(N, M, j) = (N - M + j)\Delta G^\circ + (N - M)(M - 2j)U \tag{5.26}$$

and reduces to eq (5.22) in the absence of interactions. The free energy of binding is closely related to the transition modes of surface B and the results derived in the previous section can be exploited in the analysis of the A–B interaction. It follows from the results in Section 5.1 that $\Delta G(N, M, j)$ is symmetric around $\Delta G(N, M, M/2)$ for fixed M and that the $\Delta G(N, M, j)$ values are uniformly distributed with a spacing of $\Delta G^\circ - 2(N - M)U$. The subdistribution for M fixed can be approximated by a Gaussian with mean and variance

$$\mu(N, M) = \left(N - \frac{M}{2}\right)\Delta G^\circ \tag{5.27a}$$

$$\sigma^2(N, M) = \frac{M}{4}[\Delta G° - 2(N - M)U]^2 \qquad (5.27b)$$

and the overall distribution of $\Delta G(N, M, j)$ values can be written as the superposition (5.24). The relationship between σ^2 and μ is

$$\sigma^2 = (N\Delta G° - \mu)\frac{(\Delta G°^2 + 2NU\Delta G° - 4\mu U)^2}{2\Delta G°^3} \qquad (5.28)$$

The variance σ^2 is a cubic function of the mean μ, a very peculiar result. The linear dependence of σ^2 on μ in eq (5.25), observed in the case of independent donors, is recovered as expected for $U = 0$.

The dependence of σ^2 on μ predicted by eq (5.28) has been tested and confirmed experimentally in the biologically relevant case of thrombin–hirudin interaction (Keating and Di Cera, 1993). The distribution of

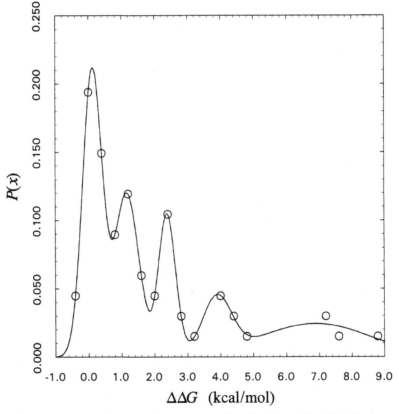

Figure 5.8 Distribution of binding free energies for thrombin–hirudin interaction relative to the wild-type (-17.93 ± 0.09 kcal/mol). The ΔG values obtained for sixty-seven mutants were grouped in bins of 0.4 kcal/mol to determine the number of mutants within each free energy interval. The curve was drawn according to eq (5.24) using five Gaussian components. The best-fit values of μ and σ^2 for each component are shown in Figure 5.9.

binding free energies is shown in Figure 5.8, as derived from the analysis of sixty-seven mutants of hirudin in their reaction with thrombin (Wallace, Dennis, Hofsteenge and Stone, 1989; Betz, Hofsteenge and Stone, 1991). These mutants have single- and multi-site substitutions of polar and apolar side chains over the entire surface of recognition, which spans about 12% of the available surface area of thrombin (Rydel, Tulinsky, Bode and Huber, 1991). The landscape of $\Delta\Delta G$ values spans a range of nearly 10 kcal/mol, as shown in Figure 5.8, and is distinctly multimodal. Individual Gaussian components can easily be detected. Hirudin is modeled as surface B and thrombin as surface A. The continuous line depicts the best-fit of the experimental data to a superposition of Gaussian terms, whose mean and variance are given in Figure 5.9. The cubic dependence of σ^2 on μ predicted by the model is indeed seen experimentally. All independent parameters in eqs (5.27a) and (5.27b) can be resolved from nonlinear least squares. In particular, the number of recognition subsites,

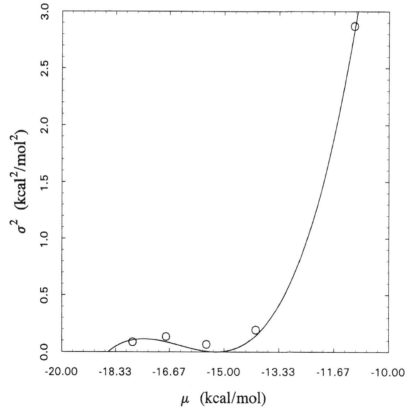

Figure 5.9   Dependence of σ^2 on μ for the five Gaussian components shown in Figure 5.8. The prediction drawn from eq (5.28) is confirmed experimentally.

N, is found to be 12 ± 3, a value very close to the number of ion pairs and hydrogen bonds revealed by the crystal structure of the thrombin–hirudin complex (Rydel *et al.*, 1991). Each subsite contributes -1.5 ± 0.3 kcal/mol to the binding free energy. The interaction energy $U = -0.10 \pm 0.04$ kcal/mol indicates that neighboring subsites are negatively linked and provides a measure of the energetic cost of switching one donor of the contraction domain to state 1. For example, if a mutation causes $M = 1$ and $j = 1$, then the free energy cost that translates in a loss of binding free energy is $-(N - 1)U = 1.1$ kcal/mol.

The example dealt with above should be considered paradigmatic for the development of more elaborate treatments of molecular recognition using Ising networks. It is quite interesting that a simple framework is capable of encapsulating the salient features of a protein–protein inter-action and its perturbation by site-directed mutations. Clearly, the treat-ment used for surface B can be extended to surface A and both surfaces can be assumed to have their own combinatorial switching mechanism upon binding. Arbitrary networks of interaction among recognition sub-sites can be introduced, consistent with information about the detailed crystal structure of the complex. Also, different contacts can be assigned different energetics. In general, the two surfaces should be modeled as heterogeneous two-dimensional Ising networks and the combinatorics should be exhausted by computer for each particular case. Alternatively, an analytical treatment of such networks for systems of biological relev-ance should be pursued.

5.3 Site-specific effects in Ising networks

Homogeneous networks provide useful modes of cooperative transitions, but heterogeneous networks embody most of the fundamental features of biological macromolecules, as they allow for intrinsic differences among individual sites and asymmetric or arbitrary patterns of interactions. An important aspect of heterogeneous networks is that they lend themselves to the analysis of properties arising locally at the level of each individual site. This feature stands in contrast to homogeneous networks where symmetry renders the global properties of the system a scaled replica of those arising at any site in the network. Obviously, the mathematical analysis of heterogeneous networks is far more complex. There are essentially two ways of approaching the study of Ising networks. The classical approach is based on the 'transfer matrix' method and allows for calculation of the partition function of the network in many cases of interest (Onsager, 1944; Newell and Montroll, 1953; Poland and Scheraga, 1970; Stanley, 1971). The other approach is somewhat more general and is based on consideration of important topological properties of the network. We will start our discussion with the former approach, which is more 'technical' and then consider aspects of the latter approach, which is more 'intuitive'.

Consider the one-dimensional Ising ring in Figure 5.1. If all sites are equivalent, the ring is a homogeneous network for which global and local properties are derived by simple rescaling. We assume the lattice-gas rule of interaction (Mayer and Mayer, 1940; Yand and Lee, 1952; Hill, 1960) because it simplifies the mathematics significantly, and define the inter-action constant between two neighboring sites in state 1 as σ. Positive or negative coupling between neighboring sites correspond to $\sigma > 1$ or $\sigma < 1$ respectively. The reference system is obtained for $\sigma = 1$. We start with the analysis of the ring to illustrate the basic properties of the transfer matrix. Consider two neighboring sites on the ring, as in Figure 5.10, and their four allowable configurations. The transfer matrix is an operator that encapsulates the allowable states as follows

$$\mathbf{W} = \begin{pmatrix} 1 & 1 \\ \omega & \sigma\omega \end{pmatrix} \tag{5.29}$$

The meaning of each term of $\mathbf{W}$ can be understood as follows. If the two sites in Figure 5.10 are considered in terms of the reference cycle dealt with in Section 3.1, the partition functions of site j when site $j + 1$ is unligated or ligated are respectively

$$^0\Psi_{j+1}(x) = 1 + K_j x = 1 + Kx = 1 + \omega \tag{5.30a}$$

$$^1\Psi_{j+1}(x) = 1 + \sigma K_j x = 1 + \sigma Kx = 1 + \sigma\omega \tag{5.30b}$$

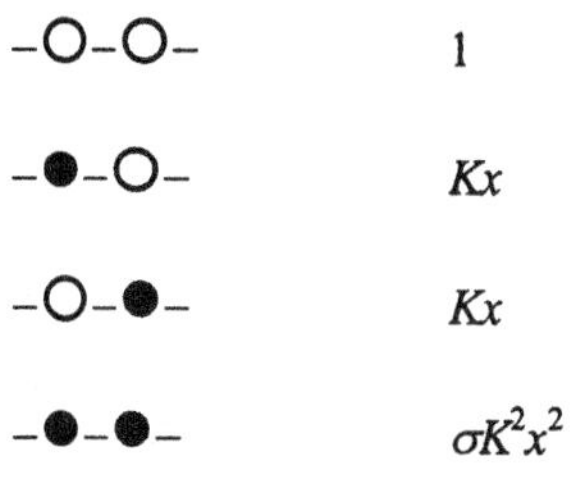

$$
\begin{array}{ccc}
\text{–O–O–} & & 1 \\[1em]
\text{–●–O–} & & Kx \\[1em]
\text{–O–●–} & & Kx \\[1em]
\text{–●–●–} & & \sigma K^2 x^2
\end{array}
$$

Figure 5.10 Possible configurations of a nearest-neighbor pair including sites j (left) and $j + 1$ (right) in the Ising ring shown in Figure 5.1. Open and filled circles depict free and bound states respectively. Contacts between sites are indicated by a connecting line. The statistical term weighting each configuration is also indicated, assuming the lattice-gas rule of interaction. K is the site-specific binding constant, x the ligand activity and σ the interaction parameter.

The terms in eqs (5.30a) and (5.30b) involve the site-specific binding constant for site j, K_j, the interaction constant, σ, and the ligand activity, x. Since the sites are identical, $K_j = K$ for all sites and, for the sake of simplicity, the product Kx can be used to define a dimensionless activity variable ω. The meaning of ω is that of a ligand activity measured in K^{-1} units. The terms defining the contracted partition functions (5.30a) and (5.30b) are the same as those entering the definition of the transfer matrix $\mathbf{W}$. The first column of $\mathbf{W}$ is constructed from the terms of ${}^{0}\Psi_{j+1}$ and labels the configurations of site j with site $j + 1$ unligated. Likewise, the second column of $\mathbf{W}$ is constructed from the terms of ${}^{1}\Psi_{j+1}$ and labels the configurations of site j with site $j + 1$ ligated. Hence, the elements of the transfer matrix $\mathbf{W}$ can be mapped into terms defining specific configurations in the contracted partition functions derived for each neighboring pair.

The partition function of the ring is obtained by multiplying together the transfer matrices for all possible neighboring pairs around the ring and taking the trace of the product (Hill, 1960; Poland and Scheraga, 1970; Stanley, 1971), i.e.,

$$\Psi(\omega) = \text{Tr}\,(\mathbf{W}_{12}\mathbf{W}_{23} \ldots \mathbf{W}_{N-1N}\mathbf{W}_{N1}) \tag{5.31}$$

This is the general solution for a ring of N sites, with N distinct transfer matrices containing different interaction constants for each pair and different binding constants for each site. In the specific case of a homogeneous ring, the sites bind with the same affinity and interact equally, so that there are N equivalent pairs and eq (5.31) simplifies to

$$\Psi(\omega) = \text{Tr}\,\mathbf{W}^N \tag{5.32}$$

The explicit form of Ψ is obtained by switching to an orthogonal basis set. This is equivalent to diagonalizing $\mathbf{W}$, a transformation that leaves the trace unchanged (Gantmacher, 1959). The similarity transformation is well known in matrix algebra and takes on the familiar form

$$\Psi(\omega) = \text{Tr}\,\mathbf{W}^N = \text{Tr}\,\mathbf{\Lambda}^N = \lambda_1^N + \lambda_2^N \tag{5.33}$$

where $\mathbf{\Lambda}$ is the diagonal form of $\mathbf{W}$. The eigenvalues of $\mathbf{W}$ are the solutions of the secular equation

$$\begin{vmatrix} 1 - \lambda & 1 \\ \omega & \sigma\omega - \lambda \end{vmatrix} = (1 - \lambda)(\sigma\omega - \lambda) - \omega = 0 \tag{5.34}$$

Hence,

$$\lambda_{1,2} = \frac{1 + \sigma\omega \pm \sqrt{(1 - \sigma\omega)^2 + 4\omega}}{2} \tag{5.35}$$

Once Ψ is known, all response functions of interest are derived from differentiation as dealt with extensively in Chapter 2. The number of ligated sites is $X = \mathrm{d}\ln\Psi/\mathrm{d}\ln\omega$ and the binding capacity is $B = \mathrm{d}X/\mathrm{d}\ln\omega$. The function X is shown in Figure 5.11 for various values of the interaction parameter σ. The case $\sigma = 1$ yields the properties of a reference system as already dealt with in Chapter 2.

We now consider a simple modification of the Ising ring that is of particular biological relevance and that leads to consideration of site-specific effects. If the ring is 'perturbed' by eliminating the contact between two sites, say 1 and N, a linear chain is obtained as shown in Figure 5.1. Although the sites bind with the same affinity K, the perturbation introduces an asymmetry in the network. As a result, the sites are no longer equivalent and the network as a whole is heterogeneous. The

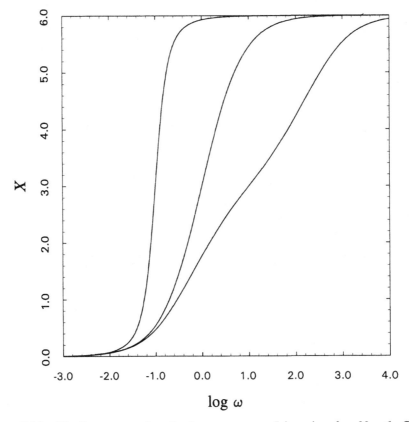

Figure 5.11 Binding curves for the homogeneous Ising ring for $N = 6$. Curves were drawn using the partition function in eq (5.33) with values of $\sigma = 10$ (left), $\sigma = 1$ (middle) and $\sigma = 0.1$ (right). The values of σ correspond to positive, no and negative cooperativity respectively.

partition function in this case involves $N - 1$ transfer matrices of the same form as $\mathbf{W}$, but also two end vectors such that (Poland and Scheraga, 1970)

$$\Psi(\omega) = \mathbf{L}_+\mathbf{W}^{N-1}\mathbf{L}_- \tag{5.36}$$

where

$$\mathbf{L}_+ = (1 \quad 1) \tag{5.37a}$$

$$\mathbf{L}_- = \begin{pmatrix} 1 \\ \omega \end{pmatrix} \tag{5.37b}$$

Calculation of the partition function is lengthy, but straightforward. However, instead of pursuing the matrix approach, which is somewhat tedious in the general case, we shall use a different approach to derive Ψ for the linear chain. The advantage of this alternative approach, which exploits contracted partition functions, will soon be evident. We use the fact that in the lattice-gas rule the interaction constant σ between neighboring sites only applies when both sites are in state 1. Hence, keeping a site in its unligated state is equivalent to deleting the site from the network. In fact, neighboring sites will not 'feel' the presence of the site being frozen in a particular state, unless it is in state 1. Freezing a site in state 0 among N sites leaves the properties of the remaining $N - 1$ sites unaltered. This fact has far-reaching consequences.

Consider any of the N sites of a homogeneous ring. The site-specific binding curve of this site is derived from the phenomenological theory as

$$X_i = 1 - \frac{{}^0\Psi_i(\omega)}{\Psi(\omega)} \tag{5.38}$$

where Ψ is the partition function of the ring. Since the network is homogeneous, it makes no difference which site is considered and $X_i = \theta = X/N = (1/N)\,\mathrm{d}\ln\Psi/\mathrm{d}\ln\omega$ for all sites. The contracted partition function in the numerator is obtained from that of the ring when site i is frozen in its unligated form. This is equivalent to dropping site i from the ring and, since all units are equivalent, to generating a linear chain of $N - 1$ sites from a ring of N sites. The contracted partition function in the numerator of eq (5.38) is therefore the partition function for a linear chain of $N - 1$ sites. Rearranging eq (5.38) and using the foregoing definitions leads to

$$^{\mathrm{L}}\Psi_N(\omega) = {}^{\mathrm{R}}\Psi_{N+1}(\omega) - \frac{1}{N+1}\frac{\mathrm{d}^{\mathrm{R}}\Psi_{N+1}(\omega)}{\mathrm{d}\ln\omega} \tag{5.39}$$

Knowledge of the partition function of a ring of $N + 1$ sites, $^{\mathrm{R}}\Psi_{N+1}$, defines the partition function of a linear chain of N sites, $^L\Psi_N$, through the transformation in eq (5.39). Using the eigenvalues of the transfer matrix and eq (5.33) yields

$$^L\Psi_N(\omega) = \lambda_1^N(\lambda_1 - \lambda_1') + \lambda_2^N(\lambda_2 - \lambda_2') \qquad (5.40)$$

where $\lambda_{1,2}' = d\lambda_{1,2}/d\ln\omega$ is given by

$$\lambda_{1,2}' = \frac{\sigma\omega\sqrt{(1 - \sigma\omega)^2 + 4\omega} \pm \omega[2 - \sigma(1 - \sigma\omega)]}{2\sqrt{(1 - \sigma\omega)^2 + 4\omega}} \qquad (5.41)$$

Consideration of contracted partition functions yields a straightforward derivation of the properties of the linear chain from those of the Ising ring.

The foregoing method can be extended to the analysis of site-specific effects in the linear chain. Unlike the ring, the linear chain is heterogeneous and lends itself to consideration of site-specific effects. Quite interestingly, analysis of these effects can be parameterized in terms of the interaction constant σ alone. The site-specific properties of the linear chain can be derived by consideration of the basic eq (5.38), which applies regardless of the particular model being considered. In the case of a linear chain, the contracted partition function $^0\Psi_i$ has a very simple interpretation, as shown in Figure 5.12. When site i is kept in its unligated state, the chain is cut into two smaller chains containing $i - 1$ and $N - i$ sites. Hence,

$$^0\Psi_i(\omega) = {}^L\Psi_{i-1}(\omega)^L\Psi_{N-i}(\omega) \qquad (5.42)$$

and each term on the right-hand side of (5.42) is of the form (5.40). The partition function of the linear chain can be rewritten in a more compact form as follows

$$^L\Psi_N(\omega) = \lambda_1^{N+1}[1 - \alpha_1 + (1 - \alpha_2)\Delta^{N+1}] = \lambda_1^{N+1}(\alpha_2 + \alpha_1\Delta^{N+1}) \quad (5.43)$$

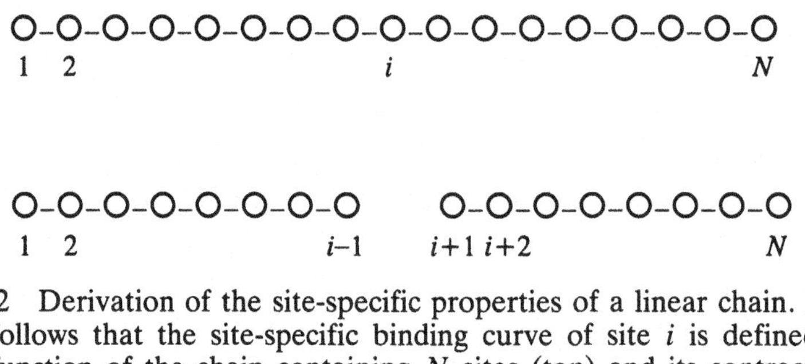

O–O–O–O–O–O–O–O–O–O–O–O–O–O–O–O–O
1 2 i N

O–O–O–O–O–O–O–O   O–O–O–O–O–O–O–O
1 2 i–1 i+1 i+2 N

Figure 5.12 Derivation of the site-specific properties of a linear chain. From eq (5.38) it follows that the site-specific binding curve of site i is defined by the partition function of the chain containing N sites (top) and its contracted form with site i kept unligated. This contraction generates two smaller chains of $i - 1$ and $N - i$ sites (bottom).

using the definitions $\alpha_1 = \lambda_1'/\lambda_1$, $\alpha_2 = \lambda_2'/\lambda_2$, $\Delta = \lambda_2/\lambda_1$. It is easy to verify that $\alpha_1 + \alpha_2 = 1$ and $-1 \leqslant \Delta \leqslant 1$. In particular, Δ is a parameter of smallness, since it gives the ratio between the minimum and the maximum eigenvalues of **W**. The site-specific binding isotherm written in terms of the foregoing definitions is given by

$$X_i = \frac{\alpha_1\alpha_2(1 - \Delta^i)(1 - \Delta^{N-i+1})}{\alpha_2 + \alpha_1\Delta^{N+1}} \tag{5.44}$$

and is plotted in Figure 5.13 for $N = 9$. As expected on intuitive grounds, the sites in the middle of the chain behave more cooperatively than those toward the ends. Symmetry around the mid-point $M = (N + 1)/2$ is also evident from eq (5.44). Sites equidistant from M have the same binding curve, since the replacement $i \rightarrow N - i + 1$ leaves eq (5.44) unchanged.

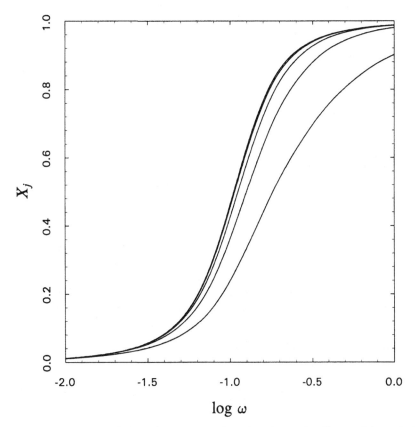

Figure 5.13   Site-specific binding curves for the sites of a linear Ising chain for $N = 9$. Curves were drawn using eq (5.44) with values of $\sigma = 10$ and refer to sites 1–5 from right to left. The curves relative to sites 6–10 are identical to those of sites 1–4 because of the symmetry of the chain around the site in the middle (site 5).

Particularly interesting is to express X_i relative to the value of X_M, i.e., the site-specific binding curve of the site in the middle of the chain. For the sake of simplicity we assume that N is odd. From eq (5.44) it follows that

$$X_M = \frac{\alpha_1\alpha_2\left(1 - \Delta^{\frac{N+1}{2}}\right)^2}{\alpha_2 + \alpha_1\Delta^{N+1}} \tag{5.45}$$

Hence,

$$X_j = X_M\frac{(1 - \Delta^j)(1 - \Delta^{N-j+1})}{\left(1 - \Delta^{\frac{N+1}{2}}\right)^2} \tag{5.46}$$

For large N eq (5.46) can be approximated by

$$X_j = X_M(1 - \Delta^j) = X_M[1 - \exp(j\ln\Delta)]. \tag{5.47}$$

For fixed ω, the difference $X_M - X_j$ decays exponentially from the end of the chain to the middle, as illustrated in Figure 5.14 for different values of the interaction parameter σ. The case $\sigma = 1$ gives $X_M - X_j = $ constant throughout, as expected. The case $\sigma > 1$, corresponding to positive nearest-neighbor interactions, follows the envelope of the exponential decay. The case $\sigma < 1$, corresponding to negative nearest-neighbor interactions, is characterized by damped oscillations of the function $X_M - X_j$, with the damping term given by the exponential decay in eq (5.47).

A basic property of the decay is the 'relaxation length'

$$l = -\frac{1}{\ln|\Delta|} \tag{5.48}$$

which follows directly from eq (5.47). This parameter provides a measure of the communication between sites in the linear chain. The absolute sign in eq (5.48) is necessary, because $\Delta < 0$ for $\sigma < 1$. The relaxation length is independent of the length of the chain and is uniquely defined by the eigenvalues of the transfer matrix. The value of l as a function of ω is plotted in Figure 5.15. The relaxation length is peaked at a point such that $\sigma\omega_{max} = 1$, as can be proved easily from the definition of Δ. Memory effects in the linear chain reach their maximum for this value of ω. The value of l at the peak is

$$l_{max} = -\frac{1}{\ln\left|\frac{\sqrt{\sigma} + 1}{\sqrt{\sigma} - 1}\right|} \tag{5.49}$$

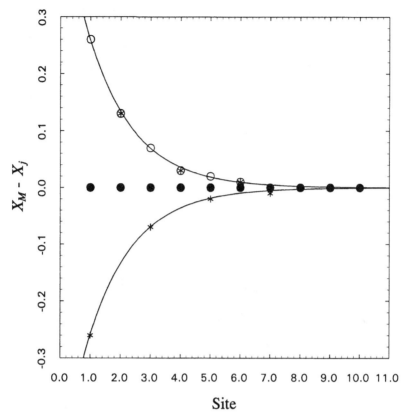

Figure 5.14 Relaxation properties of the linear chain shown as the difference $X_M - X_j$ versus $j = 1, 2, \ldots (N + 1)/2$, for a fixed value of ω and $N = 19$. The continuous lines depict the envelope of the exponential decay embodied by eq (5.47). Points refer to: (open circles) $\sigma = 10$, $\omega = 0.1$; (filled circles) $\sigma = 1$, $\omega = 1$; (stars) $\sigma = 0.1$, $\omega = 10$.

A value of $l_{max} = 0$ is obtained in the absence of interactions ($\sigma = 1$), thereby denoting the absence of 'memory' along the chain. In this limit all sites behave identically. On the other hand, $l_{max} \to \infty$ for $\sigma \to 0$, or $\sigma \to \infty$, thereby indicating the presence of long-range correlation and communication among the sites of the chain in the case of strong coupling. Note that the replacement $\sigma \to \sigma^{-1}$ leaves l_{max} unchanged. Values of $1.5 \leqslant l_{max} \leqslant 5$ are obtained for $10 \leqslant \sigma \leqslant 100$, which are typically seen in practical applications.

The foregoing analysis allows us to tackle the rather interesting problem of how information arising from site–site coupling is communicated in an Ising network. Here we further exploit the simplicity of the lattice-gas type of interaction to understand how perturbations of a homogeneous Ising

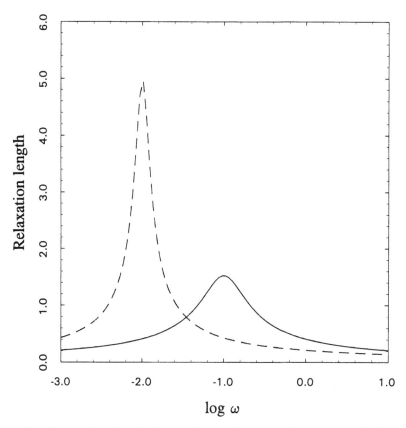

Figure 5.15   Relaxation length for the linear chain as a function of the logarithm of ω. The value of l peaks at $\omega_{max} = \sigma^{-1}$. The continuous and dashed lines refer to $\sigma = 10$ and $\sigma = 100$ respectively.

ring produce the heterogeneous linear chain. Consider the Ising ring in Figure 5.16. All neighboring pairs are coupled with an interaction constant σ and have a site-specific binding constant K, with the exception of site 1, for which the binding constant is K'. The purpose of our analysis is to study the transition for a homogeneous network, obtained for $K' = K$, to a heterogeneous linear chain, obtained for $K' = 0$. Also, all cases obtained for $K' \neq K$ deserve much consideration in a site-specific description, since they refer as well to heterogeneous networks. We shall refer to the Ising network in Figure 5.16 as the 'perturbed ring' and we seek to find the analytical expression for binding to site $i \neq 1$ in the ring.

The basic eq (5.38) demands the definition of the partition function of the perturbed ring, Ψ, along with its contracted form $^0\Psi_i$. Using the definition of contracted partition functions we have

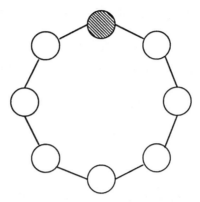

Figure 5.16 Heterogeneous Ising ring. All sites interact equally, but site 1 (hatched circle) has a different site-specific binding constant, K'. In the limit of $K' \to 0$, the ring turns into a linear chain of $N - 1$ sites.

$$\Psi(\omega) = {}^0\Psi_1(\omega) + {}^1\Psi_1(\omega)\omega' \qquad (5.50)$$

where $\omega' = K'x$. The partition function ${}^0\Psi_1$ is that of a linear chain of $N - 1$ identical sites, ${}^L\Psi_{N-1}$, whose solution is given by eq (5.40). The partition function ${}^1\Psi_1$ is the same as that of a homogeneous ring of N sites, with one site kept in its ligated state. For a homogeneous ring the partition function is the same as eq (5.50), except that ω' should be substituted with ω. Hence,

$$ {}^1\Psi_1(\omega) = \frac{{}^R\Psi_N(\omega) - {}^L\Psi_{N-1}(\omega)}{\omega} \qquad (5.51) $$

and the partition function of the perturbed ring is given by

$$\Psi(\omega) = v{}^R\Psi_N(\omega) + (1 - v){}^L\Psi_{N-1}(\omega) \qquad (5.52)$$

where $v = \omega'/\omega = K'/K$. As expected, Ψ is the weighted sum of the partition functions of the homogeneous ring of N sites and the heterogeneous linear chain of $N - 1$ sites, with the ratio between K' and K being the weighting factor. The two limiting cases are obtained for $v = 0$ and $v = 1$.

The contracted partition function ${}^0\Psi_i$ in the perturbed ring can be computed as follows. Contraction with respect to site i yields a linear chain containing $N - 1$ sites, $N - 2$ of which bind with a site-specific constant K, and the perturbed site 1 binding with a constant K'. To obtain the partition function we further contract ${}^0\Psi_i$, so that

$$ {}^0\Psi_i(\omega) = {}^{00}\Psi_{i1}(\omega) + {}^{01}\Psi_{i1}(\omega)\omega' \qquad (5.53) $$

The significance of the various terms is obvious. The partition function

$^{00}\Psi_{i1}$ is the product of two partition functions referring to linear chains of $N - i$ and $i - 2$ identical sites. The partition function $^{01}\Psi_{i1}$ is the same as that of a linear chain of $N - 1$ identical sites, with one site kept in the ligated state. For such a chain the partition function is the same as eq (5.53), with ω' substituted with ω. Hence,

$$^{01}\Psi_{i1}(\omega) = \frac{^{L}\Psi_{N-1}(\omega) - {}^{L}\Psi_{N-i}(\omega){}^{L}\Psi_{i-2}(\omega)}{\omega} \tag{5.54}$$

The contracted partition function $^{0}\Psi_{i}$ is given by

$$^{0}\Psi_{i}(\omega) = v\,{}^{L}\Psi_{N-1}(\omega) + (1 - v)\,{}^{L}\Psi_{N-i}(\omega){}^{L}\Psi_{i-2}(\omega) \tag{5.55}$$

As expected, $^{0}\Psi_{i}$ is the weighted sum of the partition functions of a linear chain of $N - 1$ identical sites and the products of linear chains of $N - i$ and $i - 2$ identical sites. The solution for X_i is then

$$X_i = 1 - \frac{v\,{}^{L}\Psi_{N-1}(\omega) + (1 - v)\,{}^{L}\Psi_{N-i}(\omega){}^{L}\Psi_{i-2}(\omega)}{v\,{}^{R}\Psi_{N}(\omega) + (1 - v)\,{}^{L}\Psi_{N-1}(\omega)} \tag{5.56}$$

X_i is different along the perturbed ring due to the presence of the terms depending on i.

The properties of X_i can be understood by simplifying eq (5.56). Let $\theta_{(N)}$ be the binding isotherm of a site in a homogeneous ring of N sites. Then

$$\theta_{(N)} = 1 - \frac{^{L}\Psi_{N-1}(\omega)}{^{R}\Psi_{N}(\omega)} \tag{5.57}$$

and eq (5.57) applies to all sites of the ring. The binding isotherm of site i in the absence of site 1, or for $K' = 0$, is

$$X_i^{\circ} = 1 - \frac{^{L}\Psi_{N-i}(\omega){}^{L}\Psi_{i-2}(\omega)}{^{L}\Psi_{N-1}(\omega)} \tag{5.58}$$

The function X_i° has the same form as eq (5.44), with N replaced by $N - 1$. Hence,

$$X_i = \frac{v\theta_{(N)} + (1 - v)(1 - \theta_{(N)})X_i^{\circ}}{v\theta_{(N)} + 1 - \theta_{(N)}} \tag{5.59}$$

The site-specific information of X_i is contained in the function X_i°, which provides a convenient reference state. There are essentially two ways of looking at the perturbed ring. One is to consider it as a homogeneous ring with a perturbed site, while the other is to see it as a heterogeneous linear chain of identical sites (2 through N) perturbed by the presence of site 1 bridging the end sites. The latter interpretation is adopted in what follows. We already know from the properties of the 'unperturbed' linear chain that site-specific effects are dominated by the parameter Δ, which also sets the relaxation length for the different binding properties of any site relative to that in the middle. If the chain is perturbed as shown in Figure 5.16, what kind of properties are to be expected for the individual sites? How does site 1 affect the other $N - 1$ sites and the communication among them? If we define the properties of the perturbed chain relative to the site in the middle, like the case dealt with in the foregoing analysis of the unperturbed chain, we obtain the following expression

$$\Delta X_{M,i} = X_M - X_i = \frac{(1 - \nu)(1 - \theta_{(N)})}{\nu\theta_{(N)} + 1 - \theta_{(N)}} \Delta X^\circ_{M,i} \tag{5.60}$$

Except for a scaling term that does not depend on i, $X_M - X_i$ is identical to the value in the unperturbed linear chain. The scaling factor does not affect the relaxation length of the chain that retains the same properties. Hence, the presence of site 1 does not affect the way other sites in the linear chain communicate with each other.

Quite interestingly, the result embodied by eq (5.60) has a more general validity. Consider the arbitrary perturbation of a linear chain as shown in Figure 5.17. Now an arbitrary Ising network, A, perturbs the linear chain composed of identical sites, from 2 to N, through sites connected to site 1. Again, we seek to find the analytical expression for the binding isotherm X_i of a given site in the perturbed linear chain ($i = 2, 3, \dots N$). The definition of X_i demands knowledge of Ψ for the perturbed chain and its contracted form $^0\Psi_i$. Using the contracted partition functions we have again

$$\Psi(\omega) = {}^0\Psi_1(\omega) + {}^1\Psi_1(\omega)\omega' \tag{5.61}$$

where we recall that $\omega' = K'x$, and K' is the binding constant of site 1. The partition function $^0\Psi_1$ refers to a linear chain of $N - 1$ sites with

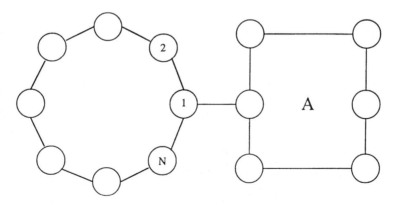

Figure 5.17 Heterogeneous Ising network representing an arbitrary perturbation of the linear chain through a network A connected to the end sites, 2 and N.

identical binding constants, $^L\Psi_{N-1}$, times the contracted partition function $^0\Psi_{1(A)}$ of network A. The partition function $^1\Psi_1$ is the product of two terms. One is the partition function of a homogeneous ring of N sites, with one site kept in its ligated state, and is given by eq (5.51). The second term is the contracted partition function $^1\Psi_{1(A)}$ of network A. Simple rearrangements yield the partition function of the perturbed chain as follows

$$\Psi(\omega) = {}^0\Psi_{1(A)}(\omega)[{}^\omega v^R\Psi_N(\omega) + (1 - {}^\omega v)^L\Psi_{N-1}(\omega)] \qquad (5.62)$$

where $^\omega v = {}^\omega K_{1(A)}/K$, and

$$^\omega K_{1(A)} = K'\frac{{}^1\Psi_{1(A)}(\omega)}{{}^0\Psi_{1(A)}(\omega)} \qquad (5.63)$$

is the site-specific affinity function of site 1 in network A. The similarity between eqs (5.52) and (5.62) is evident. The site-specific affinity function, $^\omega K_{1(A)}$, replaces the site-specific affinity constant of site 1 and v becomes a function of ω. The term $^0\Psi_{1(A)}$ factors out in the definition of Ψ. The simple case embodied by eq (5.52) is obtained as a special case when the perturbing network A is composed of site 1 alone. The contracted partition function $^0\Psi_i$ in the perturbed chain is derived as before from the contraction

$$^0\Psi_i(\omega) = {}^{00}\Psi_{i1}(\omega) + {}^{01}\Psi_{i1}(\omega)\omega' \qquad (5.64)$$

The partition function $^{00}\Psi_{i1}$ is the product of three partition functions, $^0\Psi_{1(A)}$, $^L\Psi_{N-i}$ and $^L\Psi_{i-2}$. The partition function $^{01}\Psi_{i1}$ is the same as that of a linear chain of $N-1$ identical sites, with one site kept in the ligated state, times $^1\Psi_{1(A)}$. Hence, the contracted partition function $^0\Psi_i$ is given by

$$^0\Psi_i(\omega) = {}^0\Psi_1(\omega)[{}^\omega v^L\Psi_{N-1}(\omega) + (1 - {}^\omega v)^L\Psi_{N-i}(\omega)^L\Psi_{i-2}(\omega)] \quad (5.65)$$

The solution for X_i is then

$$X_i = 1 - \frac{{}^\omega v^L\Psi_{N-1}(\omega) + (1 - {}^\omega v)^L\Psi_{N-i}(\omega)^L\Psi_{i-2}(\omega)}{{}^\omega v^R\Psi_N(\omega) + (1 - {}^\omega v)^L\Psi_{N-1}(\omega)} \qquad (5.66)$$

Except for the fact that v is now a function of ω, eq (5.66) is identical to eq (5.56). The conclusion drawn from the foregoing analysis is that the perturbation does not affect the communication among sites in the linear chain, regardless of the form of network A. In the general case, X_i depends solely on the affinity function of site 1 in network A and the binding properties of the unperturbed linear chain. In fact, carrying over the analysis as in the case dealt with previously we have the expression analogous to eq (5.59)

$$\Delta X_{M,i} = \frac{(1 - {}^{\omega}v)(1 - \theta_{(N)})}{{}^{\omega}v\theta_{(N)} + 1 - \theta_{(N)}} \Delta X^0_{M,i} \tag{5.67}$$

as expected.

5.4 The probe theorem

We now address the problem of handling perturbations in an Ising network in more general terms. Consider a heterogeneous Ising network, as schematically illustrated in Figure 5.18. Here we make no assumptions about the binding properties of the sites in the network, or about the nearest-neighbor interaction parameters. These site-specific parameters can be assigned arbitrary values and the dimensional embedding of the network is also arbitrary. Consider the section of the network in Figure 5.18 where sites j and j' are connected. If this connection is removed, the network splits into two smaller networks, A and B. Let A be the network containing site j and B the network containing site j'. Then the following 'probe' theorem holds (Di Cera and Keating, 1994):

Theorem: If an Ising network can be split into two independent networks by deleting the contact between two sites, then these sites uniquely specify each other's site-specific properties.

If the site-specific binding isotherm of one site is known for the separate

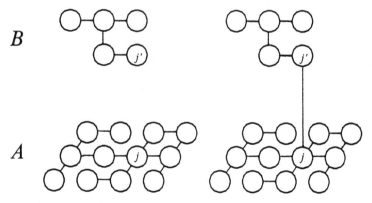

Figure 5.18   Heterogeneous Ising network that can be split into the independent subnetworks, A and B, by deleting the contact between sites j and j'. The probe theorem states that if the properties of site j' in the 'probe' network B are known when the network is isolated (left), or coupled to network A, then the properties of site j can be derived.

network and the network as a whole, then the same quantity can be derived for the other site.

To prove this theorem we consider the affinity function $^x K_{j'(AB)}$ for site j' in the network as a whole

$$^x K_{j'(AB)} = K_{j'} \frac{^1\Psi_{j'(AB)}(x)}{^0\Psi_{j'(AB)}(x)} \tag{5.68}$$

The contracted partition function at the denominator is given by

$$^0\Psi_{j'(AB)}(x) = {}^0\Psi_{j'(B)}(x)\Psi_{(A)}(x) \tag{5.69}$$

where $^0\Psi_{j'(B)}$ is the contracted partition function of the network B isolated with site j' unligated, while $\Psi_{(A)}$ is the partition function of network A isolated. The contracted partition function $^1\Psi_{j'(AB)}$ can itself be expressed as a product. Freezing site j' in its ligated form decouples networks A and B. Network B has a partition function $^1\Psi_{j'(B)}$. The properties of network A are equivalent to those in the absence of B, except for the fact that all configurations with site j in state 1 are to be multiplied by $\sigma_{jj'}$, the interaction constant between sites j and j' in state 1. Hence,

$$^1\Psi_{j'(AB)}(x) = {}^1\Psi_{j'(B)}(x)[^0\Psi_{j(A)}(x) + {}^1\Psi_{j(A)}(x)\sigma_{jj'}K_j x] \tag{5.70}$$

and eq (5.68) becomes

$$^x K_{j'(AB)} = {}^x K_{j'(B)}[1 + (\sigma_{jj'} - 1)X_{j(A)}] \tag{5.71}$$

which proves the theorem. Measurements of the site-specific properties of site j' in the isolated network B and in the network as a whole yield information on $X_{j(A)}$ and hence $X_{j(AB)}$ that reflects the site-specific properties of site j in the isolated network A and the network as a whole. The interaction parameter $\sigma_{jj'}$ can easily be derived in the limit $x \to \infty$ for which $X_{j(A)} \to 1$, since the ratio $^\infty K_{j'(AB)}/^\infty K_{j'(B)}$ is accessible experimentally. If the ratio $^x f_{j'} = {}^x K_{j'(AB)}/^x K_{j'(B)}$ is defined as a function of x, then $X_{j(A)}$ is given by

$$X_{j(A)} = \frac{^x f_j - {}^0 f_j}{^\infty f_j - {}^0 f_j} \tag{5.72}$$

Once $X_{j(A)}$ is known, then $X_{j(AB)}$ can be determined by interchanging j and j' in eq (5.71), so that

$$X_{j(AB)} = \frac{X_{j(A)}[1 + (\sigma_{jj'} - 1)X_{j'(B)}]}{1 + (\sigma_{jj'} - 1)X_{j'(B)}X_{j(A)}} \tag{5.73}$$

which is a useful alternative form. It also follows from the foregoing expressions that

$$X_{j(AB)} - X_{j'(AB)} = \frac{X_{j(A)} - X_{j'(B)}}{1 + (\sigma_{jj'} - 1)X_{j'(B)}X_{j(A)}} \tag{5.74}$$

The sign of the difference between the site-specific binding curves of sites j and j' is not affected by the presence of the contact between the sites. If the probability of binding to site j in network A, for a given x, is higher than that of binding to site j' in network B, the preferential interaction is still observed when the two sites are coupled in the network as a whole.

The significance of the probe theorem stems from the possibility of relating the properties of perturbed and unperturbed networks. Network B can be seen as a perturbation acting on network A, whereby information on the site-specific properties of site j in the unperturbed network A can be derived. More generally, network B functions as a probe for site j in the network A being probed. If the properties of the probe are known when isolated and coupled to A, then the properties of the site in the network being probed can be determined. This is completely independent of the nature of the networks, whether homogeneous or heterogeneous, and is also independent of the particular dimensional embedding. The theorem requires only that the two networks be linearly connected through a single contact.

We now derive a rather counterintuitive property of the linear Ising chain, which is a direct consequence of the probe theorem. A linear chain composed of N distinct sites interacting pairwise in an arbitrary fashion is shown in Figure 5.19. The chain has the remarkable property that its energetics in the site-specific components can be uniquely decoupled if the site-specific binding curve of the end site is known. To prove this, it is sufficient to consider the end site of the linear chain of N sites as the probe of the remaining $N - 1$ sites. Application of the probe theorem yields

$$^{x}K_{N(N)} = K_N[1 + (\sigma_{N-1N} - 1)X_{N-1(N-1)}] \tag{5.75}$$

The first $N - 1$ sites of the chain are to be identified with network A in Figure 5.18 and site N is network B. Here $^{x}K_{N(N)}$ is the affinity function of site N in the chain as a whole, $X_{N-1(N-1)}$ is the binding curve of site $N - 1$ in a chain of $N - 1$ sites (network A), while σ_{N-1N} is the inter-

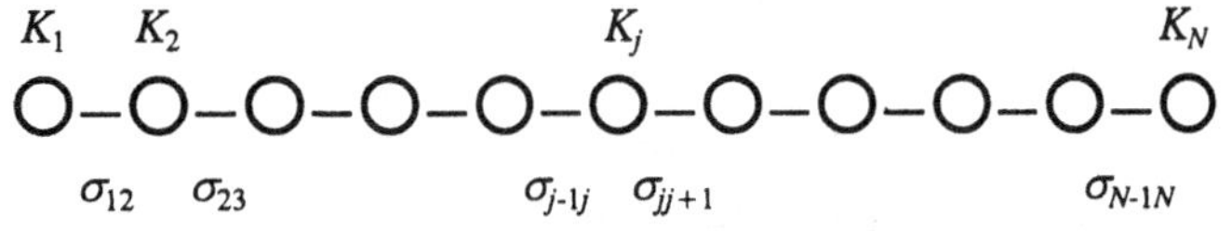

$$K_1 \quad K_2 \qquad\qquad\qquad K_j \qquad\qquad\qquad\qquad K_N$$

$$O{-}O{-}O{-}O{-}O{-}O{-}O{-}O{-}O{-}O{-}O$$

$$\sigma_{12} \quad \sigma_{23} \qquad\qquad \sigma_{j-1j} \quad \sigma_{jj+1} \qquad\qquad \sigma_{N-1N}$$

Figure 5.19 Heterogeneous Ising chain composed of N different sites interacting in an arbitrary way. The entire set of $2N - 1$ independent site-specific parameters can be resolved from knowledge of the binding properties of site N, or site 1, using the probe theorem.

action parameter between sites N (network B) and $N - 1$ (from network A). The value of K_N gives the affinity function of site N isolated, which is clearly independent of x and simply reflects the equilibrium binding constant of this site. In the general case, the chain as a whole is character- ized by N site-specific binding constants such as K_N, and $N - 1$ inter- action parameters $\sigma_{12}, \sigma_{23}, \ldots \sigma_{N-1N}$, for the coupling between sites 1 and 2, 2 and 3, $\ldots N - 1$ and N. These $2N - 1$ independent parameters define the partition function and can be resolved from eq (5.71), if $^xK_{N(N)}$ is known. In fact, the limits

$$^0K_{N(N)} = K_N \tag{5.76a}$$

$$^\infty K_{N(N)} = \sigma_{N-1N}K_N \tag{5.76b}$$

yield K_N and σ_{N-1N}. When these parameters are known, $X_{N-1(N-1)}$ can be constructed for all values of x from eq (5.75). Since $X_{N-1(N-1)}$ is the binding curve of the end site in a chain of $N - 1$ sites, this site can, in turn, be treated as the probe of a chain of $N - 2$ sites. The affinity function of this end site is $^xK_{N-1(N-1)} = X_{N-1(N-1)}/[(1 - X_{N-1(N-1)})x]$ and application of eq (5.75) yields

$$^xK_{N-1(N-1)} = K_{N-1}[1 + (\sigma_{N-2N-1} - 1)X_{N-2(N-2)}] \tag{5.77}$$

Again, the limits of $^xK_{N-1(N-1)}$ yield K_{N-1} and σ_{N-2N-1} and hence $X_{N-2(N-2)}$ for all values of x. The entire linear chain can be energetically 'sequenced' by iterating this procedure. At each step the site-specific properties of the end site are used to resolve a site-specific equilibrium constant and an interaction parameter. This information allows for con- struction of the site-specific properties of the end site of a linear chain one unit shorter and so forth. Hence, all Ks and σs can be resolved from knowledge of the site-specific properties of the end site of the original chain.

This peculiar result only applies to the end site, as follows directly from the probe theorem. Consider the case where $X_{j(N)}$ and hence $^xK_{j(N)}$ is known for a site other than the end site ($j \neq 1, N$). Then application of eq (5.71) yields

$$^xK_{j(N)} = {}^xK_{j(j)}[1 + (\sigma_{jj+1} - 1)X_{j+1(N-j)}] \tag{5.78}$$

Here the network from site 1 through j is taken as the probe. The function $^xK_{j(j)}$ describes the properties of site j in a linear chain of length j and is therefore subject to the probe theorem, i.e.,

$$^xK_{j(j)} = K_j[1 + (\sigma_{j-1j} - 1)X_{j-1(j-1)}] \tag{5.79}$$

Hence,

$$^xK_{j(N)} = K_j[1 + (\sigma_{jj+1} - 1)X_{j+1(N-j)}][1 + (\sigma_{j-1j} - 1)X_{j-1(j-1)}] \tag{5.80}$$

The equation above reduces to eq (5.75) for $j = N$, or equivalently for $j = 1$. In general, for $j \neq 1, N$, knowledge of ${}^x K_{j(N)}$ does not provide sufficient information to sequence the entire chain, because the number of variables involved in eq (5.80) exceeds the number of constraints. In fact, the limiting values of ${}^x K_{j(N)}$ are

$$ {}^0 K_{j(N)} = K_j \tag{5.81a} $$

$$ {}^\infty K_{j(N)} = \sigma_{j-1j} \sigma_{jj+1} K_j \tag{5.81b} $$

and although the value of K_j can easily be resolved, the interaction constants cannot be decoupled. Only in the case in which all sites interact equally does it make no difference which site-specific binding isotherm is used to resolve the Ks and σs.

The validity of the probe theorem is not limited to the lattice-gas type of interaction for neighboring sites. If the spin-lattice type of interaction is assumed, with $\sigma_{jj'}$ being the interaction constant between sites j and j', straightforward calculations lead to

$$ {}^0\Psi_{j'(AB)}(x) = {}^0\Psi_{j'(B)}(x)[{}^0\Psi_{j(A)}(x) + \sigma_{jj'}{}^1\Psi_{j(A)}(x)K_j x] \tag{5.82} $$

$$ {}^1\Psi_{j'(AB)}(x) = {}^1\Psi_{j'(B)}(x)[\sigma_{jj'}{}^0\Psi_{j(A)}(x) + {}^1\Psi_{j(A)}(x)K_j x] \tag{5.83} $$

Hence,

$$ {}^x K_{j'(AB)} = {}^x K_{j'(B)} \frac{\sigma_{jj'} + (1 - \sigma_{jj'})X_{j(A)}}{1 - (1 - \sigma_{jj'})X_{j(A)}} \tag{5.84} $$

and the theorem is proved. Application of the spin-lattice type of interaction to the linear Ising chain gives

$$ {}^x K_{N(N)} = K_N \frac{\sigma_{N-1N} + (1 - \sigma_{N-1N})X_{N-1(N-1)}}{1 - (1 - \sigma_{N-1N})X_{N-1(N-1)}} \tag{5.85} $$

and again, K_N and σ_{N-1N} are derived from the asymptotic values of ${}^x K_{N(N)}$ thereby allowing a complete energetic sequencing of the chain as in the case of the lattice-gas type of interaction.

The probe theorem echoes a well-known principle in statistical thermodynamics whereby information on the state of a system can be obtained from the response of the system to a perturbation (de Groot and Mazur, 1984). The probe represents the perturbation acting 'adiabatically' on a network whose site-specific properties are to be dissected in energetic terms. Information on site-specific properties of the system is derived by measuring the properties of the probe isolated and coupled to the system. It is the perturbation caused by the probe that generates information about the system. The response of the system to the perturbation is measured as the change in the properties of the probe upon its coupling to the system. This suggests that the properties of a specific site can be derived from those of a suitable probe attached to it. The probe theorem suggests a simple strategy for deriving site-specific information from *ad hoc* perturbations of the system of interest. This provides a connection with arguments discussed in Section 4.5.

Extension of this theorem to the case where the probe is attached to two or more sites in the network will provide a key to unraveling the properties of heterogeneous Ising networks in two and higher dimensions. Site-specific thermodynamics promises to be an exciting new approach to the Ising problem in arbitrary dimensions, which still remains one of the toughest unsolved problems of statistical thermodynamics.

References

Ackers, G. K. (1990) *Biophys. Chem.* **37**, 371–82.

Ackers, G. K. & Hazzard, J. H. (1993) *Trends Biochem. Sci.* **18**, 385–90.

Ackers, G. K. & Smith, F. R. (1985) *Ann. Rev. Biochem.* **54**, 597–629.

Ackers, G. K. & Smith, F. R. (1987) *Ann. Rev. Biophys. Biophys. Chem.* **16**, 583–609.

Ackers, G. K., Shea, M. A. & Smith, F. R. (1983) *J. Mol. Biol.* **170**, 223–42.

Ackers, G. K., Doyle, M. L., Myers, D. & Daugherty, M. A. (1992) *Science* **255**, 54–63.

Adams, E. Q. (1916) *J. Am. Chem. Soc.* **38**, 1503–10.

Ahlström, P., Teleman, O., Kördel, J., Forsén, S. & Jönsson, B. (1989) *Biochemistry* **28**, 3205–11.

Akke, M., Forsén, S. & Chazin, W. J. (1991) *J. Mol. Biol.* **220**, 173–89.

Anderson, C. F. & Record, M. T., Jr. (1993) *J. Phys. Chem.* **97**, 7116–26.

Aurora, R., Srinivasan, R. & Rose, G. D. (1994) *Science* **264**, 1126–30.

Avnir, D., Farin, D. & Pfeifer, P. (1984) *Nature* **308**, 261–3.

Ayala, Y. & Di Cera, E. (1994) *J. Mol. Biol.* **235**, 733–46.

Bai, Y., Milne, J. S., Mayne, L. & Englander, S. W. (1994) *Proteins* **20**, 4–14.

Baldwin, J. & Chothia, C. (1979) *J. Mol. Biol.* **129**, 175–220.

Bard, Y. (1974) *Nonlinear Parameter Estimation* (Academic, New York).

Bashford, D. & Karplus, M. (1991) *J. Phys. Chem.* **95**, 9556–61.

Betz, A. J., Hofsteenge, J. & Stone, S. R. (1991) *Biochem. J.* **275**, 801–3.

Bode, W., Turk, D. & Karshikov, A. (1992) *Protein Sci.* **1**, 426–71.

Bondon, A. & Simonneaux, G. (1990) *Biophys. Chem.* **37**, 407–11.

Bondon, A., Petrinko, P., Sodano, P. & Simonneaux, G. (1986) *Biochem. Biophys. Acta* **872**, 163–6.

Box, G. E. P. & Jenkins, G. M. (1976) *Time Series Analysis* (Holden-Day, Oakland, CA).

Briggs, W. E. (1983) *Biophys. Chem.* **18**, 67–71.

Briggs, W. E. (1984) *J. Theor. Biol.* **108**, 77–83.

Briggs, W. E. (1985) *J. Theor. Biol.* **114**, 605–14.

Brodin, P., Johansson, C., Forsén, S., Drakenberg, T. & Grundström, T. (1990) *J. Biol. Chem.* **265**, 11125–30.

Brown, I. D. & Wu, K. K. (1976) *Acta Crystallogr. B* **32**, 1957–9.

Bucci, E. & Fronticelli, C. (1965) *J. Biol. Chem.* **240**, 551–2.

Bujalowski, W. & Lohman, T. M. (1989) *J. Mol. Biol.* **207**, 249–68.

285

Bunn, H. F. & Forget, B. G. (1986) *Hemoglobin: Molecular, Genetic and Clinical Aspects* (Saunders, Philadelphia, PA).

Cantor, C. R. & Schimmel, P. R. (1980) *Biophysical Chemistry* (Freeman and Co., New York).

Carlström, G. & Chazin, W. J. (1993) *J. Mol. Biol.* **231**, 415–30.

Carter, P. & Wells, J. A. (1988) *Nature* **332**, 564–8.

Chandrasekhar, S. (1943) *Rev. Mod. Phys.* **15**, 1–89.

Chu, A. H., Turner, B. W. & Ackers, G. K. (1982) *Biochemistry* **23**, 604–17.

Cohen, S. S. & Penrose, O. (1970) *J. Chem. Phys.* **52**, 5018–21.

Cohen, E. J. & Edsall, J. T. (1943) *Proteins, Amino Acids and Peptides* (Reinhold, New York).

Colombo, M. F., Rau, D. C. & Parsegian, V. A. (1992) *Science* **256**, 655–9.

Cornish, V. W. & Schultz, P. G. (1994) *Curr. Op. Struct. Biol.* **4**, 601–7.

Crawford, F. H. (1950) *Proc. Am. Acad. Arts Sci.* **78**, 165–84.

Crawford, F. H. (1955) *Proc. Am. Acad. Arts Sci.* **83**, 191–222.

d'A. Heck, H. (1971) *J. Am. Chem. Soc.* **93**, 23–9.

Dang, Q. D., Vindigni, A. & Di Cera, E. (1995) *Proc. Natl. Acad. Sci. USA* **92**, 5977–81.

Daugherty, M. A., Shea, M. A. & Ackers, G. K. (1994) *Biochemistry* **33**, 10345–57.

Daugherty, M. A., Shea, M. A., Johnson, J. A., LiCata, V. J., Turner, G. T. & Ackers, G. K. (1991) *Proc. Natl. Acad. Sci. USA* **88**, 1110–4.

de Groot, S. R. & Mazur, P. (1984) *Non-Equilibrium Thermodynamics* (Dover, New York).

de Heer, J. (1986) *Phenomenological Thermodynamics* (Prentice-Hall, London, UK).

Di Cera, E. (1989) *J. Theor. Biol.* **136**, 467–74.

Di Cera, E. (1990) *Biophys. Chem.* **37**, 147–64.

Di Cera, E. (1992a) *Methods Enzymol.* **210**, 68–87.

Di Cera, E. (1992b) *J. Chem. Phys.* **96**, 6515–22.

Di Cera, E. (1994a) *Methods Enzymol.* **232**, 655–83.

Di Cera, E. (1994b) *Biopolymers* **34**, 1001–5.

Di Cera, E. & Chen, Z.-Q. (1993) *Biophys. J.* **65**, 164–70.

Di Cera, E. & Keating, S. (1994) *Biopolymers* **34**, 673–8.

Di Cera, E., Doyle, M. L., Connelly, P. R. & Gill, S. J. (1987a) *Biochemistry* **26**, 6494–502.

Di Cera, E., Doyle, M. L. & Gill, S. J. (1988b) *J. Mol. Biol.* **200**, 593–9.

Di Cera, E., Gill, S. J. & Wyman, J. (1988a) *Proc. Natl. Acad. Sci. USA* **85**, 449–52.

Di Cera, E., Hopfner, K.-P. & Wyman, J. (1992) *Proc. Natl. Acad. Sci. USA* **89**, 2727–31.

Di Cera, E., Robert, C. H. & Gill, S. J. (1987b) *Biochemistry* **26**, 4003–8.

Di Cera, E., Doyle, M. L., Morgan, M. S., De Cristofaro, R., Landolfi, R., Bizzi, B., Castagnola, M. & Gill, S. J. (1989) *Biochemistry* **28**, 2631–8.

Dill, K. A., Bromberg, S., Yue, K., Friebig, K. M., Yee, D. P., Thomas, P. D. & Chan, H. S. (1995) *Protein Sci.* **4**, 561–602.

Dolman, D. & Gill, S. J. (1978) *Anal. Biochem.* **87**, 127–34.

Domb, C. & Green, M. S. (1972) *Phase Transitions and Critical Phenomena* (Academic, New York).

Douglas, C. G., Haldane, J. S. & Haldane, J. B. S. (1912) *J. Physiol.* **44**, 275–304.

Doyle, M. L. & Ackers, G. K. (1992) *Biochemistry* **31**, 11182–95.

Draper, D. E. (1993) *Proc. Natl. Acad. Sci. USA* **90**, 7429–30.

Ebright, R. H. (1986) *Proc. Natl. Acad. Sci. USA* **83**, 303–7.

Edsall, J. T. & Blanchard, M. H. (1933) *J. Am. Chem. Soc.* **55**, 2337–53.

Edsall, J. T. & Wyman, J. (1958) *Biophysical Chemistry* (Academic, New York).

Englander, S. W. & Kallenbach, N. R. (1984) *Q. Rev. Biophys.* **16**, 521–655.

Feller, W. (1950) *An Introduction to Probability Theory and its Applications* (Wiley, New Yok).

Fermi, E. (1936) *Thermodynamics* (Dover, New York).

Fermi, G., Perutz, M. F., Shaanan, B. (1984) *J. Mol. Biol.* **175**, 159–74.

Fleischmann, M., Tildesley, D. J. & Ball, R. C. (1990) *Fractals in the Natural Sciences* (Princeton University Press, Princeton, NJ).

Fong, T. M., Huang, R.-R. C. & Strader, C. D. (1992) *J. Biol. Chem.* **267**, 25664–7.

Gantmacher, F. R. (1959) *The Theory of Matrices* (Chelsea, New York).

Gibbs, J. W. (1875) *Trans. Conn. Acad.* **3**, 108–248.

Gibbs, J. W. (1902) *Elementary Principles in Statistical Mechanics Developed in Especial Reference to the Rational Foundation of Thermodynamics* (C. Scribner's Sons, New York).

Gibbs, J. W. (1928) *Collected Works* (Longman, Green and Co., New York).

Gill, S. J., Di Cera, E., Doyle, M. L., Bishop, G. A. & Robert, C. H. (1987) *Biochemistry* **26**, 3995–4002.

Glauber, R. J. (1963) *J. Math. Phys.* **4**, 294–307.

Gould, R. (1988) *Graph Theory* (Benjamin/Cummings Publ. Co., Menlo Park, CA).

Gradshtein, S. & Ryzhik, I. M. (1980) *Tables of Integrals, Series and Products* (Academic, New York).

Hamming, R. W. (1986) *Coding and Information Theory* (Prentice-Hall, Englewood Cliffs, NJ).

Hankey, A. & Stanley, H. E. (1972) *Phys. Rev. B* **6**, 3515–42.

Harper, E. T. & Rose, G. D. (1993) *Biochemistry* **32**, 7605–9.

Hendler, R. W., Subba Reddy, K. V., Shrager, R. I. & Caughey, W. S. (1986) *Biophys. J.* **49**, 717–29.

Hermans, J. & Premilat, S. (1975) *J. Phys. Chem.* **79**, 1169–75.

Herzfeld, J. & Stanley, H. E. (1974) *J. Mol. Biol.* **82**, 231–65.

Hill, A. V. (1910) *J. Physiol. (London)* **40**, iv–vii.

Hill, T. L. (1944) *J. Chem. Phys.* **12**, 56–61.

Hill, T. L. (1960) *Introduction to Statistical Thermodynamcis* (Dover, New York).

Hill, T. L. (1977) *Free Energy Transduction in Biology* (Academic, New York).

Hill, T. L. (1984) *Cooperativity Theory in Biochemistry* (Springer-Verlag, New York).

Ho, C. (1992) *Adv. Protein Chem.* **43**, 153–312.

Horovitz, A. & Fersht, A. R. (1990) *J. Mol. Biol.* **214**, 613–17.

Horovitz, A. & Fersht, A. R. (1992) *J. Mol. Biol.* **224**, 733–40.

Horovitz, A., Serrano, L. & Fersht, A. R. (1991) *J. Mol. Biol.* **219**, 5–9.

Horovitz, A., Serrano, L., Avron, B., Bycroft, M. & Fersht, A. R. (1990) *J. Mol. Biol.* **216**, 1031–44.

Huang, Y. & Ackers, G. K. (1995) *Biochemistry* **34**, 6316–27.

Ikeda-Saito, M. & Yonetani, T. (1980) *J. Mol. Biol.* **138**, 845–58.

Imai, K. (1974) *J. Biol. Chem.* **249**, 7607–12.

Imai, K. (1982) *Allosteric Effects in Hemoglobin* (Cambridge University Press,

Cambridge, UK).

Inubushi, T., D'Ambrosio, C., Ikeda-Saito, M. & Yonetani, T. (1986) *J. Am. Chem. Soc.* **108**, 3799–803.

Itzykson, C. & Drouffe, J.-M. (1980) *Statistical Field Theory* (Cambridge University Press, Cambridge, UK).

Janin, J. & Wodak, S. J. (1993) *Proteins* **15**, 1–4.

Jencks, W. P. (1981) *Proc. Natl. Acad. Sci. USA* **78**, 4046–50.

Jensen, C. J., Gcrard, N. P., Schwartz, T. W. & Gether, U. (1994) *Mol. Pharmacol.* **45**, 294–9.

Johnson, M. L., Turner, B. W. & Ackers, G. K. (1984) *Proc. Natl. Acad. Sci. USA* **81**, 1093–7.

Keating, S. & Di Cera, E. (1993) *Biophys. J.* **65**, 253–69.

Khinchin, A. I. (1949) *Mathematical Foundations of Statistical Mechanics* (Dover, New York).

Klotz, I. M. (1985) *Q. Rev. Biophys.* **18**, 227–59.

Klotz, I. M. (1993) *Proc. Natl. Acad. Sci. USA* **90**, 7191–4.

Klotz, I. M. & Hunston, D. L. (1975) *J. Biol. Chem.* **250**, 3001–9.

Knowles J. R. (1987) *Science* **236**, 1252–8.

Kojima, N. & Palmer, G. (1983) *J. Biol. Chem.* **258**, 14908–13.

Koshland, D. E., Nemethy, G. & Filmer, D. (1966) *Biochemistry* **5**, 365–85.

Kretsinger, R. H. (1980) *C.R.C. Crit. Rev. Biochem.* **8**, 119–74.

Kubo, R. (1965) *Statistical Mechanics* (North-Holland, Amsterdam, The Netherlands).

Lattman, E. E. & Rose, G. D. (1993) *Proc. Natl. Acad. Sci. USA* **90**, 493–7.

Lee, T. D. & Yang, C. N. (1952) *Phys. Rev.* **87**, 410–9.

Lesk, A. M., Janin, J., Wodak, S. & Chothia, C. (1985) *J. Mol. Biol.* **183**, 267–70.

Lesser, D. R., Kurpiewski, M. R. & Jen-Jacobson, L. (1990) *Science* **250**, 776–86.

LiCata, V. J., Dalessio, P. M. & Ackers, G. K. (1993) *Proteins* **17**, 279–96.

Linderstrøm-Lang, K. (1924) *Compt. Rend. Lab. Carlsberg* **15**, 1–29.

Linse, S., Brodin, P., Drakenberg, T., Thulin, E., Sellers, P., Elmdén K., Grundström, T. & Forsén, S. (1987) *Biochemistry* **26**, 6723–35.

Linse, S., Brodin, P., Johansson, C., Thulin, E., Grundström, T. & Forsén, S. (1988) *Nature* **335**, 651–2.

Linse, S., Johansson, C., Brodin, P., Grundström, T., Drakenberg, T. & Forsén, S. (1991) *Biochemistry* **30**, 154–62.

Löwdin, P.-O. (1964) *J. Mol. Spectroscopy* **14**, 112–8.

Lyu, P. C., Liff, M. I., Marky, L. A. & Kallenbach, N. R. (1990) *Science* **250**, 669–73.

Martin, S. R., Linse, S., Johansson, C., Bayley, P. M. & Forsén, S. (1990) *Biochemistry* **29**, 4188–93.

Matthews, B. W. (1993) *Ann. Rev. Biochem.* **62**, 139–60.

Mayer, J. E. & Mayer, M. G. (1940) *Statistical Mechanics* (Wiley, New York).

Monod, J., Changeux, J. P. & Jacob, F. (1963) *J. Mol. Biol.* **6**, 306–29.

Monod, J., Wyman, J. & Changeux, J. P. (1965) *J. Mol. Biol.* **12**, 88–118.

Mountcastle, D. B., Freire, E. & Biltonen, R. L. (1976) *Biopolymers* **15**, 355–71.

Nayal, M. & Di Cera, E. (1994) *Proc. Natl. Acad. Sci. USA* **91**, 817–21.

Neuberger, A. (1936) *Biochem. J.* **30**, 2085–94.

Newell, G. F. & Montroll, E. W. (1953) *Rev. Mod. Phys.* **25**, 353–94.

Onsager, L. (1944) *Phys. Rev.* **65**, 117–28.

Padmanabhan, S., Marqusee, S., Ridgeway, T., Laue, T. M. & Baldwin, R. L. (1990) *Nature* **344**, 268–70.

Pauli, W. (1927) *Zeits. Phys.* **41**, 81–92.

Pauling, L. (1929) *J. Am. Chem. Soc.* **51**, 1010–26.

Pauling, L. (1935) *Proc. Natl. Acad. Sci. USA* **21**, 186–91.

Pauling, L. (1947) *J. Am. Chem. Soc.* **69**, 542–53.

Pearson, C. E. (1990) *Handbook of Applied Mathematics* (Van Nostrand Reinhold, New York).

Peller, L. (1982) *Nature* **300**, 661–2.

Perrella, M. (1995) in preparation.

Perrella, M. & Rossi-Bernardi, L. (1981) *Methods Enzymol.* **76**, 133–43.

Perrella, M., Davids, N. & Rossi-Bernardi, L. (1992) *J. Biol. Chem.* **267**, 8744–51.

Perrella, M., Benazzi, L., Cremonesi, L., Vesely, S., Viggiano, G. & Rossi-Bernardi, L. (1983) *J. Biol. Chem.* **258**, 4511–7.

Perrella, M., Sabbioneda, L., Samaja, M. & Rossi-Bernardi, L. (1986) *J. Biol. Chem.* **261**, 8391–6.

Perrella, M., Colosimo, A., Benazzi, L., Ripamonti, M. & Rossi-Bernardi, L. (1990) *Biophys. Chem.* **37**, 211–23.

Perrella, M., Shrager, R. I., Ripamonti, M., Manfredi, G., Berger, R. L. & Rossi-Bernardi, L. (1993) *Biochemistry* **32**, 5233–8.

Perrella, M., Benazzi, L., Ripamonti, M. & Rossi-Bernardi, L. (1994) *Biochemistry* **33**, 10358–66.

Perutz, M. F. (1970) *Nature* **228**, 726–39.

Perutz, M. F. (1989) *Q. Rev. Biophys.* **22**, 139–236.

Poland, D. & Scheraga, H. A. (1970) *Theory of Helix-Coil Transitions in Biopolymers* (Academic Press, New York).

Pradier, L., Menager, J., Le Guern, J., Bock, M.-D., Heuillet, E., Fardin, V., Garret, C., Doble, A. & Mayaux, J.-F. (1994) *Mol. Pharmacol.* **45**, 287–93.

Presta, L. G. & Rose, G. D. (1988) *Science* **240**, 1632–41.

Prigogine, I. & Defay, R. (1954) *Chemical Thermodynamics* (Longman, New York).

Ptashne, M. (1986) *A Genetic Switch: Gene Control and Phage λ* (Cell Press, Cambridge, MA).

Record, M. T., Jr. & Anderson, C. F. (1995) *Biophys. J.,* **68**, 786–94.

Record, M. T., Jr., Anderson, C. F. & Lohman, T. M. (1978) *Q. Rev. Biophys.* **11**, 103–78.

Roughton, F. J. W., Otis, A. B. & Lyster, R. J. L. (1955) *Proc. R. Soc. London B* **144**, 29–54.

Rydel, T. J., Tulinsky, A., Bode, W. & Huber, R. (1991) *J. Mol. Biol.* **221**, 583–601.

Scatchard, G. (1949) *Ann. N.Y. Acad. Sci.* **51**, 660–72.

Schechter, I. & Berger, A. (1967) *Biochem. Biophys. Res. Commun.* **27**, 157–62.

Schellman, J. A. (1975) *Biopolymers* **14**, 999–1018.

Schellman, J. A. (1978) *Biopolymers* **17**, 1305–22.

Schellman, J. A. (1987) *Biopolymers* **26**, 549–59.

Schellman, J. A. (1990a) *Biopolymers* **29**, 215–24.

Schellman, J. A. (1990b) *Biophys. Chem.* **37**, 121–40.

Schellman, J. A. (1994) *Biopolymers,* **34**, 1015–25.

Schrödinger, E. (1946) *Statistical Thermodynamics* (Dover, New York).

Scrutton, N. S., Berry, A. & Perham, R. N. (1990) *Nature* **343**, 38–43.

Senear, D. F. & Ackers, G. K. (1990) *Biochemistry* **29**, 6568–77.

Shaanan, B. (1983) *J. Mol. Biol.* **171**, 31–59.

Shaw, A. N. (1935) *Philos. Trans. R. Soc. London Ser. A* **234**, 299–328.

Shibayama, N., Inubushi, T., Morimoto, H. & Yonetani, T. (1987) *Biochemistry* **26**, 2194–201.

Silva, M. M., Rogers, P. H. & Arnone, A. (1994) *J. Biol. Chem.* **267**, 17248–56.

Simms, H. S. (1926) *J. Am. Chem. Soc.* **48**, 1239–61.

Simonneaux, G., Bondon, C., Brunel, C. & Sodano, P. (1988) *J. Am. Chem. Soc.* **110**, 7637–41.

Skelton, N. J., Kördel, J., Akke, M., Forsén, S. & Chazin, W. J. (1994) *Nature Struc. Biol.* **1**, 239–45.

Smith, F. R. & Ackers, G. K. (1985) *Proc. Natl. Acad. Sci. USA* **82**, 5347–51.

Smith, F. R. & Simmons, K. C. (1994) *Proteins* **18**, 295–300.

Smith, F. R., Lattman, E. E. & Carter, C. W. (1991) *Proteins* **10**, 81–91.

Smith, F. R., Gingrich, D., Hoffman, B. & Ackers, G. K. (1987) *Proc. Natl. Acad. Sci. USA* **84**, 7089–93.

Spiros, P. C., LiCata, V. J., Yonetani, T. & Ackers, G. K. (1991) *Biochemistry* **30**, 7254–62.

Srinivasan, R. & Rose, G. D. (1994) *Proc. Natl. Acad. Sci. USA* **91**, 11113–17.

Stanley, H. E. (1971) *Introduction to Phase Transitions and Critical Phenomena* (Oxford University Press, Oxford, UK).

Stockmayer, W. H. (1950) *J. Chem. Phys.* **18**, 58–61.

Sturgill, T. W. (1978) *Biopolymers* **17**, 1793–810.

Sturgill, T. W. & Biltonen, R. L. (1976) *Biopolymers* **15**, 337–54.

Sturgill, T. W., Johnson, R. E. & Biltonen, R. L. (1978) *Biopolymers* **17**, 1773–92.

Szabo, A. & Karplus, M. (1972) *J. Mol. Biol.* **72**, 126–97.

Szebenyi, D. M. E. & Moffat, J. (1986) *J. Biol. Chem.* **261**, 8761–77.

Tanford, C. (1969) *J. Mol. Biol.* **39**, 539–44.

Tanford, C. (1970) *Adv. Protein Chem.* **24**, 2–95.

Tanford, C. & Kirkwood, J. G. (1957) *J. Am. Chem. Soc.* **79**, 5333–9.

Timasheff, S. (1990) *Biochemistry* **31**, 9857–64.

Tolman, R. C. (1938) *The Principles of Statistical Mechanics* (Oxford University Press, Oxford, UK).

Viggiano, G. & Ho, C. (1979) *Proc. Natl. Acad. Sci. USA* **76**, 3673–7.

von Hippel, P. H. & Schleich, T. (1969) *Acc. Chem. Res.* **2**, 257–65.

Wallace, A., Dennis, S., Hofsteenge, J. & Stone, S. R. (1989) *Biochemistry* **28**, 10079–84.

Weaver, L. H. & Matthews, B. W. (1987) *J. Mol. Biol.* **193**, 189–99.

Weber, G. (1975) *Adv. Protein Chem.* **29**, 1–83.

Weber, G. (1982) *Nature* **300**, 603–7.

Weber, G. (1992) *Protein Interactions* (Chapman & Hall, New York).

Wegscheider, R. (1895) *Monatsch. Chem.* **16**, 153–8.

Weinhold, F. (1975a) *J. Chem. Phys.* **63**, 2479–83.

Weinhold, F. (1975b) *J. Chem. Phys.* **63** 2484–7.

Weinhold, F. (1975c) *J. Chem. Phys.* **63**, 2488–95.

Wells, C. M. & Di Cera, E. (1992) *Biochemistry* **31**, 11721–730.

Wells, J. A. (1990) *Biochemistry* **29**, 8509–17.

Whitehead, E. P. (1980a) *J. Theor. Biol.* **86** 45–82.

Whitehead, E. P. (1980b) *J. Theor. Biol.* **87**, 153–70.

Whitehead, E. P. (1981) *J. Theor. Biol.* **93**, 547–83.

Wilson, E. B. (1911) *Advanced Calculus* (McGraw-Hill, New York).

Wyman, J. (1948) *Adv. Protein Chem.* **4**, 407–531.

Wyman, J. (1964) *Adv. Protein Chem.* **19**, 223–86.

Wyman, J. (1967) *J. Am. Chem. Soc.* **89**, 2202–18.

Wyman, J. & Allen, D. W. (1951) *J. Polymer. Sci.* **7**, 499–518.

Wyman, J. & Gill, S. J. (1990) *Binding and Linkage: The Physical Chemistry of Biological Macromolecules* (University Science Books, Mill Valley, CA).

Yang, C. N. & Lee, T. D. (1952) *Phys. Rev.* **87**, 404–9.

Zhou, H. X., Hull, L. A., Kallenbach, N. R., Mayne, L., Bai, Y. & Englander, S. W. (1994) *J. Am. Chem. Soc.* **116**, 6482–3.

Zimm, B. H. & Bragg, J. K. (1959) *J. Chem. Phys.* **31**, 526–35.

Index

Printed in the United States
By Bookmasters